ADVANCES IN AMINO ACID MIMETICS AND PEPTIDOMIMETICS

Volume 1 • 1997

ADVANCES IN AMINO ACID MIMETICS AND PEPTIDOMIMETICS

Editor: ANDREW ABELL
Department of Chemistry
University of Canterbury
Christchurch, New Zealand

VOLUME 1 • 1997

JAI PRESS INC.

Greenwich, Connecticut *London, England*

55 Old Post Road No. 2
Greenwich, Connecticut 06836

JAI PRESS LTD.
38 Tavistock Street
Covent Garden
London WC2E 7PB
England

ISBN: 0-7623-0200-3

Printed and bound by CPI Antony Rowe, Eastbourne
Transferred to digital printing 2006

CONTENTS

LIST OF CONTRIBUTORS

Jeffrey Aubé	Department of Medicinal Chemistry University of Kansas Lawrence, Kansas
Colin James Barrow	School of Chemistry University of Melbourne Parkville, Victoria, Australia
Sylvie E. Blondelle	Torrey Pines Institute for Molecular Studies San Diego, California
Chong-Hwan Chang	Experimental Station The DuPont Merck Pharmaceutical Company Wilmington, Delaware
Colette T. Dooley	Torrey Pines Institute for Molecular Studies San Diego, California
Barbara Dörner	Torrey Pines Institute for Molecular Studies San Diego, California
Jean-Philippe Dumas	Department of Chemistry University of Houston Houston, Texas
Charles J. Eyermann	Experimental Station The DuPont Merck Pharmaceutical Company Wilmington, Delaware
David P. Fairlie	Centre for Drug Design and Development University of Queensland Brisbane, Queensland, Australia

Juris Paul Germanas	Department of Chemistry University of Houston Houston, Texas
Michael N. Greco	Drug Discovery The R.W. Johnson Pharmaceutical Research Institute Spring House, Pennsylvania
Andrew D. Hamilton	Departments of Chemistry and Pharmacology University of Pittsburgh Pittsburgh, Pennsylvania
C. Nicholas Hodge	Experimental Station The DuPont Merck Pharmaceutical Company Wilmington, Delaware
Richard A. Houghten	Torrey Pines Institute for Molecular Studies San Diego, California
P.K. Jadhav	Experimental Station The DuPont Merck Pharmaceutical Company Wilmington, Delaware
Kyonghee Kim	Department of Chemistry University of Houston Houston, Texas
Patrick Y.S. Lam	Experimental Station The DuPont Merck Pharmaceutical Company Wilmington, Delaware
Bruce E. Maryanoff	Drug Discovery The R.W. Johnson Pharmaceutical Research Institute Spring House, Pennsylvania
John M. Ostresh	Torrey Pines Institute for Molecular Studies San Diego, California

Mark S. Plummer	Department of Chemistry Parke-Davis Pharmaceutical Research Division of Warner-Lambert Company Ann Arbor, Michigan
Yimin Qian	Departments of Chemistry and Pharmacology University of Pittsburgh Pittsburgh, Pennsylvania
Robert C. Reid	Centre for Drug Design and Development University of Queensland Brisbane, Queensland, Australia
James D. Rodgers	Experimental Station The DuPont Merck Pharmaceutical Company Wilmington, Delaware
Tomi K. Sawyer	Department of Chemistry Parke-Davis Pharmaceutical Research Division of Warner-Lambert Company Ann Arbor, Michigan
Saïd M. Sebti	H. Lee Moffitt Cancer Center and Department of Biochemistry and Molecular Biology University of South Florida Tampa, Florida
Charles J. Stankovic	Department of Chemistry Parke-Davis Pharmaceutical Research Division of Warner-Lambert Company Ann Arbor, Michigan
Philip Evan Thompson	Centre for Bioprocess Technology Monash University Clayton, Victoria, Australia

PREFACE

Peptidomimetics are compounds which mimic the biological activity of peptides while offering the advantages of increased bioavailability, biostability, bioefficiency, and bioselectivity against the natural biological target of the parent peptide. Examples of peptidomimetics have been isolated as natural products, synthesized as libraries from novel subunits, and designed on the basis of X-ray crystallographic studies and through an intricate knowledge of the biological mode of action of natural peptides. They offer challenging synthetic targets and are increasingly important medicinal agents and biological probes. As a consequence, peptidomimetics embrace much of what is modern medicinal and organic chemistry. This volume highlights some recent and exciting developments in the area.

Andrew Abell
Editor

THE ROLE OF COMPUTER-AIDED AND STRUCTURE-BASED DESIGN TECHNIQUES IN THE DISCOVERY AND OPTIMIZATION OF CYCLIC UREA INHIBITORS OF HIV PROTEASE

Charles J. Eyermann, P. K. Jadhav, C. Nicholas Hodge,
Chong-Hwan Chang, James D. Rodgers,
and Patrick Y. S. Lam

Advances in Amino Acid Mimetics and Peptidomimetics
Volume 1, pages 1-40

ISBN: 0-7623-0200-3

ABSTRACT

A series of cyclic urea HIV protease inhibitors have been discovered using computer-aided *de novo* design and structure-based design methods. Extensive medicinal chemistry and additional structure-based design on this series has led to the clinical candidates DMP323 and DMP450. DMP450 shows good pharmacokinetics in man, but does not achieve the plasma levels necessary to be effective against known HIV mutants. Recent work on this series has focused on identifying HIV protease inhibitors with improved antiviral efficacy against wild-type and mutant strains of HIV. This work has primarily focused on nonsymmetrical compounds, heterocyclic substituents, and incorporating hydrogen-bonding interactions with the protease backbone for an improved resistance profile. Additional computer-aided and structure-based design studies have explored the interaction of these inhibitors with HIV protease and whether desolvation energies can aid in understanding plasma levels after oral dosing.

1. INTRODUCTION

Increasingly, the search for new drugs is driven by an intimate knowledge of the molecular interactions involved in a pharmacologically important mechanism of action. The case where the mechanism of action involves a proteolytic process is particularly exciting because there is a good probability that a medicinal chemist, even without a detailed structural model of the enzyme, can quickly design and synthesize very potent inhibitors by incorporating a transition-state mimetic into a substrate-like inhibitor.

However, substrate-derived protease inhibitors usually have poor *in vivo* pharmacokinetics, most notably poor oral bioavailability. This is often attributed to their relatively high molecular mass (> 600 daltons), poor solubility, and pseudopeptide character. Attempts to improve oral bioavailability have traditionally focused on keeping the core transition state mimetic, replacing the amide bonds, and analoguing the inhibitor substituents until the desired combination of potency, specificity, and solubility is obtained. While this process is very time consuming and was of limited effectiveness in obtaining orally bioavailable renin inhibitors,[1] it can be successful as demonstrated by recent FDA approval of three orally bioavailable substrate-derived HIV protease (HIVPR) inhibitors.[2-4] There are many reviews[5] and additional papers[6,7] on how substrate-derived inhibitors have been designed and developed into useful AIDS therapies.

In this chapter, we present an example of an alternative approach to analoguing around a substrate-based inhibitor, namely the use of structure–activity relationships (SAR) on substrate-based inhibitors and computer-aided *de novo* design techniques to identify a novel non-peptide HIVPR inhibitor scaffold. The role of structure-based design techniques in identifying inhibitors for clinical study as well as identifying potent inhibitors against mutant forms of HIVPR is also summarized.

An additional goal of this chapter is to illustrate how simple initial potency and pharmacokinetic criteria for selecting a compound for clinical study has evolved into a complex set of criteria which requires an inhibitor to occupy a very small region of physiochemical property space. Efforts to satisfy these current criteria are reviewed.

2. HIV PROTEASE

HIV protease (HIVPR) is an aspartyl protease critical to processing HIV's gag and gag-pol gene products.[8] Since inhibition of HIVPR prevents replication of HIV, the causative agent of AIDS, identifying potent HIVPR inhibitors emerged as an important strategy for developing therapeutically useful treatments for AIDS.[9]

Many laboratories had already developed potent inhibitors for aspartyl proteases in their search for renin inhibitors.[1c] While renin inhibitors turned out to be weak to poor HIVPR inhibitors, many of the lessons learned from renin inhibitor research did provide useful insights into how to quickly design potent HIVPR inhibitors. For example, a number of different transition-state mimetics (Figure 1) were available for aspartyl proteases and almost all of the early attempts to design HIVPR inhibitors incorporated one of these mimetics into peptides with residues similar to HIVPR substrates.

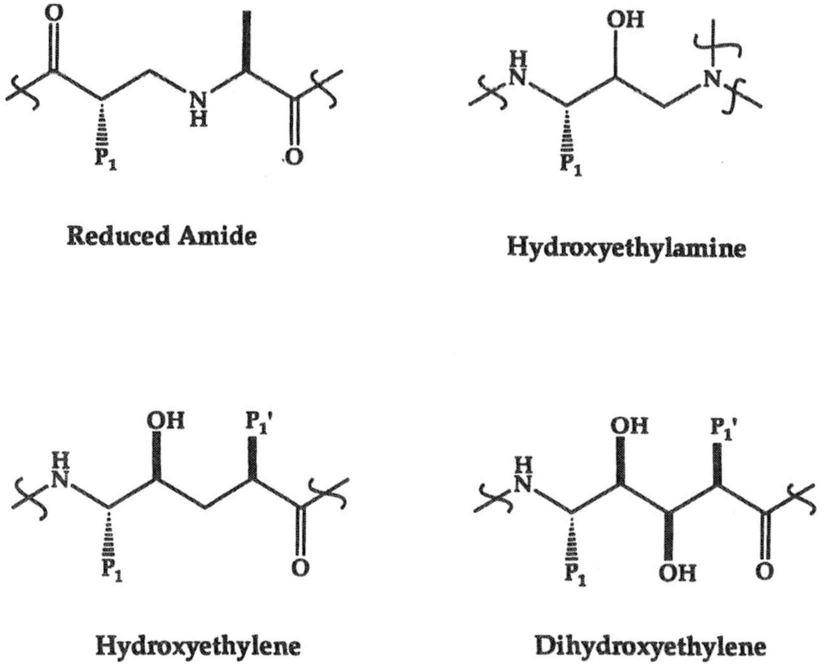

Reduced Amide

Hydroxyethylamine

Hydroxyethylene

Dihydroxyethylene

Figure 1. Examples of transition-state mimetics developed for use in aspartyl protease inhibitors.

3. SUBSTRATE-ANALOGUE HIVPR INHIBITORS

3.1. Acyclic C_2-Symmetric Diol HIVPR Inhibitors

In late 1988 Wlodawer postulated that like the *rous sarcoma* virus, HIVPR may also have a C_2 axis of symmetry. Based on this information, it was thought that an inhibitor which is C_2 symmetric and uses a diol as the transition-state mimetic would be more complimentary to a C_2 symmetric enzyme than an inhibitor which was asymmetric and used a mono-ol transition-state mimetic. Indeed C_2 symmetric diols were found to be micromolar inhibitors of HIVPR.[10] Our initial lead compound,[1] P9695, was found to be active against HIVPR with IC50 = 500 nM.

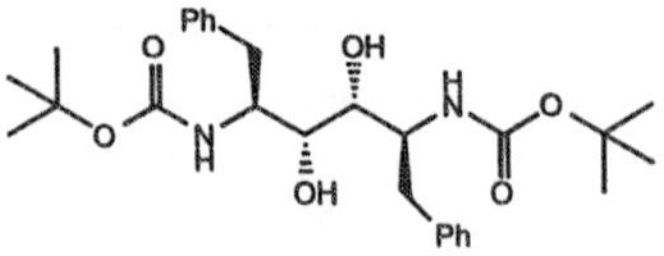

P9695

While this work was ongoing, the first X-ray crystal structure of HIVPR–inhibitor complex revealed that HIVPR is indeed C_2 symmetric and that the active site has six subsites all of which are occupied by an inhibitor binding in an extended conformation.[12] It became apparent from this first HIVPR–inhibitor crystal structure that P9695 was capable of occupying only four subsites. Based on this analysis, P9941 was designed and synthesized as our first low nanomolar HIVPR inhibitor. P9941 contains isoleucine at P2/P2' and 2-pyridylacetate as the capping groups. The choice of isoleucine was based on the observation that HIVPR's natural substrate has relatively small groups at the P2/P2' positions. 2-Pyridylacetic acid was used as a water-solubilizing group at the P3/P3' position.

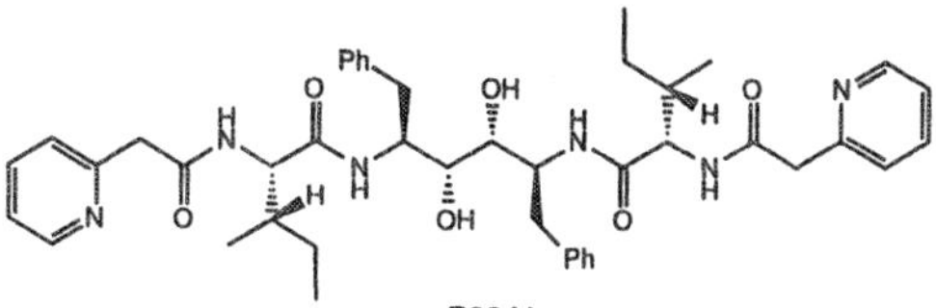

P9941

Many analogues of P9941 were synthesized to generate a SAR (Table 1). The benzyl group at P1 and P1' was found to give optimum potency. The diols with SRRS absolute stereochemistry at the diol and P1/P1' chiral centers were more active than diols with RSSR, SRSS, or SSSS stereochemistry. Small P2 or P2' residues were found to be optimum at the P2/P2' positions. The hydrogen bond donor and acceptor functionality of the amide or carbamate was very important for interactions with Asp29/29', Asp30/Asp30', and Gly48/48'. Groups of diverse steric and electronic character could be accommodated at P3/P3'. Q8024 was one of our most potent C_2 symmetric diols[13] both in an *in vitro* enzyme inhibition assay ($K_i = 0.24$ nM) and anti-infectivity assay (50 nM).

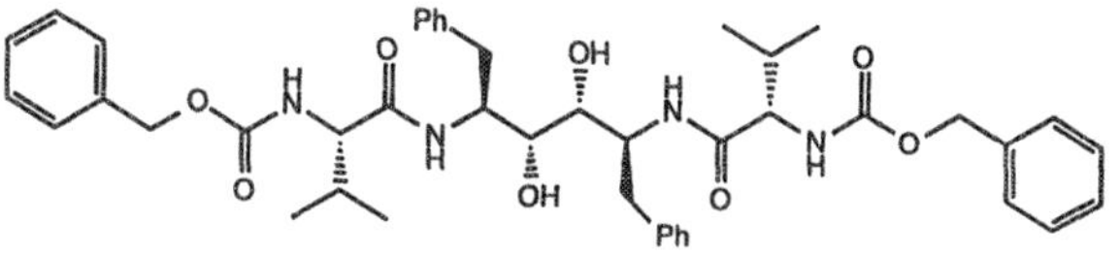

Q8024

Unfortunately, most of our C_2 symmetric diols are highly insoluble in aqueous formulations as well as common organic solvents. For example, the oral bioavailability of P9941 in rats is <10F%.

After the synthesis of Q8024, we learned that Abbott, Smith Kline Beecham, Hoechst and other laboratories had also made significant discoveries using the C_2-symmetric diol approach. Abbott has published very elegant work in this area.[14]

Given the undesirable physical properties and molecular mass (>700 daltons) of our diol series, the limited synthetic resources we had available for analoguing P9941 and Q8024, and the intense competitive pressures from other research labo-

Table 1. HIVPR Inhibition Activity of C_2-Symmetric Diols

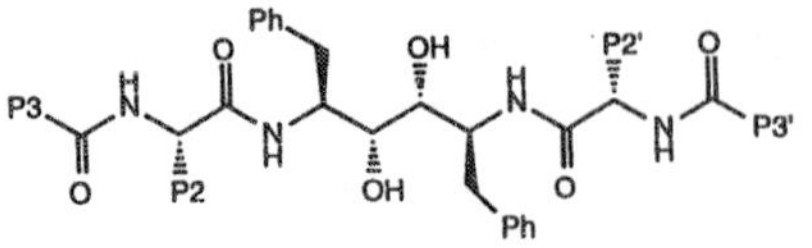

P2/P2'	*P3/P3'*	*K_i (nM)*[a]
2-butyl	2-pyridylmethyl	6.6
2-methyl-1-propyl	2-pyridylmethyl	220
2-butyl	3-pyridylmethyl	9
2-butyl	4-pyridylmethyl	13.5
2-butyl	phenylmethyl	4
2-butyl	cyclohexylmethyl	>200
2-butyl	*t*-butylmethyl	486
2-butyl	2-quinolinyl	26
2-butyl	4-morpholinyl	88
2-butyl	(2,4-dihydroxypyrimidine) methyl	1.5
2-propyl	benzyloxy	0.24
methyl	benzyloxy	31
2-butyl	benzyloxy	1
$NH_2COCH_2CH_2$-	benzyloxy	257
NH_2COCH_2-	benzyloxy	38
2-propyl	methyl	4.6
2-propyl	hydrogen	6.4
1-hydroxy-1-ethyl	benzyloxy	54

Note: [a] The K_i values were generated using spectrophotometry assay.

ratories, we felt it would be very difficult to bring a pharmaceutically useful compound from our C_2 symmetric diol series to market. We began to consider alternative approaches to identifying HIVPR inhibitors.

3.2. First Structures of HIVPR–Inhibitor Complexes

Wlodawer's first high resolution X-ray crystal structures of HIVPR–inhibitor complexes were of substrate-like inhibitors synthesized in the Rich (JG365)[15] and Marshall (MVT-101)[12] laboratories. These inhibitors (Figures 2a and 2b) contain a reduced scissile bond and hydrophobic sidechains at P1 and P1'. They bind to C_2-symmetric HIVPR (Figure 2c) in an extended conformation with an extensive network of hydrogen-bonding interactions with the protease's backbone amides (Figure 2d).

In addition to defining enzyme-inhibitor hydrogen-bonding and subsite interactions, the X-ray structures revealed that HIVPR uses a tetra-coordinated structural

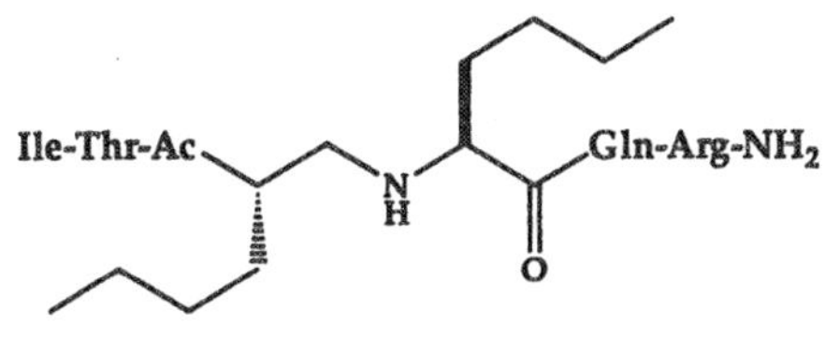

Ile-Thr-Ac
N
H
Gln-Arg-NH2
O

MVT-101

Figure 2a. HIVPR inhibitor MVT-101 was developed in Marshall's laboratory and contains a reduced amide transition-state mimetic.

OH
Ac-Ser-Leu-Asn-NH
N
Ph
O
Ile-Val-OCH3

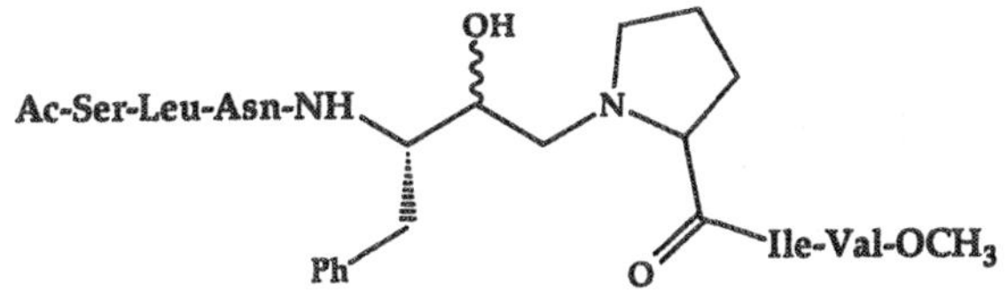

JG365

Figure 2b. HIVPR inhibitor JG365 was developed in the Rich laboratory and contains a hydroxyethylamine transition-state mimetic.

"Flaps"
Structural
Water
D25
D25'
C2 Axis of Symmetry

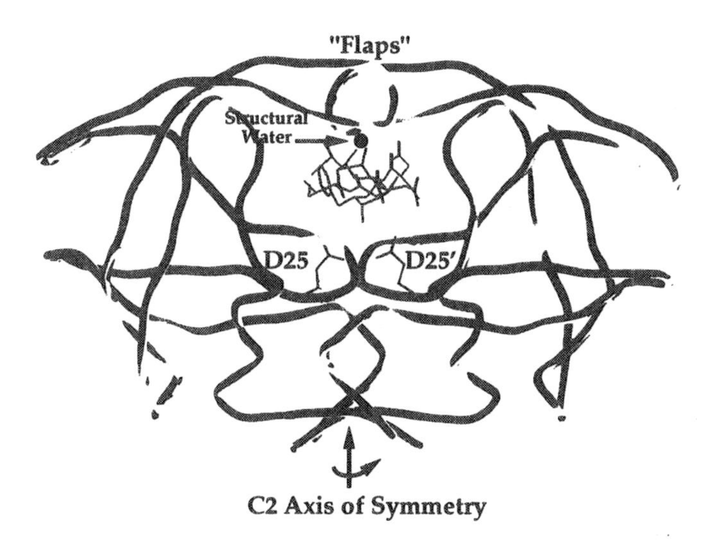

Figure 2c. X-ray structure of HIVPR–JG365 illustrating C_2 axis of enzyme symmetry, the location of the structural water molecule and the catalytic aspartates.

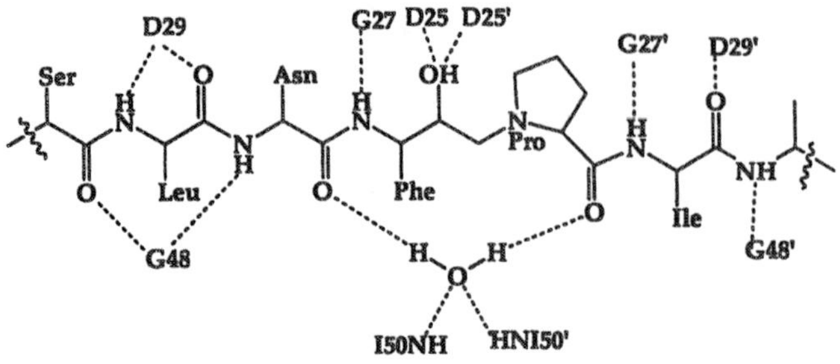

Figure 2d. Hydrogen-bonding observed in the crystal structue of JG365 complexed with HIVPR.

water molecule to link the bound inhibitor to the flexible and glycine-rich ß-strands or "flaps" of the HIVPR dimer. This structural water molecule accepts two hydrogen bonds from backbone amide hydrogens of symmetry-related residues Ile50 and Ile50', and donates two hydrogen bonds to the carbonyl oxygens flanking the transition-state mimetic of the inhibitor molecule. The presence of the structural water molecule in this HIV–PR inhibitor complex is a unique feature of the retrovirus protease.

The X-ray structure of the HIVPR–JG365 complex was particularly interesting because it defined the interactions of the hydroxyethylene (mono-ol) transition-state mimetic with the catalytic Asp25 and Asp25' residues.

4. DISCOVERY OF CYCLIC UREA Inhibitors

4.1. Design Strategy

Wlodawer's X-ray crystal structures of HIVPR–inhibitor complexes made it possible to initiate an effort to use *de novo* design tools to identify a new class of inhibitors. Our goal was to design a novel series of *non-peptide* inhibitors having the enzyme inhibitory and antiviral potency, selectivity, and *oral bioavailability* required to be therapeutically useful. To achieve this goal we started out with two very simple design criteria; first, the inhibitor should contain no, or very few, amide bonds, and second, it should have a molecular mass below 600 Daltons. If such an inhibitor could achieve nanomolar enzyme inhibition and micromolar antiviral activity we believed we would be well on our way to a clinical candidate.

4.2. *De Novo* Design Tools

In the last half of the 1980s it was becoming clear that one of the limiting steps in applying structure-based design was generating novel scaffolds which could properly position functional groups for interaction with key receptor sites. Only a few researchers were successful in designing scaffolds using only molecular graphics and chemical intuition.[16]

The desire to overcome this difficulty in identifying synthetically useful scaffolds inspired several laboratories to develop computer methods for *de novo* design[17]. Two of the primary *de novo* design tools which emerged were shape-based docking[18] of 3D database molecules into a receptor active site and pharmacophore searching of 3D databases.[19]

Pharmacophore searching of 3D databases involves searching a database of three-dimensional molecular structures for those molecules which satisfy some combination of distances, angles, dihedrals, and other geometric constraints between a set of functional groups necessary for a desired biological activity. The hits (molecules) from these 3D database searches can then be analyzed for scaffolds which are synthetically attractive.

Our approach was to use pharmacophore searching instead of shape-based searching of 3D databases to identify scaffolds, primarily because with shape-based methods interesting scaffolds could be missed due to bad van der Waals contacts with the receptor. Our first successful application of 3D pharmacophore searching had been in designing PLA2 inhibitors.[20]

4.3. Pharmacophore Models

Docking of Acyclic Diol to HIVPR

As noted above, using pharmacophore searching of 3D databases as a *de novo* inhibitor design tool requires a 3D pharmacophore model; that is, the 3D arrangement of the functional groups required for biological activity. While a good understanding of the functional groups required for potent HIVPR inhibitors was available from the SAR of our C_2 symmetric diols and other known inhibitors, an X-ray structure of a C_2 symmetric diol bound to HIVPR was unavailable.

An understanding at the molecular level was sought for why C_2 symmetric diol inhibitors have a 10-fold greater potency over the potency observed for comparable mono-ols. Is the potency due to the diol transition-state mimetic or is it due to better hydrogen bonding and van der Waals interactions between the diol's substituents and HIVPR? Having access to Wlodawer's crystal structures of the linear substrate-based inhibitors gave some insights into how the diol's substituents might interact with HIVPR, but a very important question remained. Does one or both of the diol's hydroxy groups interact with both catalytic aspartates; i.e. is binding symmetrical or asymmetrical?

Distance geometry is a powerful method for building complex 3D models, including models where a number of contraints are to be applied.[21] Using distance geometry, P9941, which was one of our key linear C_2 symmetric diol inhibitors, was docked into the HIVPR active site to test whether the diol transition-state mimetic could achieve a symmetrical binding mode. Each distance geometry model included the following constraints: (1) each diol oxygen must be within hydrogen-bonding distance of the catalytic D25 or D25' residues; (2) hydrogen bonding between inhibitor amide bonds and protease amide bonds must be similar to the MVT-101–HIVPR complex; (3) two atoms from the side chains of P9941 and MVT-101 were forced to overlap (the overlapping atoms were usually CB and CG);(4) the amide carbonyl oxygen between P1 and P2, and P1' and P2' were forced to be in hydrogen-bonding distance of the tetra-coordinated structural water molecule; (5) the chiralities of residues P1, P1', P2, and P2' were maintained; and (6) no bad van der Waals contacts were allowed between P9941 and HIVPR.

One hundred dockings generated 14 unique conformations of P9941 (Figure 3a). The hydrogen-bonding constraints between the amide backbone of the peptide and protease generates a very specific backbone conformation for P9941. However, there is a reasonable degree of rotational freedom for the inhibitor sidechains, especially for P3 and P3'. A low-energy minimized structure of P9941 docked in the active site of HIVPR is shown in Figure 3b. Note the hydrogen-bonding interactions for the structural water molecule and the dol group of P9941. While no con-

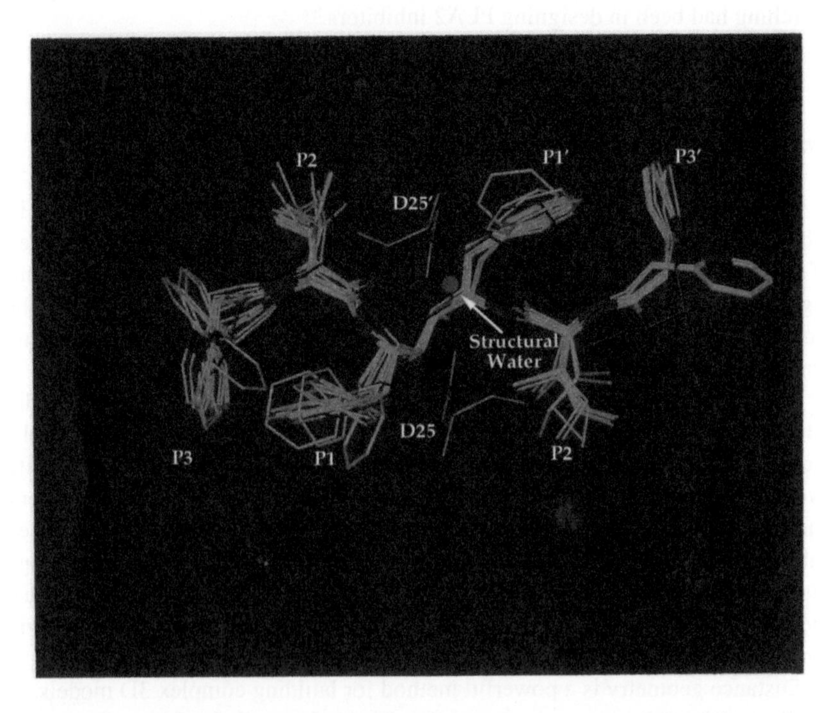

Figure 3a. Distance geometry models of the C_2-symmetric diol inhibitor P9941 docked into HIVPR active site. Rich inhibitor is shown in magenta and the catalytic aspartates D25 and D25' are shown in cyan.

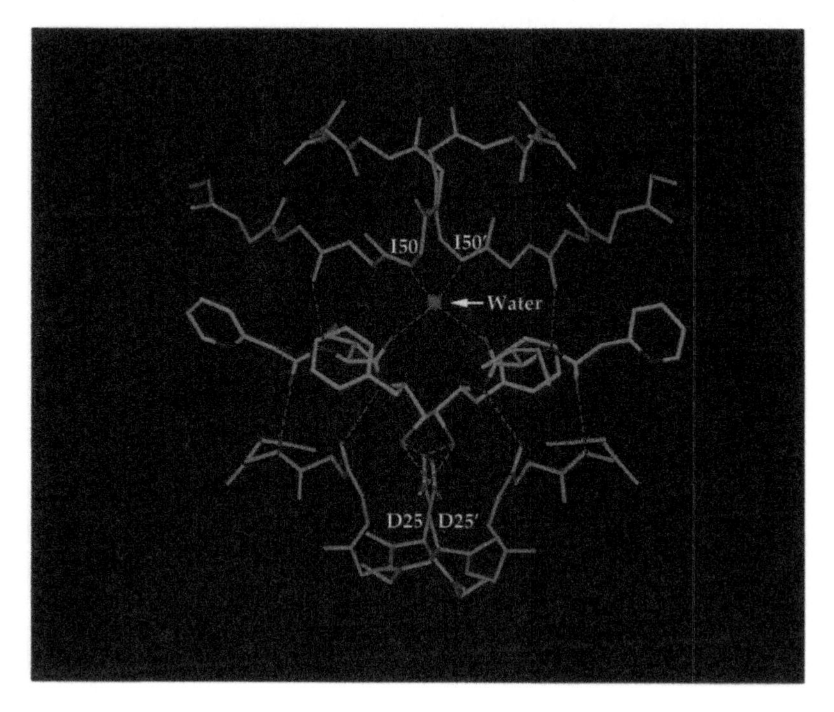

Figure 3b. Distance geometry model of P9941 docked to HIVPR. Hydrogen-bonding interactions between the structural water molecule, HIVPR and P9941 are shown as dashed magenta lines. The hydrogen bonds defining the extended binding conformation of P9941 are also similarly indicated.

formational analysis was performed on unbound P9941, the distance geometry-generated conformations contained no obvious flaws and it was assumed that the docked conformations were within 20 kcal/mole of a global energy minimum. Figure 3c compares the distance geometry-derived model of P9941 and a subsequently published X-ray structure of HIVPR complexed with a related diol inhibitor.[22] The distance geometry model for the diol, P1, and P1' region of P9941 are very similar to the crystal structure.

GRID Calculations on HIVPR Active Site

Another approach to developing 3D pharmacophore models is the GRID method developed by Goodford.[23] GRID calculations "probe" the active site using hydrophobic and hydrophilic functional groups to determine regions within the active site favorable for a given functional group. Energetically favorable regions are called "hot spots" and their type and position can directly generate a 3D pharmacophore model.

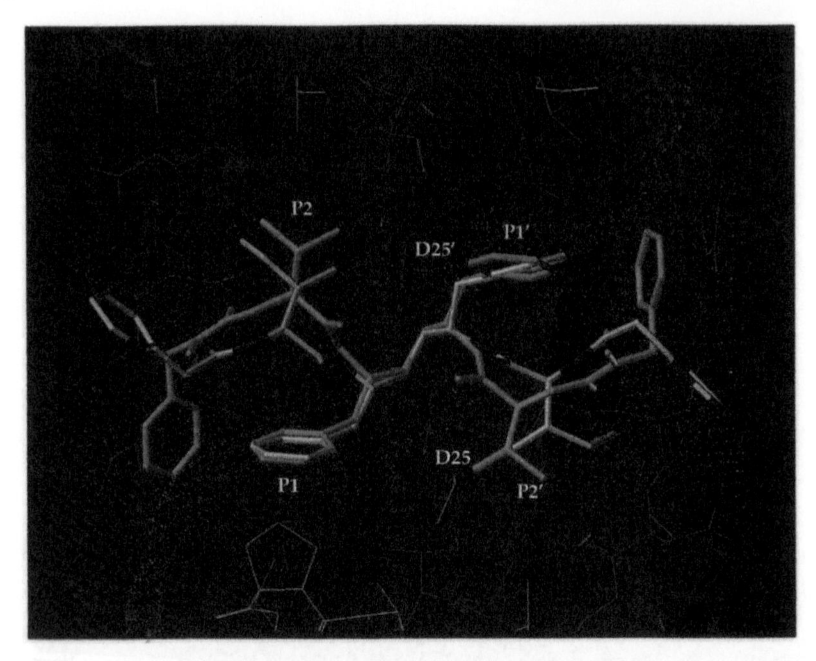

Figure 3c. Comparison of a distance geometry model of P9941 (colored by atom) with the X-ray structure (green) of an Abbott diol with *(R,R)* stereochemistry at the diols. The geometry of the P9941 model was optimized using molecular mechanics.

Interaction energies between methyl, amino, proton, and carboxy oxygen probes and the HIVPR active-site residues were calculated.[24] The proton probe's most favorable position is near D25 and D25', the carboxy probe's most favorable position represents the structural water molecule, and the most energetically favorable methyl contours represent the strong hydrophobic nature of the S1, S1', S2, and S2' sites. The water contours are the least energetically significant among the four probes used. Finally, significant proton contours were found near the S3 and S3' subsites where water molecules are observed in the HIVPR–MVT-101 crystal structure.

Pharmacophore Refinement and Searching

Several 2D pharmacophore models were proposed based on the SAR of our diol series. An integral component of each of the pharmacophore models was a hydrogen-bond donor/acceptor to interact with the catalytic D25 and D25' residues and groups which would interact with several different combinations of the S1, S2, S3, S1', S2', and S3' subsites.

In addition, the X-ray structures of the HIVPR–inhibitor complexes suggested that incorporation of the binding features of the structural water molecule into an inhibitor might be sufficient to provide high-affinity binding without the need for multiple interactions at the specificity subsites. Part of the improved binding affinity would result from the increased entropy of the displaced structural water molecule.[25] This approach might also permit synthesis of low molecular weight, yet highly potent inhibitors with significant oral bioavailability. Finally, incorporating the functional equivalents of the structural water molecule into an inhibitor might also ensure specificity for HIVPR over other cellular aspartic acid proteases because only in the retroviral aspartic acid proteases is this structural water molecule known to play a role in substrate and inhibitor binding.

Using the crystal structures of the protease with the JG635 and MVT-101, the model of the P9941–HIVPR complex, and the results of the probe interaction calculations, the 2D pharmacophores were refined into three complete 3D pharmacophore models.

Pharmacophore model A (Figure 4a) was used to search[26] for scaffolds which would connect a transition-state mimetic, a carbonyl for hydrogen bonding to the structural water, and a P1 group. Pharmacophore model B (Figure 4b) was used to search for scaffolds which would connect a transition-state mimetic, P1, and a P1' group. Pharmacophore model C (Figure 4c) was used to search for scaffolds which would connect a transition-state mimetic, P1, P2, and P3 groups.

Hits from 3D searches of the Cambridge Crystallographic Database[27] using pharmacophore models A and C yielded a number of scaffolds and a number of synthetic targets were designed. However, actual syntheses of these targets were never

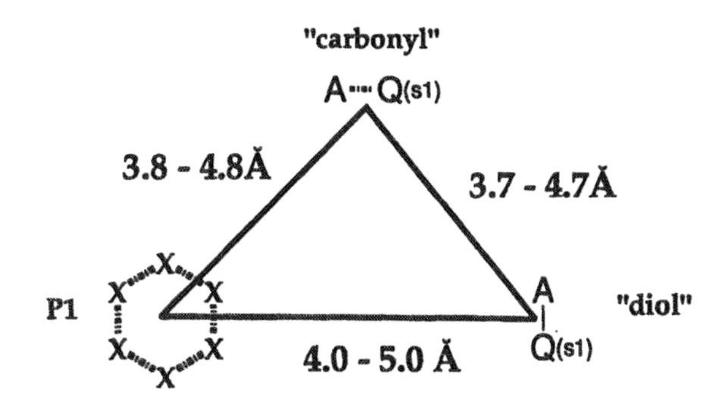

Figure 4a. Pharmacophore model A used in searching for scaffolds which connect the "diol" and "carbonyl" groups, and the P1 substituent. Distances shown are between the centroids of the pharmacophore groups. Dashed lines indicate any bond type is allowed. The P1 substituent can be an aromatic or non-aromatic six-membered ring containing any element. Q(s1) represents any atom other than carbon with one connection. Additional geometric constraints include a dihedral angle of 130°–170° for the "diol" and "ketone" vectors, and the angle between the plane of the P1 substituent and a plane containing the "diol" and ketone groups of 25°–30°.

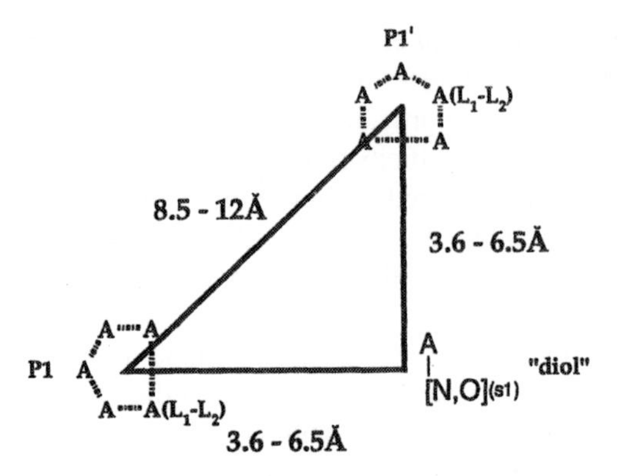

Figure 4b. Pharmacophore model B used in searching for scaffolds which connect the "diol" group and the P1 and P1' substituents. Distances shown are between the centroids of the pharmacophore groups. Dashed lines indicate any bond type is allowed. The A(L_1-L_2) indicates that the P1/P1' substituents can be either a five- or six-membered ring. The P1/P1' rings can be either aromatic or nonaromatic and contain any element. The "diol" group must have a single bond, not be in a ring, and have a singly connected nitrogen or oxygen atom. An additional geometric constraint is an angle of 150°–180° between the P1 centroid, "diol", and P1' centroid.

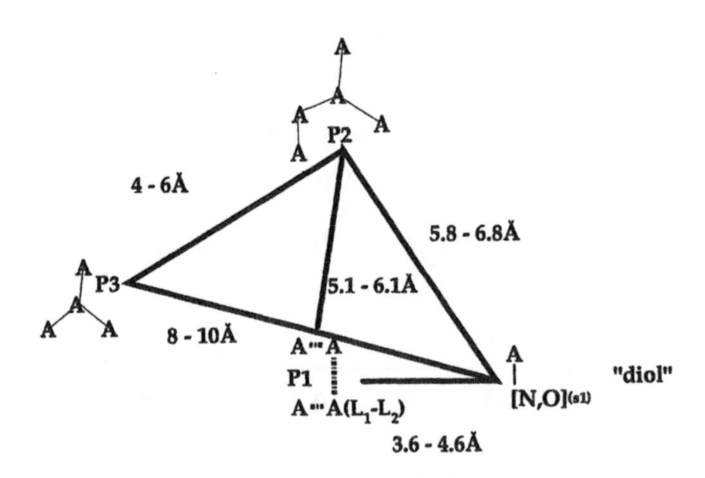

Figure 4c. Pharmacophore model C used in searching for scaffolds which connect the "diol", P1, P2, and P3 substituents. Distances shown are between the centroids of the pharmacophore groups. Dashed lines indicate any bond type is allowed. Special query atoms are A for any atom and (s1) indicates the atom must have a single connection. The "diol" group must have a single bond, not be in a ring, and have a singly connected nitrogen or oxygen atom. The bonds for the P2 substituent must also be in a chain.

initiated because of the excitement created by a hit from a 3D database search using pharmacophore model B.

4.4. Design of Cyclic Urea (CU) Scaffold

Pharmacophore model B was intended to find novel scaffolds which would connect a transition-state mimetic capable of interacting with the active site aspartic acid residues, and hydrophobic rings designated P1 and P1' that would occupy the symmetrically arranged S1 and S1' subsites of the enzyme. A 3D pharmacophore (Figure 5a) search of the Cambridge crystal structure database yielded the structure[28] shown in Figure 5b, which not only met the initial search criteria, but also included an oxygen which matched well with the structural water molecule found in HIVPR complexes (Figure 5c). Discovering that a phenyl ring could properly position a hydrogen bond donor/acceptor group capable of interacting with the catalytic aspartic acid residues *and* a functional equivalent of the structural water molecule led us to choose a six-membered ring as our initial non-peptide scaffold (Figure 5d). The saturated six-membered ring scaffold was chosen because it would be very difficult to properly orient substituents on a phenyl ring into all of the S1, S1', S2, and S2' subsites of HIVPR. The six-membered ring was subsequently modified to a seven-membered ring system (Figure 5e) in order to incorporate the diol functionality. The synthetic target was then modified to the cyclic urea (Figure 5f) based on two independent lines of reasoning. First, it was realized that the cyclic urea was synthetically accessible by carbonyldiimidazole cyclization of the diaminodiol precursor used in the linear C_2-symmetric diol series. And second, cyclic ureas have precedent as excellent hydrogen-bond acceptors both in nature (e.g. biotin/streptavidin interactions[29]) and in synthetic systems.[30]

Seven-membered ring cyclic ureas can exist in two pseudo chair conformations (Figure 6). When the nitrogens are unsubstituted, 1,3-diaxial strain[31] dominates and conformer **2** with pseudo diequatorial benzyl groups and pseudo diaxial hydroxyl groups is preferred. In contrast, when the two nitrogens are substituted with P2/P2' groups, the converse is true even for a substituent as small as a methyl group. The partial double bond character of the urea C–N bond introduces severe allylic 1,2-strain[32] between the benzylic groups and the nitrogen substituents. This allylic 1,2-strain overcomes the 1,3-diaxial strain, and conformer **3** with pseudo diaxial benzyl groups and pseudo diequatorial hydroxyl groups is preferred. This conformational prediction by first principles was subsequently confirmed by small molecule X-ray crystallography.[33]

Based on the above conformational analysis, modeling of a cyclic urea *unsubstituted* on nitrogen derived from either natural (L) or unnatural (D) phenylalanine revealed that both cyclic urea enantiomers can be docked equally well into the active site. More importantly, and unexpectedly, a *nitrogen-substituted* cyclic urea derived from unnatural D-phenylalanine was found to be more complementary to the active site than one derived from natural L-phenylalanine (Figure 5g).

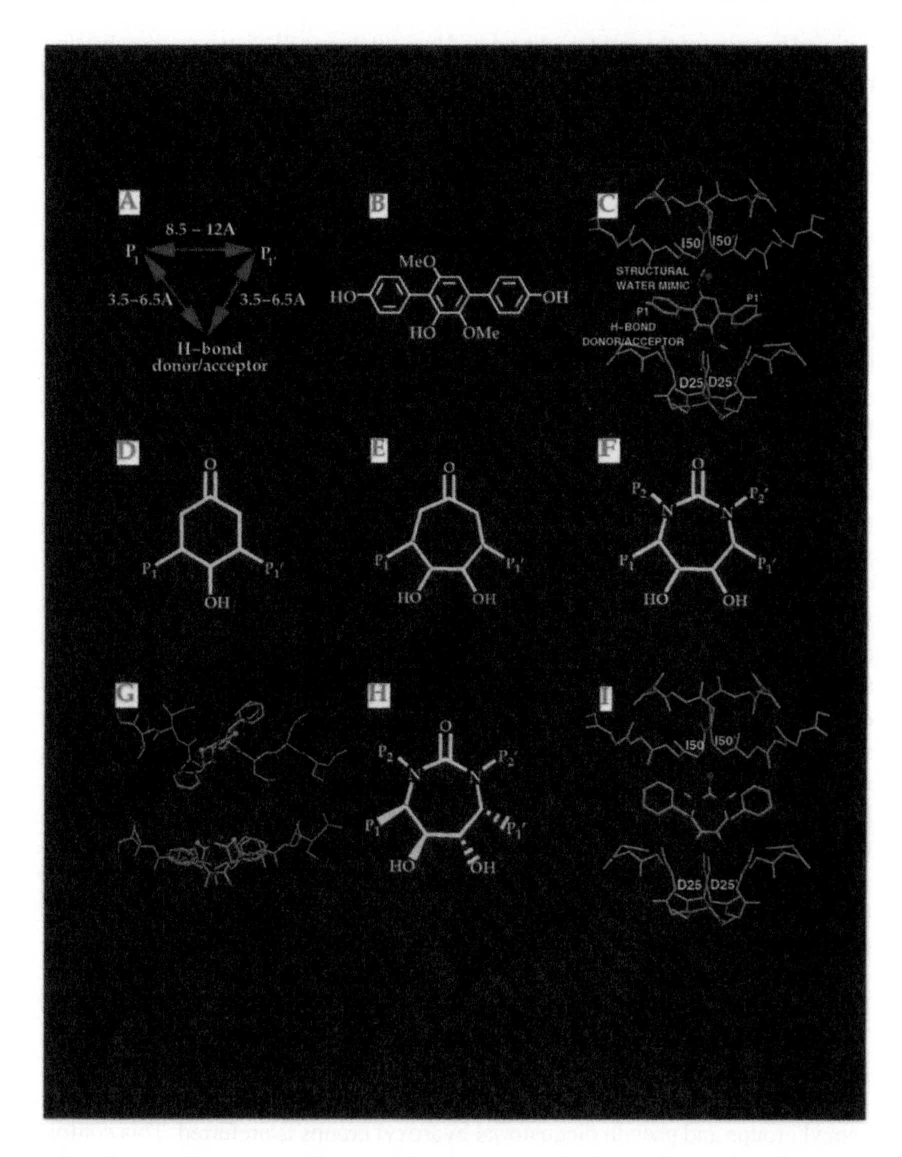

Figure 5. Steps in the design of the cyclic urea scaffold. In 5g the green colored structure represents the cyclic urea derived from D-phenylalanine, the red colored structure represents the cyclic urea derived from L-phenylalanine, and the blue structure is JG365 from the HIVPR crystal structure. Note that the "spheres" in 5G represent the P2/P2' attachment atoms and that the green attachment spheres for D-phenylalanine derived cyclic urea are better positioned to place substituents into the S2/S2' subsites than the red attachment spheres for L-phenylalanine derived cyclic ureas.

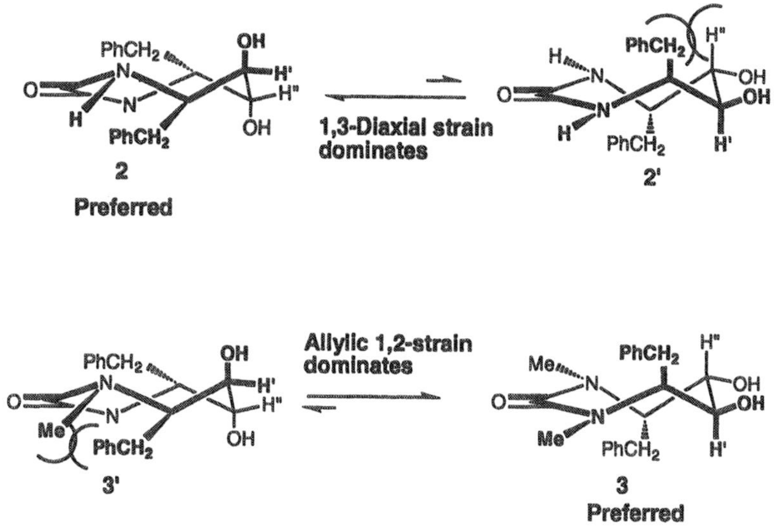

Figure 6. Conformational analysis of designed cyclic ureas predicting that **2** is preferred when the nitrogens are not substituted, whereas due to A_{12} strain, conformation **3** is preferred when the nitrogens are substituted.

Thus, for inhibiting HIVPR, the preferred stereochemistry of N-substituted seven-membered CUs was predicted to be 4*R*,5*S*,6*S*,7*R* as shown in Figure 5h. This sterochemical preference is opposite to the stereochemistry which we were using in our linear C_2-symmetric diols. Figure 5i shows our proposed model for the binding of the cyclic urea scaffold to HIVPR.

4.5. Testing of CU Scaffold

The CU synthesized from natural L- phenylalanine and with no substituents at nitrogen has a K_i against peptide substrate of 3.0 μM while the unsubstituted CU (**1** in Table 2) based on unnatural D-phenylalanine has a K_i value of 4.5 μM.[34,35] The CU enantiomer **2**, derived from L-phenylalanine with allyl groups on nitrogen, is only a 4.5uM inhibitor. In contrast, **3**, the CU with 4*R*,5*S*,6*S*,7*R* stereochemistry derived from D-phenylalanine and with allyl groups on nitrogen, is a 5.2 nM inhibitor of HIVPR! Compound **3**, XK216, was the first indication that our design was a success.

XK216's discovery mostly depends on the careful conformational analysis of substituted seven-membered CUs which led us to synthesize D-phenylalanine-based CUs. Other laboratories which cyclized L-phenylalanine based linear diol inhibitors found only micromolar HIVPR inhibitors.

Synthesis and testing of other active CUs with simple hydrocarbon P2/P2' substituents quickly followed the discovery of XK216. Cyclopropylmethyl CU **4** was

Table 2. P2/P2' SAR of Symmetric Cyclic Urea HIVPR Inhibitors

Compound	*P2/P2'*	*Ki* [a] (μM)	IC_{90} [b] (μM)
1	H	4,500	>100
2	allyl (enantiomer of **3**)	4,500	>100
3	**allyl (XK216)**	**5.2**	**4.7**
4	**cyclopropylmethyl**	**2.1**	**1.8**
5	benzyl	3.0	0.83
6	α-naphthylmethyl	86	16
7	β-naphthylmethyl	0.31	3.9
8	***p*-hydroxymethylbenzyl (DMP323)**	**0.34**	**0.057**
9	m-hydroxymethylbenzyl	0.14	0.038
10	p-hydroxybenzyl	0.12	0.032
11	m-hydroxybenzyl	0.12	0.054
12	***m*-aminobenzyl$\cdot 2CH_3SO_3H$ (DMP450)**	**0.28**	**0.13**
13	3-(1-hydroxy-1-methylethyl)benzyl	0.067	0.034
14	*p*-acetoxybenzyl	0.90	0.73
15	*m*-acetylaminobenzyl	0.42	0.14
16	*m*-methylaminobenzyl	0.28	0.034
17	*m*-*N,N*-dimethylaminobenzyl	0.06	0.038
18	*N*-gylcinamidobenzyl	0.04	0.16
19	*m,p*-dihydroxybenzyl	0.038	0.74
20	*m*-carboxybenzyl	0.43	34
21	*m*-carbomethoxybenzyl	1.3	0.30
22	*m*-carboxamidobenzyl	0.060	0.71
23	*m*-methylcarboxamidobenzyl	0.060	0.081
24	*m*-acetylbenzyl	0.07	0.04
25	*m*-trifuoroacetylbenzyl	0.037	0.040
26	***m*-aldoximebenzyl**	**0.01**	**0.005**
27	***m*-acetoximebenzyl**	0.01	**0.002**

Notes: [a] K_i values were measured as described.[35]
[b] Inhibition of viral replication was quantified in HIV-1 (RF)-infected MT2 cells by measurement of viral RNA with an oligonucleotide-based sandwich hybridization assay; *n*=1 except as indicated.

found to have one of the two best K_is among the cycloalkylmethyl CUs, and benzyl CU 5 showed a K_i of 3.0 nM.[36] According to X-ray structures[12,15], the S2/S2' subsites of HIVPR are very large. An α-naphthylmethyl group was introduced at

P2/P2' as in **6** and found to be a poor binder. On the other hand, β-naphthylmethyl CU **7** was found to be a subnanomolar inhibitor with a K_i of 0.3 nM.

Cyclic ureas XK216 and **4** are very low molecular weight compounds which achieve nanomolar HIVPR inhibition. We attribute a significant portion of this potency to preorganization[30a,37]; namely, the unbound solution conformation of these (and almost all other CUs) is very similar to the bound conformation. The energy required to adopt the bound conformation is therefore "prepaid" in synthesis. While it is impossible to truly know how much preorganization contributes to the binding energy of cyclic ureas, a *seco* compound which has benzyl groups at P2/P2' is only a 6700 nM inhibitor of HIVPR compared to CU **5**. This corresponds to a gain of almost 5 kcal/mol in binding affinity for preorganizing the diols and substituents to be complementary to the active site!

In addition to demonstrating that the CU scaffold resulted in potent HIVPR inhibition, it was necessary to demonstrate good antiviral activity in terms of IC_{90}, the concentration of inhibitor resulting in 90% inhibition of viral RNA production in HIV-1 infected MT-2 cells.[38] XK216's IC_{90} is 4.7 μM and XK234's IC_{90} is 1.8 μM. However, the IC_{90} for **7** is rather poor relative to its K_i, probably due to the extremely high lipophilicity of the molecule (clogP[39] 9.2).

Gratifyingly, in contrast to the enzyme potency shown by these CUs toward HIVPR, no inhibition of the cellular aspartic acid proteases—renin, pepsin, and cathepsin D—was observed at concentrations at least 3000 times higher than the K_i value for HIVPR. This specificity for the retroviral HIVPR is a consequence, at least in part, of the design of the CUs, since only in the retroviral aspartic proteases has the structural water molecule been shown to be involved in enzyme–substrate interactions.

In addition to having good potency and selectivity, XK216 was found to be 49% orally bioavailable (F%) in rats.[40] Presumably XK216's excellent rat bioavailability is partly due to its low molecular mass of 407 amu. The oral bioavailability of cyclopropylmethyl CU **4** in rat is 100% with a C_{max} of 4.3 μM at 10 mg/kg dose. In dog, the oral bioavailability of **4** is 48% with a C_{max} of 9.2 μM at a similar dose.

With these results a new scaffold with an exciting profile had clearly been identified and efforts to convert this discovery into a clinical candidate were quickly accelerated.

5. OPTIMIZATION OF CYCLIC UREAS

Examination of the potency, selectivity, and oral bioavailability results for the early CUs indicated that IC_{90}, the antiviral potency, was the critical property needing optimization. For example, CU **4** with cyclopropylmethyl P2/P2' groups has excellent oral bioavailability but its IC_{90} is only 1.8 μM. A 10-fold improvement in IC_{90} while maintaining the excellent oral bioavailability was desired.

However, the early results for hydrophobic P2/P2' substituents also hinted that optimizing antiviral potency was not equivalent to optimizing enzyme potency. In fact, most inhibitors were 1000 times weaker in the RNA IC_{90} assay than in the K_i

assay. We quantify this loss in potency using the ratio of the K_i (nanomolar) to the RNA IC_{90} (μM) and call the ratio the inhibitor's *translation* value:

$$\text{Translation} = K_i/IC_{90}$$

While the observed translation values were not completely surprising since antiviral potency is dependent on an inhibitor's ability to penetrate cell membranes, the size of the reduction in potency was daunting. Nevertheless, it was expected that the CU's physiochemical properties could be modified to improve cell membrane permeability and consequently significantly improve antiviral potency.

The initial strategy was to reduce the lipophilicity of the CUs by using more polar P2/P2' substituents. Examination of the HIVPR active site reveals that there are a few hydrogen bond donors/acceptors, namely the side chains and backbone amides of Asp29/29' and Asp30/30', close to the *meta* and *para* positions of an *N*-benzyl--substituted CU. Hydroxy and hydroxymethyl groups were incorporated as in **8–11**.[36] These compounds have K_i values in the subnanomolar range. Moreover, because of its favorable lipophilicity, the translation from K_i to IC_{90} is greatly improved. For example, CU **8**, DMP323, (clog P and HPLC log P are 4.8 and 3.6, respectively[41]) translates two orders of magnitude better than other subnanomolar inhibitors like **7** (clogP 9.2). The IC_{90}'s of these CUs, **8–11**, are in the range of 0.032–0.057 μM.

Table 3 lists some of the different P2/P2' substitutents which were made.[36,42] In addition to hydroxy groups, amino groups were also introduced as in **12**, resulting

Table 3. P1/P1' SAR of Symmetric Cyclic Urea Inhibitors of HIV

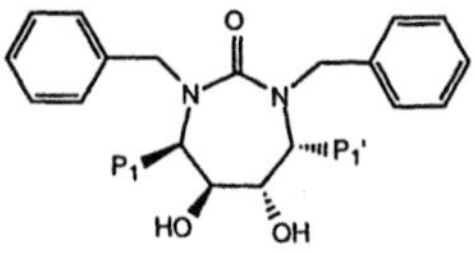

Compound	*P1/P1'*	*Ki*[a] (μM)	IC_{90}[b] (μM)
5	**benzyl**	**3.0**	**0.83**
28	methyl	5,000	>10,000
29	isobutyl	1,700	>10,000
30	2-(methylthio)ethyl	1,100	>10,000
31	cyclohexylmethyl	28	>10,000
32	phenethyl	320	>10,000
33	2-naphthylmethyl	9.8	3,300
34	3-aminobenzyl	2.8	1,300
35	4-pyridylmethyl	22	5,700
36	3,4-(methylenedioxy)benzyl	1.3	500

Note: [a-b] See Table 2.

in a potent compound (IC_{90} 130 nM) with weakly basic anilino P2/P2' substituents. Many of the functional groups with multiple hydrogen donors and acceptors as in **13–27** are an order of magnitude better binder than DMP323. However, many of them are too polar and the translation to antiviral assay is poor. Oximes **26** and **27** are exceptions with antiviral potency down to 5 nM. Unfortunately, the oral bioavailability in rats for many of these compounds is worse than DMP323.

Table 3 shows a partial SAR of P1/P1' modifications.[43] Aliphatic substitutions as in **28–33** give poorer K_i compared with DMP323. Modelling shows that CUs with benzyl P1/P1' interact very effectively with the S1/S1' subsites. Any change in atom hybridization at the β-carbon of P1/P1' from sp^2 to sp^3 increases the negative steric interactions. Aromatic rings of various sizes and polarity as in **34–36** are generally acceptable. Benzyl P1/P1' is one of the best P1/P1' in terms of affinity and translation (Table 3).

6. CLINICAL STUDIES

6.1. DMP323

As a class, the cyclic ureas are potent antivirals and are significantly bioavailable. Table 4 summarizes the pharmacokinetic parameters in various species for a number of substituted benzylic CUs.[34,36,40]

During the time period the first CU was being selected for clinical studies it was still unclear whether a HIVPR inhibitor would be therapeutically useful for treating AIDS. Consequently, there was tremendous interest in getting a molecule from this class of inhibitors into the clinic as quickly as possible. Considering the overall pharmacokinetic and safety parameters in animals, DMP323 was selected shortly after its initial design and synthesis.

Unfortunately, poor solubility (6 ug/ml at pH 7), first pass metabolism,[44] and limited formulation possibilities resulted in poor performance in Phase I studies.

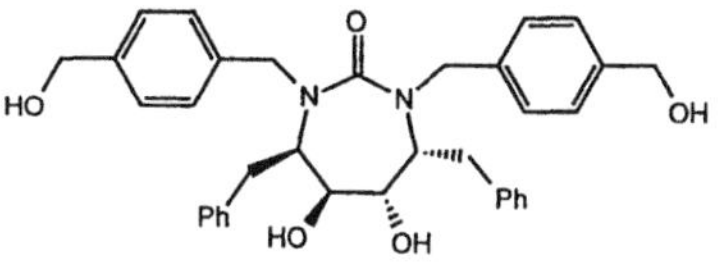

DMP323

6.2. DMP450

Based on the performance of DMP323 in the clinic, significant attention was given to the metabolic and solubility profiles of CUs being considered for our second clinical candidate. In addition to meeting metabolism and solubility requirements, two new requirements also became quite important in selecting a clinical candidate.

First, the propensity of CUs, and HIVPR inhibitors in general, to be highly bound to plasma protein[45] was discovered. In the clinical setting, it is the intracellu-

Table 4. Pharmacokinetics of Symmetric Cyclic Urea HIVPR Inhibitors

Compound	*P2/P2'*	*Rat P.O. Bioavail[a] C_{max}(µmM)*	*Rat P.O. Biovail[a] F%*
3	**allyl (XK216)**	**2.7**	**49**
4	**cyclopropylmethyl**	**4.3**	**100**
		9.2 (dog)	**48 (dog)**
5	benzyl	1.3	—
7	β-naphthylmethyl	0.38	—
8	***p*-hydroxymethylbenzyl**	**0.78**	**27**
	(DMP323)	**2.8 (dog)**	**37 (dog)**
9	*m*-hydroxymethylbenzyl	0.83	18
10	*p*-hydroxybenzyl	0.39	22
11	*m*-hydroxybenzyl	0.81	30
		2.0 (dog)	16 (dog)
12	***m*-aminobenzyl•$2CH_3SO_3H$**	**2.25**	**71**
	(DMP450)	**11.2 (dog)**	**79 (dog)**
13	3-(1-hydroxy-1-methylethyl)	0.63	—
	benzyl	0.24(dog)	<4%(dog)
16	*m*-methylaminobenzyl	3.8	—
		1.3(dog, @5mg/kg)	
17	*m-N,N*-dimethylaminobenzyl	0.07	
18	*N*-gylcinamidobenzyl	2.0	
23	*m*-methylcarboxamidobenzyl	0.12	—
24	*m*-acetylbenzyl	0.28	—
25	*m*-trifuoroacetylbenzyl	1.0	—
26	***m*-aldoximebenzyl**	**0.35**	—
27	***m*-acetoximebenzyl**	**0.08**	—
		0.51(gelucire)	

Note: [a] Bioavailability was determined in groups of rats, unless otherwise indicated (n = 4 per group), dosed with compound in formulations containing propylene glycol, polyethylene glycol 400, water at 10mg/kg. The maximun plasma concentration (C_{max}) is the observed peak plasma concentration after an oral dose. Oral bioavailability (F%) was determined by the ratio AUC PO/AUC/ IV, where AUC is the area under the plasma concentration-time curve from time zero to infinity and is normalized for dose.

lar free or "unbound" compound concentration that will determine the level of inhibition observed and the resultant level of efficacy. Indeed, excessive plasma protein binding can lead to clinical failure as recently observed with the HIVPR inhibitor Sc52151.[46]

Second, resistance to HIVPR inhibitors[47] which were undergoing clinical trials was also reported, so it became necessary to not only test a compound's potency against wild-type virus, but also against virus containing HIVPR with single, double, and multiple mutations.

After much work, DMP 450 was chosen as a second clinical candidate.[48] DMP 450, the bis-methanesulfonic acid salt of *m*-aminobenzyl-substituted CU, combines antiviral potency with superior physical properties. DMP450's IC_{90} is 130 nM.

DMP450

Initial concerns about the metabolic fate of the anilino P2/P2' groups were answered by an extensive set of genetic toxicity and metabolism studies. No unusual results were found in the genotoxicity studies and no evidence for the generation of undesirable reactive intermediates could be found.

DMP450's oral bioavailability in dogs is as high as 79F%, with a C_{max} of 11.2 μM at a dose of 10 mg/kg. DMP 450's binding to human plasma proteins as determined by equilibrium dialysis is in the range of 90–93%. As shown in Figure 7, DMP 450

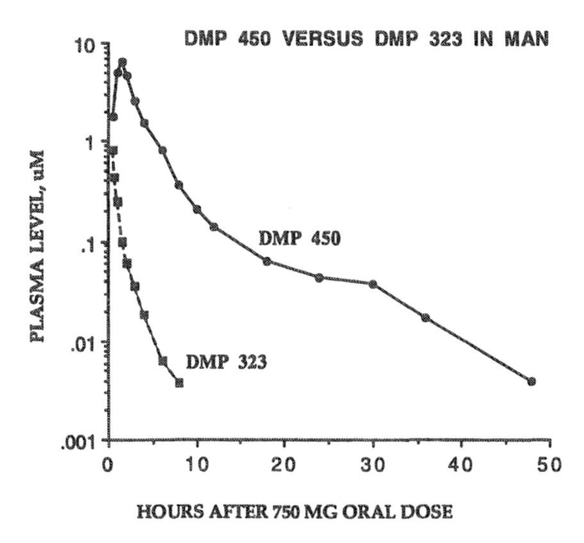

***Figure** 7.* Pharmacokinetics of DMP323 and DMP450 after oral dosing. HIV-1 sero-negative human subjects were given 750 mg of DMP323 in a liquid formulation or 750 mg of DMP450 as a neat powder in capsules. Blood samples were withdrawn at the indicated times, and quantities of compound were determined on extracts of plasma using HPLC.

Table 5. Pharmacokinetic Properties of Six HIVPR Inhibitors in Man

Inhibitor	*Dosage*	*Cmax, (μM)*	*IC_{90} (μM)*	*Trough Level, (μM)*	*Reference*
Indinavir	800 mg[a]	12.6	0.025-0.10[e]	0.25	3, 49, 50
Ritonavir	600 mg[b]	15.6	0.004-0.153[f]	4.2	4, 51, 52, 53
Saquinavir	600 mg[a]	0.079	0.009	<<0.079	54, 55
Viracept	750 mg[a]	Pending[d]	.008-0.120[g]	Pending[d]	7, 56
Vx-478	1200 mg[b,c]	18.2[c]	0.012-0.019[h]	2.06[c]	57, 58
DMP 450	750 mg single	6.49	0.144	0.3	48

Notes: [a] Administered three times daily.
[b] Administered twice daily.
[c] Single dose data. Multiple dose data have not been reported outside of oral presentations, but the long half-life reported for single doses suggests significant accumulation at trough occurs.
[d] Pharmacokinetic parameters for multiple doses have not been reported outside of oral presentations. The latter have indicated that micromolar levels are present at trough with multiple dosing.
[e] Range of IC_{95} for laboratory and clinical isolates.
[f] IC_{50} reported for a range of laboratory and clinical isolates.
[g] Range of IC_{90} for laboratory and clinical isolates.
[h] Range of IC_{50} for laboratory and clinical isolates.

yields plasma levels in man that are 6 times higher than DMP 323 at peak, and 535 times higher at 8 hours.

Although DMP 450 maintains significant antiviral activity against a variety of mutant viruses with single amino acid changes within the protease, the plasma levels at trough (Table 5) with multiple dosing regimens fall short of the concentration required to provide adequate coverage of *all* HIV variants. However, the use of DMP450 in combination with reverse transcriptase (RT) inhibitors may ultimately be of therapeutic benefit.

7. PROGRESS TOWARDS IMPROVED CU INHIBITORS

7.1. Challenges and Strategies

In recent clinical studies, several HIVPR inhibitors have been shown to reduce the viral load and increase the number of $CD4^+$ lymphocytes in HIV-infected patients. Saquinavir, Ritonovir, and Indinavir have also recently been approved by the FDA to be used as AIDS therapy in *combination* with RT inhibitors.[49-55] However, the powerful ability of the virus to rapidly generate resistant mutants[47] suggests that there will be an ongoing need for new HIVPR inhibitors with improved pharmacokinetic and efficacy profiles.

While there is still a need for better HIVPR inhibitors, identifying one which provides a significant benefit over FDA-approved therapies is a major challenge. It requires the simultaneous optimization of potency, efficacy against mutant proteases, propensity for plasma protein binding, physical properties, cost of synthesis, and pharmacokinetics. As we and others have demonstrated, computer-aided and structure-based design techniques can be of tremendous benefit in designing new and potent compounds. However, addressing pharmacokinetic issues such as plasma protein binding, oral bioavailability, and potency against wild-type and mutant forms of the protease remains extremely difficult.

Despite these sizable challenges, our interest in CUs remains high because of this series demonstrated ability to yield potent antiviral compounds at a relatively low molecular weight, DMP450's significant bioavailability in man, and examples of very potent inhibitors against highly mutated HIV strains (*vide supra*).

In the remaining part of this chapter we review additional findings from our HIVPR inhibitor discovery program which have provided a foundation for identifying a CU which we hope will provide added value to the current set of AIDS therapies.

7.2. Crystallography and NMR Studies

All X-ray crystal structures of HIVPR–CU complexes confirm our initial computer model (Figure 8) that CUs bind at the active site of HIVPR with the urea moiety oriented toward the flaps and the diol group oriented toward the catalytic aspartic acid groups.[34,48] In addition, the structural water observed in complexes with linear substrate-like inhibitors is not present in the X-ray structures of HIVPR–CU complexes. The displacement of the structural water is further confirmed by NMR ROESY experiments of the complex of isotopically enriched HIVPR with DMP323[59], whereas for the linear diol inhibitor, P9941, a long-lived water molecule in this position is observed.

As modeled, the CU diol in HIVPR–CU complexes interacts symmetrically with the catalytic aspartates, while most of the reported linear diols bind asymmetrically. Recently, structures of linear diols with systematic variations of the diol configuration have been reported.[22] It was found that (*R,R*)- and (*R,S*)-diols and (S)-mono-ol share the same binding conformation as found in the other reported linear diol structures. However, (*S,S*)-diol binds symmetrically such that the two hydroxy groups are located between the two aspartates as in CU complexes. NMR studies also indicate that when complexed with DMP323, both catalytic aspartic acids are protonated.[60] This observation may have important implications for the design of new classes of HIVPR inhibitors.

More than a dozen X-ray structures of HIVPR–CU complexes have been solved with different P2/P2' substituents. These structures have usually confirmed the modeled interactions between the P2/P2' substituents and the S2/S2' residues. The X-ray structure of HIVPR–DMP323 (Figure 8a) shows that as designed, the benzyl alcohols accept hydrogen bonds from the backbone NH of D30/30' and D29/29'.[34] The amine nitrogen of DMP450's *m*-aminobenzyl substituent is located (Figure 8b) within hydrogen-binding distance of the side chain of D30 and D30'.[48]

7.3. Other Cyclic Scaffolds

Confirmation that the CUs do indeed displace the structural water molecule has stimulated many additional ideas for cyclic HIVPR inhibitors.[61] Figure 9 summa-

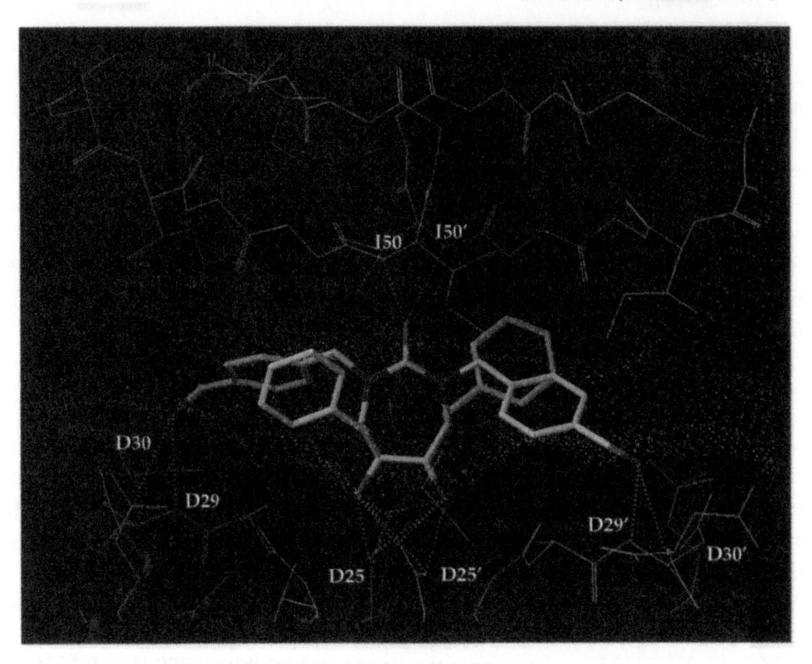

Figure 8a. X-ray crystal structure of DMP323–HIVPR. Hydrogen-bonding interactions are shown as dashed magenta lines. The contour surface represents an "extended" solvent-accessible surface generated by using twice the normal van der Waals radii to generate the surface. Inhibitor atoms which "touch" this extended surface are in van der Waals contact. Atoms which are in hydrogen-bonding distance will "penetrate" this surface.

rizes a few of these inhibitors which have been designed in our laboratories; additional cyclic structures have also been designed elsewhere.[5c,61] To date, the seven-membered CU has yielded the best combination of potency and pharmacokinetic profile. Clearly, the use of 3D database searching can serve as a powerful starting point for discovering new and diverse leads.

7.4. Benzofused CUs

One approach to overcome the metabolic instability of DMP323 and oximes **26** and **27**, and also increase the potency of DMP323, was to incorporate the P2/P2' hydrogen bond donor and acceptor functionality into various benzofused heterocycles.[62]

In varying the position of hydrogen bond donor and acceptor functionalites (Table 6), it became apparent that having a hydrogen bond acceptor at position Y (either sp^2 nitrogen or carbonyl) dramatically increases potency against HIVPR. In comparing indole **37** (K_i = 1.3 nM) with oxindole **38** (K_i = 0.14 nM) and oxazolidinone **39** (K_i = 0.18 nM), an order of magnitude improvement in K_i is observed by

D30
I84
V82
D25
D25′
D30′
V82′
I84′

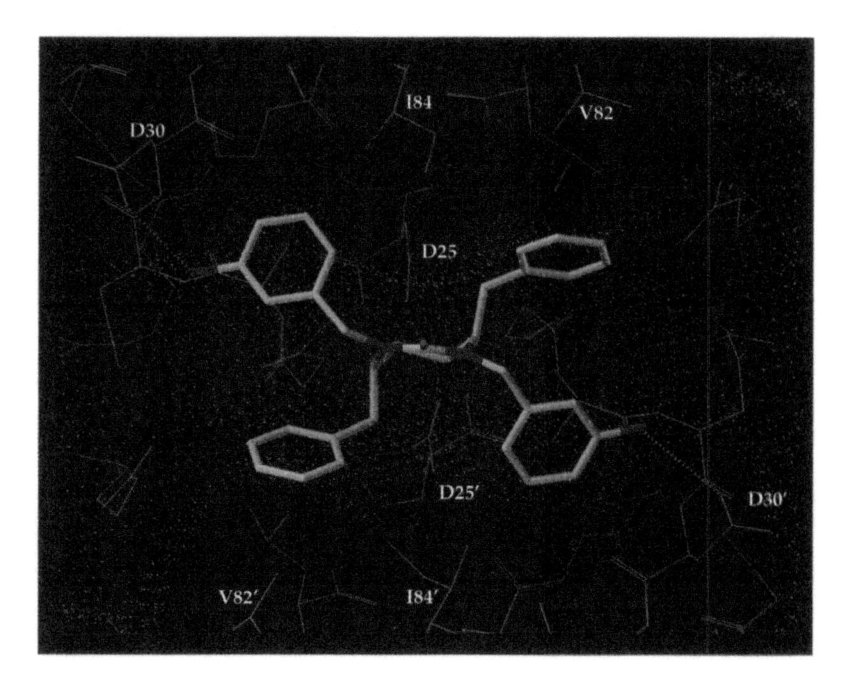

Figure 8b. X-ray crystal structure of DMP450–HIVPR. See additional notes for Figure 8a.

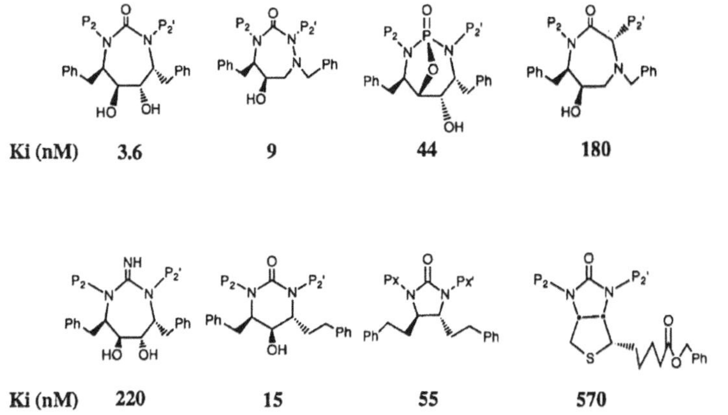

Ki (nM) 3.6 9 44 180
Ki (nM) 220 15 55 570

Figure 9. Examples of other cyclic HIVPR inhibitors. P2/P2' = benzyl; Px and Px' = 3-hydroxymethylbenzyl.

Table 6. SAR for Benzofused Cyclic Ureas

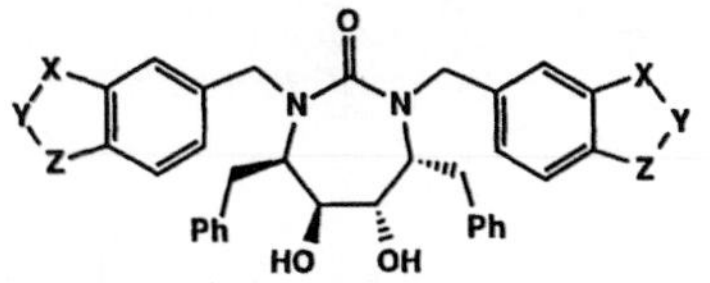

Compound	*X-Y-Z*	K_i*(nM)*[a]	IC_{90}*(nM)*[b]
37	CH=CH-NH	1.3	190
38	CH_2-C(=O)-NH	0.14	370
39	O-C(=O)-NH	0.18	40
40	CH=N-NH	0.018	8
41	NH-C(=O)-NH	<0.01	>65,000
42	O-C(NH_2)=N	0.20	84

Note: [a,b] See footnotes for Table 2.

adding the carbonyl. An even more dramatic improvement is observed when the Y-carbon of indole **37** is replaced with an sp^2 hybridized nitrogen. Indazole **40** (K_i = 0.018 nM) shows 2 orders of magnitude improvement in potency over the indole **37**. Having a hydrogen bond donor at the X or Z position is beneficial but the most potent compound, benzimidazolone **41** (K_i < 0.01 nM), has a N–H at both positions. Having an NH_2 group at position Y is detrimental except for aminobenzoxazole **42**, which has about the same K_i as benzoxazolidinone **39**.

As mentioned above, the most potent HIVPR inhibitor found in this series is benzimidazolone **41**. In order to understand the interactions responsible for this activity, an x-ray crystal structure of **41** bound to HIVPR was obtained (Figure 10). In addition to the usual binding features of the CU core, two favorable hydrogen bonds are formed between the benzimidazolone ring and the backbone of the enzyme. The N–H at the Z position forms a 2.8-Å hydrogen bond to the D30 carbonyl and the carbonyl oxygen at the Y position forms a 2.9-Å hydrogen bond to the D30 N–H. Unexpectedly, the benzimidazolone is much more potent than the benzoxazolone **39** and the oxindole **38**, which should be able to form the same hydrogen bonds. A possible explanation for these differences in potency is a bridging water molecule from the "x-position" nitrogen of **41** to the enzyme flap G48 carbonyl. Although there is an analogous water at P2', it is not within hydrogen-bonding distance of G48'.

While benzimidazolone **41** is the most potent compound against HIVPR in this series, it has very weak antiviral activity which is most likely due to its high polarity. For example, in comparing the series **41**, **38**, and **39**, the translation improves as the log P increases in going from NH, to CH_2, to O at position X. Fortunately, indazole **40** strikes a good balance and the superb potency (K_i = 0.018 nM) translates well to antiviral activity (IC_{90} = 8 nM).

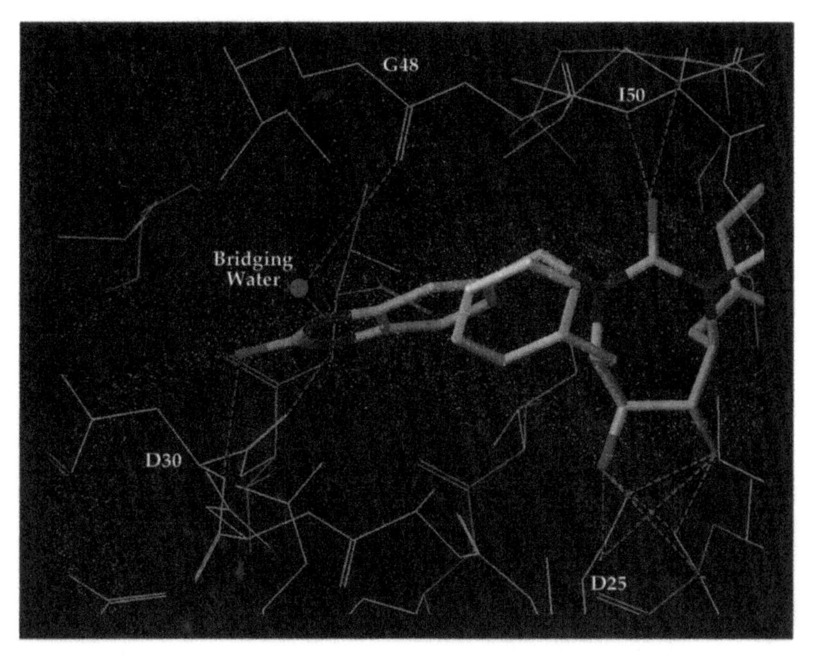

Figure 10. Interaction of the benzimidazolone P2 substituent with HIVPR's S2 subsite. Note the hydrogen-bonding interaction with G48/48' which is mediated by a water molecule.

Encouraged by the improvement in activity over DMP323, indazole **40** was selected for *in vitro* metabolism and pharmacokinetic studies. Incubation with rat liver homogenates reveals no significant metabolic liability with this indazole. Bioavailability studies in dog gave a clearance of 1.1 L/h/kg, $t_{1/2} = 2.0$ h, $C_{max} = 0.6$ μg/mL (with 10 mg/kg dose), and F% = 13%. Since metabolism and clearance are likely not the problem, it is believed that low aqueous solubility (3 ng/mL in water) results in poor intestinal permeability. Unfortunately, this low level of bioavailability is probably not sufficient for a sustained antiviral effect in man. Additional work focuses on achieving better oral bioavailability for this class of CUs by improving solubility properties, in part by making nonsymmetric CUs.[63]

7.5. Nonsymmetric CUs

The symmetric nature of the CU scaffold provides a synthetic advantage in structure–activity relationship studies since CUs with symmetric P2/P2' side chains can be prepared easily to find the best P2/P2' side chains. Subsequently, the best P2/P2' side chains can be combined to make CUs with nonsymmetric P2/P2' side chains.

Table 7. Nonsymmetrical Cyclic Urea Inhibitors of HIV Protease

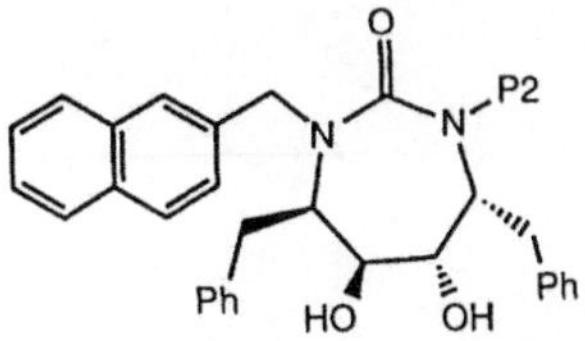

Compound	*P2*	K_{ia} *(nM)*	$IC_{90}{}^{b}$ *(μM)*	*Bioavail*^{c} C_{max}*(μM)*
7	β-naphthylmethyl	0.31 (n=3)	3.9 (n=5)	0.38
43	*n*-propyl	1.1	3.3	—
44	*n*-butyl	0.6	0.75	—
45	allyl	1.4	0.99	—
46	cyclopropylmethyl	1.5	1.3	—
47	cyclopentyl	0.28	0.66	0.32
48	benzyl	2.3	7.5	—
49	3-picolyl	5.2	0.72	—
50	4-picolyl	6.9	2.7	—
51	*p*-fluorobenzyl	3.6	1.1	—
52	*p*-hydroxymethylbenzyl	0.93	0.21	—
53	*m*-aminobenzyl	1.0	0.084	—
54	*m*-hydroxybenzyl	0.33	0.145	0.96 2.7(DOG)

Note: [a,b] Same as in Table 2.
[c] Same as in Table 4.

Table 7 shows potency and limited oral bioavailability data for nonsymmetrical CUs.[36] The K_i of the more potent P2 substituent is usually retained in the nonsymmetric CUs. This indicates that inhibitor binding to one half of the HIVPR dimer is not as important as the other half, or that the observed binding for symmetrical CUs is less than optimum. More importantly, nonsymmetric CUs have the advantage of better water solubility than symmetric CUs. Presumably, the improved solubility is partly due to the reduction in the number of *intermolecular* hydrogen bonds formed in the solid state. Recent work has therefore focused on synthesizing nonsymmetric CUs in order to improve the chances of water solubility, and by implication improve the chances of having acceptable oral bioavailability.[63]

7.6. Resistance

Cyclic urea-based HIVPR inhibitors to combat drug resistance have been designed based on the postulate that the protease of drug resistant viruses will not contain major alterations in *backbone* structure.[64] Consequently, inhibitors designed to maximize hydrogen bonds to the backbone of the wild-type enzyme might retain effectiveness against mutant strains. Furthermore, inhibitors which are capable of forming many nonbonded interactions (e.g. electrostatic, van der Waals, ionic, dipole–dipole, π–π, π–cationic, etc.), distributed throughout the active site, might have superior resistance characteristics as compared to inhibitors which only partially occupy the active site, and therefore form fewer interactions with the enzyme. For example, in the case of an inhibitor which occupies six subsites (S3, S2, S1, S1', S2,' and S3') of the enzyme, the total binding energy (DG_{total}) is a sum of binding energies of six subsites (e.g. DG_{S3}, DG_{S2} etc; Eq. 1). Similarly, Eq. 2 can be used to represent the total binding energy of an inhibitor which occupies four subsites of the enzyme.

$$DG_{total} = DG_{S3} + DG_{S2} + DG_{S1} + DG_{S1'} + DG_{S2'} + DG_{S3'}. \tag{1}$$

$$DG_{total} = DG_{S2} + DG_{S1} + DG_{S1'} + DG_{S2'} \tag{2}$$

An inhibitor which derives its binding energy by occupation of six subsites (Eq.1) is likely to have broader potency against mutant proteases than an inhibitor which occupies four subsites of the enzyme (Eq. 2) since the percent change in binding energy experienced by the former as a consequence of protease mutation will most likely be lower than in the latter.

Examination of various X-ray crystal structures of HIVPR–inhibitor complexes[12,15,5b,5c] reveals that there is an array of hydrogen-bond donors and acceptors at the S2-S3 junction of the enzyme active site comprised of D29, D30, and G48 pointing towards the inhibitor binding cleft. Indeed, the amide linkage between P2 and P3 of the linear inhibitors make critical hydrogen bonds with G48 (C–O) and D29 (N–H). These hydrogen-bonding interactions are captured in the CU-based series of inhibitors by attaching a P3 group via an amide linkage. Attachment of P3 also results in increased van der Waals interactions. Based on the above concept and design strategy, several CU amides (Figure 11) were synthesized and evaluated against a panel of drug-resistant mutant viruses (Table 8). CU amides XV652 and SD146 have exceptional resistance profiles against a panel of drug-resistant mutants. In comparison to XV652 and SD146, Q8467 and closely

Q8467 : P1 = P1' = benzyl; P2 = cyclopropylmethyl; R = 2-thiazolyl
XV651 : P1 = P1' = benzyl; P2 = cyclopropylmethyl; R = 2-imidazolyl
SD145 : P1 = P1' = benzyl; P2 = cyclopropylmethyl; R = 2-benzimidazolyl

XV638 : P1 = P1' = benzyl; R = 2-thiazolyl
XV652 : P1 = P1' = benzyl; R = 2-imidazolyl
SD146 : P1 = P1' = benzyl; R = 2-benzimidazolyl

Figure 11. Cyclic urea amides synthesized for improved potency against HIVPR mutations.

Table 8. Resistance of Reconstructed Mutant Viruses to Cyclic Urea Amides

				Resistance of Constructed Mutant Viruses[b] (IC_{90} mutant/IC_{90} WT)						
Inhibitor	*No. of Hydrogen-bonds*[a]	K_i *(nM)*[b]	IC_{90} *(nM)*[b]	*82A*	*82F*	*84V*	*ABT 538 VIRUS*	*48V/ 90M*	*MK639 VIRUS*	*46I/ 47V/ 50V*
Q8467	10	0.058	40	1.8	4.1	6.1	41	0.9	18.5	11
XV651	11	0.016	28	ND[c]	ND	ND	83	ND	12	ND
SD145	11	0.037	17	1.9	2.1	8.4	86	1.9	26	2.1
XV638	12	0.027	4.2	2.2	0.9	1.2	24	0.2	8.7	23
XV652	14	0.014	19	2.7	1.6	0.7	0.2	0.1	0.4	0.4
SD146	14	0.024	5.1	0.8	0.3	0.6	1.2	0.3	0.7	1.0

Notes: [a] Except for Q8467, XV638, and SD146, the number of hydrogen bonds were estimated from their models.
[b] Ki's , IC_{90}'s for wild type virus, and resistance of mutant viruses were measured as described in biological methods.
[c] ND = not determined.

related unsymmetric amides (Figure 11 and Table 8) exhibit relatively poor resistance profiles.

X-ray structures of HIVPR–DMP323, HIVPR–Q8467, HIVPR–XV638, and HIVPR–SD146 complexes (Figure 12) were useful in developing a correlation between the nonbonded interactions between inhibitor and enzyme and the resistance profiles of CU amides.[64] There are 10 hydrogen bonds between HIVPR and DMP323, 10 between HIVPR and Q8467, 12 between HIVPR and XV638, and 14 between HIVPR and SD146. This data indicates that the ability of an HIVPR inhibitor to form the largest number of hydrogen bonds to the backbone and the largest number of van der Waals interactions in the active site correlates strongly with the retention of high potency against drug-resistant mutant viruses.

The remarkable resistance profile of SD146 stems from its ability to occupy all six enzyme subsites, and form a large number of hydrogen bonds and extensive van der Waals contacts. SD146 is a very potent antiviral agent (IC_{90} = 5.1 nM) and has an exceptional resistance profile. Unfortunately, because of its extreme insolubility in water and oils, no formulation of SD146 has been found for oral or intravenous administration to animals. However, knowledge gained from this study could be useful in designing inhibitors of superior resistance profile while maintaining other important criteria.

Recent structural studies on the binding of DMP323 and DMP450 to wild-type and mutant forms of HIVPR indicate that mutations persist in the presence of these

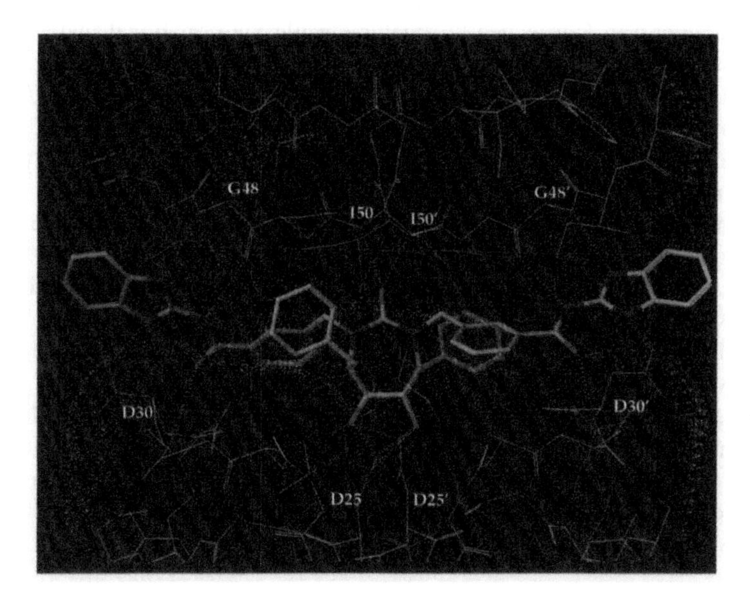

Figure 12. Hydrogen-bonding interactions observed in the X-ray structure of HIVPR–SD146 which contribute to SD146's excellent potency against mutant forms of HIVPR.

inhibitors because of a decrease in the van der Waals interactions between the mutant proteases and these CU inhibitors.[65] In addition, DMP323 and DMP450 do not contain the extensive S3 and S3' hydrogen-bonding interactions with the protease backbone that is observed in SD146–HIVPR.

Finally, it is worth noting that with respect to potency against mutant HIVPR strains, the conformational rigidity of the CU scaffold may be a liability because CU-based inhibitors can't easily adjust (optimize) their binding interactions with mutated active-site subsites. For example, the P1 and P1' benzyl groups of DMP323 and DMP450 can't easily adjust to mutations in the S1/S1' subsites.

7.7. Additional Computational Studies

Predicting Binding Affinities

As our efforts to identify a clinical candidate intensified, many new synthetic targets were being proposed. Some were based on structure-based design techniques and some were simply based on good medicinal chemistry analoguing around the known SAR from other HIVPR inhibitors. A computational technique which could provide *a priori* predictions of the binding affinity of these new synthetic targets was sought.

Unfortunately, no computational technique was available which could handle the diversity of structures being proposed and at the same time provide useful and timely predictions. Consequently, small molecule conformational analysis coupled with simple qualitative geometric and electrostatic complementarity to the HIVPR active site was sometimes used to evaluate whether to synthesize a proposed molecule. Fortunately, the rigidity of both the CU core and the HIVPR active site, along with the mostly hydrophobic nature of the active site permitted us, for the most part, to successfully use this very qualitative approach.

CoMFA (Comparative Molecular Field Analysis) was explored as a technique for quantitatively predicting affinities.[66,67] Two CoMFA models were built: one to predict *in vitro* HIVPR inhibition constant (K_i) and the other to predict antiviral activity (RNA IC_{90}).[68] The models were based on the observed K_i and IC_{90} data for 165 CUs. The molecular field analysis for each molecule was generated using both electrostatic and steric probes.

A statistical analysis of the K_i model gives a cross-validated r^2 of 0.51 using 10 components, a non-cross-validated r^2 of 0.92 with a standard error of 0.39, and a F value of 165. The predictive ability of this K_i model is about a factor of 7. All 36 compounds in a test set predicted within two log units of their actual K_i value with an average absolute error of 0.6 log units across a K_i range of 5.5 log units.

For the IC_{90} model, a cross-validated r^2 of 0.42 is calculated using 10 components, and a non-cross-validated r^2 of 0.89 with a standard error of 0.36, and a F value of 120. This model can predict IC_{90} within a factor of 10. All 30 compounds in a test set predicted within one log unit of their actual IC_{90} value with an average absolute error of 0.82 log units across a range of 5 log units.

Overall, the CoMFA models provide rough but useful predictive K_i and IC_{90} values which can help prioritize new cyclic urea targets. However, the model did not have sufficient information regarding the molecular field of the active site to predict K_i and IC_{90} values for CUs with functionalities that interacted with regions of the active site not included in the training set, or for noncyclic urea compounds.

A second approach used to help choose synthetic targets was to evaluate whether the low-energy conformations of a new cyclic scaffold positions its substituents similarly to an active CU inhibitor. Benchmark studies of the quenched dynamics technique[69] predicted CU conformations very well, and studies on other cyclic scaffolds were used to prioritize synthetic targets.

In summary, qualitative modeling and CoMFA models provided some guidance in prioritizing synthetic targets, but quantitative prediction of binding affinities remains a major challenge in the application of computer-aided techniques to drug discovery.

Predicting Membrane Permeability

Antiviral potency in cell culture and high *in vivo* plasma levels both require an inhibitor which readily permeates a membrane. Keeping a compound's molecular mass below 600 amu and its total molecular charge no more than one unit positive or negative is considered a practical heuristic for good membrane permeability. For the most part, CUs meet these broad guidelines. Quantification of a molecule's membrane permeability is usually made using physiochemical properties like molecular size and the ability to partition between a nonpolar and polar phase; for example, octanol and water partition coefficients (log P).[70]

While we found some general correlations between membrane permeability and molecular size and log P, problems in obtaining log Ps for our CUs ultimately limited the usefulness of the correlations. Obviously, measured log Ps are not available for compounds which haven't been synthesized, and we found computed log Ps using standard programs like ClogP[39] could in some cases give absolute errors of two log units.

One computational approach we explored to better understand membrane permeability was to examine if computed solvation energies could be used to predict the maximum *in vivo* plasma levels after oral dosing of a CU. In these studies the solvation energy associated with transferring a molecule from the gas phase to solution is used, not the solvation energy associated with transfer from an aqueous environment to a hydrophobic environment. Previous work on correlating membrane permeabilities of AII antogonists with solvation energies was encouraging.[71]

The calculation of an "exact" solvation energy for CUs is computationally expensive; for example, calculating the optimized conformation of DMP450 in water using AMSOL[72] requires >100 cpu hours. Therefore, several approximations to calculating CU solvation energies were tried.[73] These approximations included geometry optimization using molecular mechanics instead of AMSOL's semiempirical methods, using a single conformation instead of an ensemble of conformations, and partitioning the inhibitor into fragments. In the fragment-based calculation, a molecule such as DMP450 is partitioned into a tetramethyl cyclic urea fragment, two phenyl fragments, and two aniline fragments.

After comparing the results from exact solvation calculations on the whole inhibitor with the results using the sum of fragments approach, we concluded that for CUs, solvation energies using the sum of fragment solvation energies is an acceptable approximation.

Using 12 CUs with a C_{max} in dogs between 400 and 12,500 nM, and clearance values less than 1L/h/kg, we developed Eq. 3,

$$-\text{Log } C_{max} = -6.92 + 0.004*(\text{volume}) + 4.94*(\text{Esolv/SASA}) \tag{3}$$

where volume is the molecular volume, Esolv is the AMSOL SM2 solvation energy, and SASA is the solvent-accessible surface area.[74] The cross-validated r^2 for this model is 0.7 using two components, $r^2 = 0.85$, the standard error = 0.24 log units, and the F statistic is 24. Unlike the correlation with solvation energy observed with the AII antagonists, the correlation given by Eq. (3) for *in vivo* plasma levels for CUs is dominated by solvent-accessible surface area; solvation energy contributes only 5% to the correlation. It should be noted that a number of the AII antagonists which were used in the earlier work are charged at physiological pH.

To test the predictive power of this C_{max} model, eighteen CUs with measureable plasma level were predicted. Thirteen compounds were predicted within one-half of a log unit, but five compounds had errors in the C_{max} of 0.75 to 1.0 log unit. When binding affinities are predicted within a factor of 10 it is considered a major accomplishment. However the stakes are so high once pharmacokinetic parameters are being optimized, that predictions that are off by a factor of 10, or even a factor of 3–5 are almost meaningless.

Nevertheless, these studies, and others appearing in the literature[75] are providing important insights into some of the methods for computing physiochemical properties which may ultimately help guide optimization of pharmacokinetic properties.

8. CONCLUSIONS

Computer-aided and structure-based design techniques have led to the discovery of cyclic ureas, a novel and potent class of non-peptide HIVPR inhibitors. One compound from this class, DMP450, has the physiochemical and biological properties we initially set out to achieve, including good pharmacokinetic properties in man. Our more recent research has identified new cyclic urea inhibitors with better potencies and resistance profiles and preclinical development is underway.[63]

In reaching our goal of identifying potent nonsubstrate-like HIVPR inhibitors with good pharmacokinetic parameters, we have clearly demonstrated the feasibility of quickly "jumping" from substrate-like inhibitors to novel nonpeptide inhibitors. We anticipate there will be an ever-increasing use of this approach, including its use in conjunction with the emerging discipline of combinatorial chemistry.

ACKNOWLEDGMENTS

The authors gratefully acknowledge the tremendous dedication and contributions of the scientists we have worked with on this project. We are especially indebted to Sue Erickson-Viitanen who has enthusiastically and untiringly chaired DuPont Merck's HIVPR inhibitors working group.

REFERENCES AND NOTES

1. (a) Hamilton, H. W.; Steinbaugh, B. A.; Stewart, B. H.; Chan, O. H.; Schmid, H. L.; Schroeder, R.; Ryan, M. J.; Keiser, J.; Taylor, M. D.; Blankley, C. J.; Kaltenbronn, J. S.; Wright, J.; Hicks, J. *J. Med. Chem.* **1995**, *38*, 1446. (b) Kleinert, H. D.; Rosenberg, S. H.; Baker, W.R.; Stein, H. H.; Klinghofer, V.; Barlow, J.; Spina, K.; Polakowski, J.; Kovar, P.; Cohen J.; Denissen J. *Science* **1992**, *257*, 1940. (c) Greenlee, W. J. *Med. Res. Reviews* **1990**, *10*, 173.
2. Roberts, N. A.; Martin, J. A.; Kinchington, D.; Broadhurst, A. V.; Craig, J. C.; Duncan, I. B.; Galpin,S. A.; Handa, B. K.; Kay, J.; Krohn, A.; Lambert, R. W.; Merrett, J. H.; Mills, J. S.; Parkes, K. E. B.; Redshaw, S.; Ritchie, A. J.; Taylor, D. L.; Thomas, G. J.; Machin, P. J. *Science* **1990**, *248*, 358.
3. Vacca, J. P.; Dorsey, B.D.; Schleif, W.A.; Levin, R.B.; McDaniel, S.L.; Drake, P.L.; Zugay, J.; Quintero, J.C.; Blahy, O.M.; Roth, E.; Sardana, V.V.; Schlabach, A.J.; Graham, P.I.; Condra, J.H.; Gotlib, L.; Holloway, M.K.; Lin, J.; Chen, I.-W.; Vastag, K.; Ostovic, D.; Anderson, P.S.; Emini, E.A.; Huff, J.R. *Proc. Natl. Acad. Sci. USA* **1994**, *91*, 4096.
4. Kempf, D. J.; Marsh, K. C.; Denissen, J. F.; McDonald, E.; Vasavanonda, S.; Flentge, C. A.; Green, B. E.; Fino, L.; Park, C. H.; Kong, X.-P.; Wideburg, N. E.; Saldivar, A.; Ruiz, L.; Kati, W. M.; Sham, H. L.; Robins, T.; Stewart, K. D.; Hsu, A.; Plattner, J. J.; Leonard, J. M.; Norbeck, D. W. *Proc. Natl. Acad. Sci. USA* **1995**, *92*, 2484.
5. For recent reviews see: (a) Debouck, C. *AIDS Res. Hum. Retroviruses* **1992**, *8*, 153. (b) Wlodawer, A.; Erickson, J. W. *Annu. Rev. Biochem.* **1993**, *62*, 543. (c) Appelt, K. *Perspect. Drug Discovery Des.* **1993**, *1*, 23. (d) Katz, R. A.; Skalka, A. M. *Ann. Rev. Biochem.* **1994**, *63*, 133. (e) Kempf, D. J.; Sham, H.L. *Curr. Pharma. Design* **1996**, *2*, 225.
6. Kim, E. E.; Baker, C. T.; Dwyer, M. D.; Murcko, M. A.; Rao, B. G.; Tung, R. D.; Navia, M. A. *J. Am. Chem. Soc.* **1995**, *117*, 1181.
7. Reich, S. H.; Melnick, M.; Davies, J. F., II; Appelt, K.; Lewis, K. K.; Fuhry, M. A.; Pino, M.; Trippe, A. J.; Nguyen, D. *Proc. Natl. Acad. Sci. USA* **1995**, *92*, 3298.
8. (a) Kohl, N. E.; Emini, E. A.; Schleif, W. A.; Davis, L. J.; Heimbach, J. C.; Dixon, R. A. F.; Scolnick, E. M.; Sigal, I. S. *Proc. Natl. Acad. Sci. USA* **1988**, *85*, 4686. (b) Peng, C.; Ho, B. K.; Chang, T. W.; Chang, N. T. *J. Virol.* **1989**, *63*, 2550.
9. Drake, P. L.; Huff, J. R. *Adv. Pharmacol.* **1995**, *25*, 399.
10. Jadhav, P. K.; McGee, L. R.; Shenvi, A.; Hodge, C. N. U. S. Patent 5,294,720 issued March 15, 1994.

11. Jadhav, P. K.; Woerner, F.J. *Bioorg. Med. Chem. Lett.* **1992**, *2*, 353.
12. Miller, M.; Schneider, J.; Sathyanarayana, B. K.; Toth, M. V.; Marshall, G. R.; Clawson, L.; Selk, L.; Kent, S. B. H.; Wlodawer, A. *Science* **1989**, *246*, 1149.
13. Q8024 is same as compound **3** reported in reference 10.
14. (a) Kempf, D.J.; Norbeck, D.W.; Codacovi, L.M.; Wang, X.C.; Kohlbrenner, W.E.; Wideburg, N.E.; Paul, D.A.; Knigge, M.F.; Vasavanonda, S.; Craig-Kennard, A.; Saldivar, A.; Rosenbrook, Jr., W.; Clement, J.J.; Plattner, J.J.; Erickson, J. *J. Med. Chem.* **1990**, *33*, 2687. (b) Erickson, J.; Neidhart, D. J.; VanDrie, J.; Kempf, D. J.; Wang, X. C.; Norbeck, D. W.; Plattner, J. J.; Rittenhouse, J. W.; Turon, M.; Wideburg, N. E.; Kohlbrenner, W. E.;Simmer, R.; Helfrich, R.; Paul, D. A.; Knigge, M. F. *Science* **1990**, *249*, 527.
15. Swain, A. L.; Miller, M. M.; Green, J.; Rich, D. H.; Schneider, J.; Kent, S. B. H.; Wlodawer, A. *Proc. Natl. Acad. Sci. USA* **1990**, *87*, 8805.
16. Ripka, W. C.; Sipio, W. J.; Blaney, J. M. *Lect. Heterocycl. Chem* **1987**, *9*, 95.
17. (a) Martin, Y. C. *J. Med. Chem.* **1992**, *35*, 2145.(b) Downs, G. M.; Willett, P. *Rev. Comp. Chem.* **1996**, *7*, 1. (c) Good, A. C.; Mason, J. S. *Rev. Comp. Chem.* **1996**, *7*, 67.
18. DesJarlais, R. L.; Sheridan, R. P. Dixon, J. S.; Kuntz, I. D.; Venkataraghavan, R. *J. Med. Chem.* **1986**, *29*, 2149. (b)DesJarlais, R. L.; Sheridan, R. P.; Seibel, G. L.; Dixon, J. S.; Kuntz, I. D.; Venkataraghavan, R. *J. Med. Chem.* **1988**, 31, 722.
19. Eyermann, C. J.;Ripka, W. C. Using 3-D Similarity Searching to Develop Synthetic Targets. In *Proceedings of the 2nd International Conference on Chemical Information, Noordwijkerhout*; Warr, W., Ed.; Springer Verlag, 1990.
20. Eyermann, C. J.; Lam, P. Y. S.; Kerr, J. S.; Ripka, W. C. *Abstracts of Papers;* 202nd National Meeting of the American Chemical Society, New York, 1991.
21. (a) Blaney, J. M.;Dixon J. S. *Rev. Comput. Chem.* **1994**, *5*, 299. (b)Blaney, J. M.; Crippen, G. M.;Dearing, A.; Dixon, J. S.; *DGEOM*, Quantum Chemistry Program Exchange 590.
22. Hosur, M. V.; Bhat, T. N.; Kempf, D. J.; Baldwin, E. T.; Liu, B.; Gulnik, S.; Wideburg, N. E.; Norbeck, D. W.; Appelt, K.; Erickson, J. W. *J. Am. Chem. Soc.* **1994**, *116*, 847.
23. Goodford, P.J. A. *J. Med. Chem.* **1985**, 28, 849.
24. QUANTA, Molecular Simulations Incorporated, San Diego, CA.
25. Dunitz, J. D. *Science* **1994**, *264*, 670.
26. (a) MACCS-3D, Molecular Design Limited, San Leandro, CA. (b) 3D searches were done on a subset of the Cambridge Structural Database which has been converted for use with MACCS-3D. (Eyermann,C. J. unpublished).
27. Allen, F. H.; Kennard, O. *Chem. Des. Automation News* **1993**, *8*, 1 & 31.
28. (a) Andreetti G. D.; Bocelli, G.; Sgarabotto P. *Cryst. Struct. Comm.* **1974**, *3*, 145. (b) While the scaffold from the Cambridge database hit was not used to initiate a synthesis program, a compound from our corporate compound collection which is very similar to the original hit was found to be a 10 mM inhibitor of HIVPR (Eyermann, C. J.; Klabe, R. M. unpublished results).
29. Weber, P. C.; Ohlendorf, D. H.; Wendoloski, J. J.; Salemme, F. R. *Science* **1989**, *243*, 85.
30. (a) Cram, D. J. *Angew. Chem. Int. Ed.* **1986**, *25*, 1039.(b) Cram, D. J. *Science*, **1988**, *240*, 760.(c)Cram, D. J.; Lam, P. Y. S. *Tetrahedron* **1986**, *42*, 1607. (d) Cram, D. J.; Dicker, I. B.; Lauer, M.; Knobler, C. B.; Trueblood, K. N. *J. Am. Chem. Soc.* **1984**, 106, 7150.
31. March, J. *Advanced Organic Chemistry*, 3rd Edition; Wiley, New York, 1985, p 125.
32. (a) Johnson, F. *Chem. Rev.* **1968**, *68*, 375.(b) Hoffmann, R. W. *Angew Chem. Int. Ed. Engl.* **1992**, *31*, 1124.
33. Small molecule crystal structure studies by J. C. Calabrese of E. I. DuPont de Nemours & Co. Inc.
34. (a) Lam, P. Y. S.; Jadhav, P. K.; Eyermann, C. J.; Hodge, C. H.; Ru, Y.; Bacheler, L. T.; Meek, J. L; Otto, M. J.; Rayner, M. M.; Wong, Y. N.; Chang, C-H.; Weber, P. C.; Jackson, D.; Sharpe, T. R.; Erickson-Viitanen, S. *Science* **1994**, *263*, 380. (b) Lam, P. Y.; Jadhav, P. K.; Eyermann, C. J.; Hodge, C. N.; De Lucca, G. V.; Rodgers, J. D. U.S. Patent 5,610,294, issued March 11, 1997.

35. Erickson-Viitanen, S.; Klabe, R. M.; Cawood, P. G.; O'Neal, P. L.; Meek, J. L. *Antimicro. Agents Chemo.* **1994**, *38*, 1628.
36. Lam, P. Y. S.; Ru, Y.; Jadhav, P. K.; Aldrich, P. E.; DeLucca, G. V.; Eyermann, C. J.; Chang, C.-H.; Emmett, G.; Holler, E. R.; Daneker, W. F.; Li, L.; Confalone, P. N.; McHugh, R. J.; Han, Q.; Li, R.; Markwalder, J. A.; Seitz, S. P.; Sharpe, T. R.; Bacheler, L. T.; Rayner, M. M.; Klabe, R. M.; Shum, L.; Winslow, D. L.; Kornhauser, D. M.; Jackson, D. A.; Erickson-Viitanen, S.; Hodge, C. N. *J. Med. Chem.* **1996**, *39*, 3514.
37. Rich, D. H. Effect of Hydrophobic Collapse on Enzyme-Inhibitor Interaction. Implication for the Design of Peptidal Mimetics. In *Perspectives in Med. Chem.*; Testa, B.; Kyburz, E.; Fuhrer, W.; Giger, R., Eds.; VCH, New York, 1993, pp 15-25.
38. Bacheler, L. T.; Paul, M.; Jadhav, P. K.; Otto, M.; Stone, B.; Miller, J. *Antiviral Chem. Chemother.* **1994**, *5*, 111.
39. ClogP was calculated using Pomona's MedChem Project Software Release 3.54, Pomona College, Claremont, CA. For cyclic urea **24**, the clogP of 9.2 probably has significant amount of error.
40. Wong, N. Y.; Burcham, D. L.; Saxton, P. L., Erickson-Viitanen, S.; Grubb, M. F.; Quon, C. Y.; Huang, S.-M. *Biopharm. Drug Disp.* **1994**, *15*, 535.
41. HPLC logP was measured by J. Gerry Everlof using a C_{18} column and methanol/water eluant.
42. Wilkerson, W.W.; Akamike, E.; Cheatham, W.W.; Hollis, A.Y.; Collins, D.; DeLucca, I.; Lam, P.Y.S.; Ru, Y. *J. Med. Chem.* **1996**, *39*, 4299.
43. Nugiel, D. A.; Jacobs, K.; Worley, T.; Patel, M.; Kaltenbach, R. F. III; Meyer, D. T.; Jadhav, P. K.; De Lucca, G. V.; Smyser, T. S.; Klabe, R. M.; Bacheler, L. T.; Rayner, M. M.; Seitz, S. P. *J. Med. Chem.* **1996**, *39*, 2156.
44. Christ, D. D., Meek, J. L., Farmer, A. R., Larsen, B. S. *ISSX Proceedings* **1993**, *4*, 230.
45. Olson, R. E.; Christ D. D. *Ann. Rep. Medicin. Chem.* **1996**, *31*, 327.
46. Bryant, M.; Getman, D.; Smidt, M.; Marr, J.; Clare, M.; Dillard, R.; Lansky, D.; DeCrecenzo, G.; Heintz, R.; Houseman, K.; Reed, K.; Stoszenbach, J.; Talley, J.; Vazquez, M.; Mueller, R. *Antimicrob.Agents Chemother.* **1995**, 39, 2239.
47. (a) Perelson, A. S.; Neumann, A. U.; Markowitz, M.; Leonard, J. M.; Ho, D. D. *Science* **1996**, *271*, 1582. (b) Condra, J. H.; Schleif, W. A.; Blahy, O. M.; Gabryelski, L. J.; Graham, D. J.; Quintero, J. C.; Rhodes, A.; Robbins, H. L.; Roth, E.; Shivapraksh, M.; Titus, D.; Yang, T.; Toppler, H.; Squires, K. E.; Deutsch, P. J.; Emini, E. A. *Nature* **1995**, *374*, 569. (c) Jacobsen, H.; Yasargil, K.;Winslow, D. L.; Craig, J. C.; Krohn, A.; Duncan, I. B.; Mous, J. *Virology* **1995**, *206*, 527. (d) Markowitz, M. M.; Kempf, D. J.; Norbeck, D. W.; Bhat, T. N.; Erickson, J. W.; Ho, D. D. *J. Virol.* **1995**, *69*, 701. (e) Ridky, T.; Leis, J. *J. Biol. Chem.* **1995**, *270*, 29621.
48. Hodge, C. N.; Aldrich, P. E.; Bacheler, L. T.; Chang, C.-H., Eyermann, C. J.; Garber, S.; Grubb, M. F.; Jackson, D. A.; Jadhav, P. K.; Korant, B.; Lam, P. Y.-S.; Maurin, M. B.; Meek, J. L.; Otto, M. J.; Rayner, M. M.; Reid, C.; Sharpe, T. R.; Shum, L.; Winslow, D. L.; Erickson-Viitanen, S. *Chem. and Biol.* **1996**, *3*, 301.
49. Wei, X.; Ghosh, S. K.; Taylor, M. E.; Johnson, V. A.; Emini, E. A.; Deutsch, P.; Lifson, J. D.; Bonhoeffer, S.; Nowak, M. A.; Hahn, B. H.; Saag, M. S. and Shaw, G. M. *Nature* **1995**, *373*, 117.
50. Package insert for Indinavir.
51. Ho, D. D.; Neumann, A. U.; Perelson, A. S.; Chen, W.; Leonard, J. M.; Markowitz, M. *Nature* **1995**, *373*, 123.
52. Danner, S. A.; Carr, A.; Leonard, J. M.; Lehman, L. M.; Gudiol, F.; Gonzales, J.; Raventos, A.; Rubio, R.; Bouza, E.; Pintado, V.; Aquado, A. G.; Delomas, J. G.; Delgado, R.; Borleffs, J. C. C.; Hsu, A.; Valdes, J. M.; Boucher, C. A. B.; Cooper, D. A. ; Gimeno, C.; Clotet, B.; Tor, J.; Ferrer, E.; Martinez, P. L.; Moreno, S.; Zuncada, G.; Alcami, J.; Noriega, A. R.; Pulido, F.; Glassman, H. N. *N. Engl. J. Med.* **1995**, *333*, 1528.
53. Package insert for Ritonavir.
54. Kitchen, V. S.; Skinner, C.; Ariyoshi, K.; Lane, E. A.; Duncan, I. B.; Burckhardt, J.; Burger, H. U.; Bragman, K.; Pinching, A. J. and Weber, J. N. *The Lancet* **1995**, *345*, 952.

55. Package insert for Saquinavir.
56. Shetty, B. V.; Kosa, M. B.; Khalil, D.A.; Webber, S. *Antimicrob. Agents Chemother.* **1996**, *40*, 110.
57. Livingston, D. J.; Pazhanisamy, S.; Porter, D. J. T.; Partaledis, J. A.; Tung, R. D.; Painter, G. R. *J. Infect. Dis.* **1995**, *172*, 1238.
58. St. Clair, M. H.; Millard, J.; Rooney, J.; Tisdale, M.; Parry, N.; Sadler, B. M.; Blum, M. R.; Painter, G. *Antiviral Res.* **1996**, *29*, 53.
59. Grzesiek, S.; Bax, A.; Nicholson, L. K.; Yamazaki, T.; Wingfield, P.; Stahl, S. J.; Eyermann, C. J.; Torchia, D. A.; Hodge, C. N.; Lam, P. Y. S.; Jadhav, P. K.; Chang, C.-H. *J. Am. Chem. Soc.* **1994**, *116*, 1581.
60. Yamazaki, T.; Nicholson, L. K.; Torchia, D. A.; Wingfield, P.; Stahl, S. J.; Kaufman, J. D.; Eyermann, C. J.; Hodge, C. N.; Lam, P. Y. S.; Ru, Y.; Jadhav, P. K.; Chang, C.-H.; Weber P. C. *J. Am. Chem. Soc.* **1994**, 116, 10791.
61. (a) De Lucca, G. V.; Erickson-Viitanen, S.; Lam, P. Y. S. *Drug Discov. Today* **1997**, *2*, 6. (b) De Lucca, G. V. *Bioorg. Med. Chem. Lett.* **1997**, *7*, 501. (c) Smallheer J. M.; Seitz, S. P. *Heterocycles* **1996**, *43*, 2367. (d) Han, Q.; Lafontaine, J.; Bacheler, L. T.; Rayner, M. M.; Klabe, R. M.; Erickson-Viitanen, S.; Lam, P. Y. S. *Bioorg. Med. Chem. Lett.* **1996**, *6*, 1371.
62. Rodgers, J.D.; Johnson, B.L.; Wang, H.; Greenburg, R.A.; Erickson-Viitanen, S.; Klabe, R.M.; Cordova, B.C.; Rayner, M.M.; Lam, G.N.; Chang, C-H. *Bioorg. Med. Chem. Lett.* **1996**, *6*, 2919.
63. (a) Rodgers, J. D.; Wang, H.; Johnson, B. L.; Lam, P. Y.; Ru, Y.; De Lucca, G. V.; Kim, U. T.; Ko, S. S.; Trainor, G. L.; Klabe, R. M.; Cordova, B. C.; Bacheler, L. T.; Erickson-Viitanen, S.; Lam, G. N.; Chang, C.-H. 213th ACS National Meeting 1997, April 13-17, Div. of Med. Chem., 144 (Abstr.). (b) Lam, P. Y. S.; Rodgers, J. D.; Li, R.; Ru, Y.; Jadhav, P. K.; Chang, C.-H.; Clark, C. G.; Seitz, S. P.; Bacheler, L. T.; Lam, G. N.; Erickson-Viitanen, S.; Trainor, G. V.; Anderson, P. S. 2 13th ACS National Meeting 1997, April 13-17, Div. of Med. Chem., 278 (Abstr.).
64. Jadhav, P. K.; Ala, P.; Woerner, F. J.; Chang, C.-H.; Garber, S. S.; Anton, E. D.; Bacheler, L. T. *J. Med. Chem.* **1997**, *40*, 181.
65. Ala, P.; Huston, E.; Klabe, R.; McCabe, D.; Duke, J.; Rizzo, C.; Korant, B.; DeLoskey, R.; Lam, P. Y. S.; Hodge, C. N.; Chang, C.-H. *Biochemistry* **1997**, *36*, 1573.
66. Cramer, R.D.; Patterson, D.E.; Bunce, J.D. *J. Am. Chem. Soc.* **1988**, *110*, 5959.
67. SYBYL, Tripos Associates, St. Louis, MO, USA
68. Eyermann, C.J.; De Lucca, I. Unpublished results.
69. Hodge, C. N.; Straatsma, T. P.; McCammon J. A.; Wlodawer, A., Rational Design of HIV Protease Inhibitors. In *Structural Biology of Viruses*; Chiu, W.; Burnett, R. M.; Garcea, R., Eds.; Oxford Univ. Press, New York, 1997, pp 451-473.
70. Navia, M. A.; Chaturvedi, P. R. *Drug Discovery Today* **1996**, *1*, 179.
71. Ribadeneira, M.D.; Aungst, B.J.; Eyermann, C.J.; Huang, S.-M. *Pharm. Res.* **1996**, *13*, 227.
72. (a) Cramer, C. J.; Lynch, G. C.; Hawkins, G. D.; Truhlar, D. G.; Liotard, D. A.AMSOL-version 4.0, QCPE Bulletin **1993**, *13*, 78; based in part on AMPAC-version 2.1 by Liotard, D. A.; Healy, E. F.; Ruiz, J. M.; Dewar, M. J. S. (b) Cramer, C. J.; Truhlar, D. G. *Science* **1992**, *256*, 213.
73. Eyermann, C.J.; De Lucca, I. Unpublished results.
74. Connolly, M. L. *Science* **1983**, *221*, 709.
75. Lipophilicity in Drug Action and Toxicology. In *Methods and Principles in Medicinal Chemistry*; Pliska, V.; Testa, B; Waterbeemd, H. Eds.; VCH, New York, 1996, Vol 4.

MACROCYCLIC INHIBITORS OF SERINE PROTEASES

Michael N. Greco and Bruce E. Maryanoff

Advances in Amino Acid Mimetics and Peptidomimetics
Volume 1, pages 41-76

ISBN: 0-7623-0200-3

ABSTRACT

Macrocyclic peptides play an important role in many biological processes. In comparison to their acyclic counterparts, the restricted conformational flexibility of macrocyclic peptides offers potential advantages for binding interactions with bioreceptors. For example, Nature employs macrocyclic peptide hormones such as oxytocin, the vasopressins, and somatostatin to regulate such critical processes as lactation, uterine contraction, vasoconstriction, and growth hormone release. The serpin superfamily is a unique class of inhibitor proteins that regulate the actions of serine proteases, proteolytic enzymes involved in the regulation of physiological events such as blood coagulation, fibrinolysis, connective tissue turnover, inflammatory responses, and complement activation. Serpins operate by a mechanism whereby they present a peptide recognition epitope as a part of macrocyclic array, or loop of the enzyme. The macrocyclic peptide motif has been underexplored as a means to discover novel serine protease inhibitors. In this chapter, we review serine protease inhibitors from the perspective of our studies involving the macrocyclic peptide cyclotheonamide A (CtA), a marine natural product. CtA, itself, is a very potent inhibitor of trypsin and a potent inhibitor of thrombin. We outline our progression from fundamental studies of CtA to a focused drug discovery approach aimed at identifying novel inhibitors of thrombin, a serine protease that plays a central role in the control of thrombosis and hemostasis. Our protein structure-based approach utilized X-ray and NMR structural information to design hybrid structures that combined elements of CtA and the thrombin-recognition tripeptide, D-Phe-Pro-Arg, in an analogy with fibrinogen Aα-chain motifs. We describe synthetic chemistry, enzyme inhibition, and molecular modeling, and then rationalize thrombin versus trypsin inhibition by considering features of the CtA-bound X-ray structures of each enzyme. Our approach resulted in a class of novel macrocyclic inhibitors of thrombin and trypsin with good *in vitro* potency. Although enzyme selectivity for thrombin over trypsin was unexceptional, we managed to find some selective inhibitors of trypsin.

1. INTRODUCTION

Peptides are ubiquitous regulators of biological processes, being well represented in hormone and enzyme systems. However, their utility as therapeutic agents is limited principally because of poor pharmacodynamic properties and susceptibility to proteolytic degradation.[1] One might also suggest that their therapeutic shortcomings are, in part, a consequence of a high degree of conformational mobility. A bound bioactive peptide must attain a specific conformation out of the many random conformations that are populated by an unbound peptide in solution to operate effectively at a receptor or within an enzyme active site. From this perspective, macrocyclic peptides offer some advantages over their acyclic counterparts for interactions with biological receptors. The relatively restricted conformational flexibility of macrocylic peptides can bring about opportunities for enhanced binding interactions with receptors and enzymes. This might be expected to add a measure of selectivity to a particular binding interaction, since the availability of peptide side chains is somewhat more constrained in a macrocyclic arrangement. Additionally, conformational constraint of peptide motifs affords an entropic advantage, since the bioactive conformation of a constrained peptide is reached from a comparatively smaller population of random conformers. As a consequence, much attention has been focused on achieving a better understanding of bioactive macrocyclic peptides and their interactions with bioreceptors.[1,2]

Given the relevance of macrocyclic peptides to bioactive conformations, it is not surprising that Nature utilizes macrocyclic peptide motifs in a number of important biological processes. The peptide hormones oxytocin (**1**), arginine vasopressin (**2**), and lysine vasopressin (**3**) are macrocyclic hexapeptides that mediate such processes as lactation, uterine contraction, vasoconstriction, and naturetic functions.[3,4] Somatostatin[5] (**4**) is a macrocyclic peptide hormone that regulates the release of hormones such as growth hormone, gastrin, glucagon, and insulin. The serpin superfamily is a unique class of natural enzyme inhibitors that function as regulators of physiological events such as blood coagulation, fibrinolysis, connective tissue turnover, inflammatory responses, and complement activation.[6,7] Serpins owe their

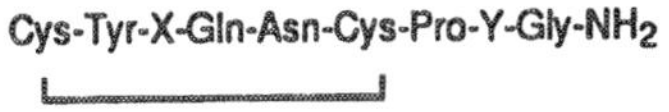

Cys-Tyr-X-Gln-Asn-Cys-Pro-Y-Gly-NH_2

1 X = Ile, Y = Leu
2 X = Phe, Y = Arg
3 X = Phe, Y = Arg

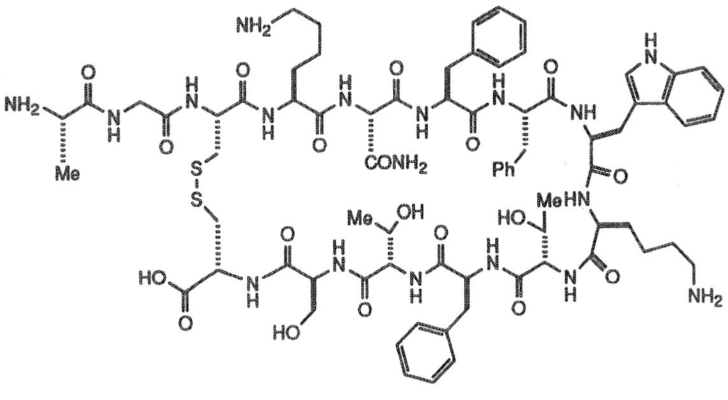

4

inhibitory activity to a mechanism (*vide infra*) whereby they present a peptide recognition epitope as a macrocylic array, or loop, to the enzyme. The macrocyclic peptide natural product cyclosporin A (CsA, **5**) has emerged as the principal immunosuppressant agent for solid organ transplantation. The wealth of studies surrounding the interaction of CsA, and the macrolide FK-506, with their corresponding immunophilin receptors has greatly advanced research in immunosupressants.[8,9] Other novel bioactive macrocyclic peptide natural products, particularly of marine origin, continue to be isolated and promise to assist the design of new therapeutic agents.[10] Indeed, macrocyclic peptides have served as starting points for major advances in the drug discovery process.

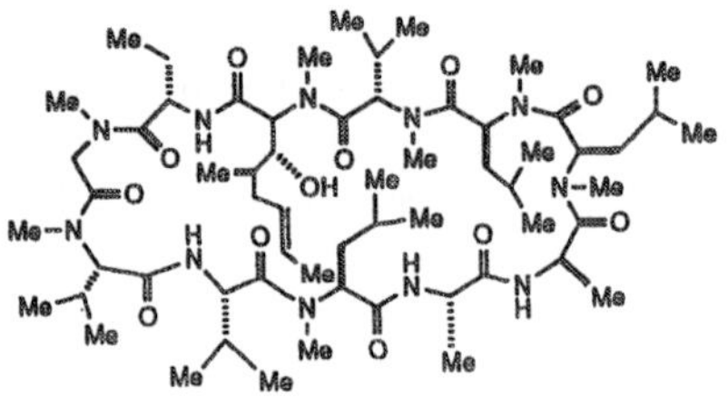

5

One of the first successful examples of peptidomimicry in drug design involved the macrocyclic peptide hormone somatostatin (**4**). Extensive structure–function and conformational studies led to the simplified cyclic hexapeptide **6**,[5,11,12] which

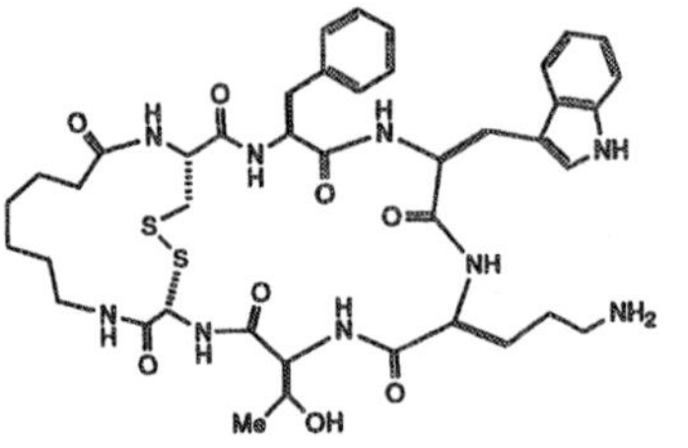

6

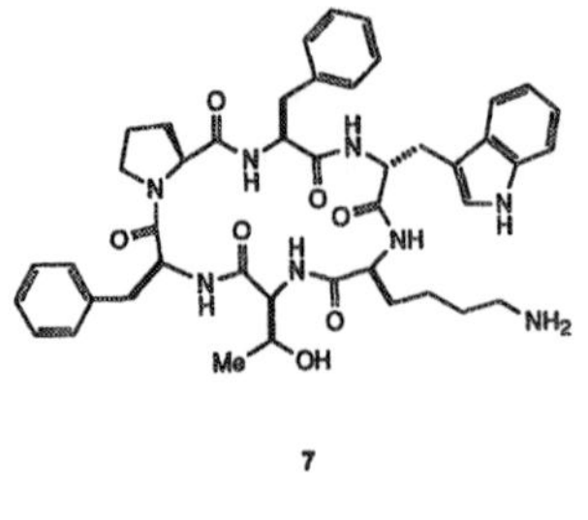

7 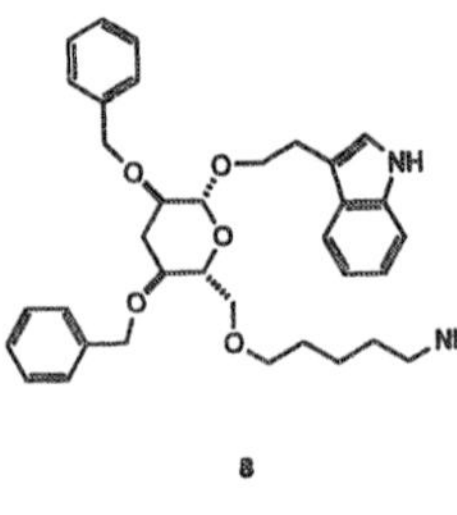 **8**

showed similar potency to somatostatin for inhibiting the release of growth hormone, glucagon, and insulin *in vitro* and *in vivo*. Continuation of this approach led to the potent cyclic hexapeptide somatostatin agonist **7**, which, in turn, led to the discovery of **8**, a non-peptidal partial somatostatin agonist containing a β-D glucose scaffold.[13,14]

Numerous additional examples of drug discovery involving peptidomimicry of bioactive macrocyclic peptides have appeared, with applications to oxytocin, vasopressin, bradykinin, opioids, substance P, renin, angiotensin, cholecystokinin, glycoprotein IIb/IIa, and HIV protease having been reported and reviewed recently.[2]

2. SERINE PROTEASES

Of the myriad biochemical processes in living organisms, proteolysis is one of the most important. Proteolysis of peptide substrates by diverse enzymes plays a central role in the regulation of physiological and pathophysiological conditions. Among the various classes of proteolytic enzymes, the serine proteases are well understood mechanistically.[15,16] They represent a large family usually distinguished by an active-site catalytic triad of Asp102/His57/Ser195, in which the serine hydroxyl is a reactive nucleophile. Serine proteases hydrolyze an amide (or ester) bond by catalyzing the transfer of a substrate acyl group (acylation) to the serine hydroxyl with departure of an amine (or alcohol) product, followed by cleavage of the acyl–enzyme (deacylation) by a water molecule (Figure 1). This process is assisted by proton transfer via the imidazole group of His57 and the carboxylate of Asp102.

Although the family members share a common catalytic apparatus, there are major differences between them relative to substrate specificity. Protein substrate alignment in the protease is denoted by the convention -P3-P2-P1-P1'-P2'-P3'- (amino-terminus of the peptide chain on the left), where Pn and P'n are amino acid

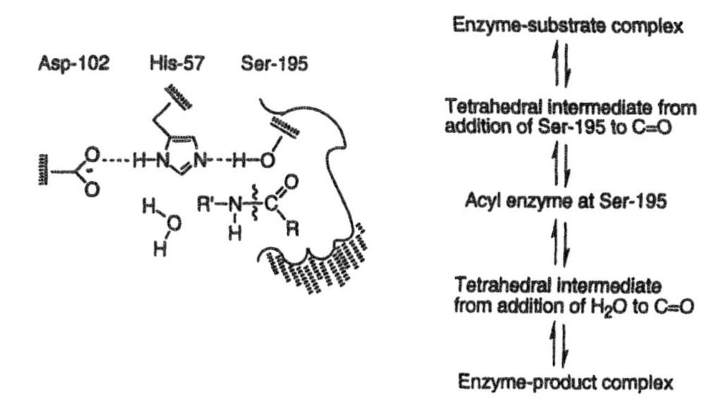

Figure 1. Drawing of an active site for a generic serine protease, with the catalytic triad poised for amide hydrolysis, along with the most discrete steps of the proteolytic process.

positions toward the N-terminus and C-terminus, respectively, relative to the scissile bond (located between P1 and P1'); the corresponding recognition subsites of the enzyme are denoted by S3-S2-S1-S1'-S2'-S3'. One key factor of specificity is the type of P1 amino acid. Trypsin and trypsin-like proteases are specific for lysine and arginine; chymotrypsin is specific for large hydrophobic amino acids, such as phenylalanine and tryptophan; and elastase is specific for small hydrophobic amino acids, such as alanine and valine. These differences result from a few key changes in the "specificity pocket" (S1) that binds the P1 amino acid side chain. In trypsin-like serine proteases, Asp189 presents a key carboxylate anchor in S1 that forms a salt-bridge with the positively charged ammonium or guanidinium groups (of Lys or Arg). In elastase, two glycines at the mouth of the pocket in chymotrypsin are replaced by the bulky residues Val216 and Thr226, thereby preventing entry of large side chains and providing a means for binding the small side chain of alanine. Besides the primary determinant of the specificity pocket, there are other critical structural differences within the active site of the serine protease such that substrate (or inhibitor) specificity will be very dependent on the steric and electrostatic properties of the exposed enzyme surface, in terms of molecular recognition.

Serine proteases are important not only in the regulation of intracellular protein turnover, but also in the control of many physiological functions such as digestion, peptide hormone synthesis and turnover,[17] blood coagulation,[18] fibrinolysis,[19] control of blood pressure,[20] fertilization,[21] ovulation,[22] and tissue remodeling.[23] Moreover, the uncontrolled destruction of intracellular proteins by these proteases is associated with many pathological conditions, such as: the breakdown of articular cartilage by elastase[24] and collagenase[25] in rheumatoid arthritis; the destruction of pulmonary elastin by elastase in emphysema;[26] the release of leukocyte elastase from polymorphonuclear cells in response to inflammation stimuli;[27] the activation of plasminogen during the invasion of healthy tissue by tumor cells;[28] and hypercoagulability in several thrombotic disorders[29,30] caused by inappropriate generation of thrombin in the final common pathway of blood coagulation, where a complex cascade of several serine proteases (factor Xa, factor VIIa, factor IXa, factor VIIIa, etc.) is involved.

2.1. Serine Protease Inhibitors

Because of the great importance of proteolysis, this process must be effectively controlled. Enter endogenous protease inhibitors, which are themselves proteins. They are found in almost all tissues where proteases are present, some showing high specificity and others showing promiscuity. When the delicate balance between an endogenous inhibitor and its target protease is not in place, proteolytic action can occur, resulting in pathological conditions such as those described above. Direct inhibitors of serine proteases have the potential to maintain therapeutic control in such disease states, making synthetic, small-molecule inhibitors with high specificity a very worthwhile goal.

Serpins are a family of natural protein-based serine protease inhibitors that tightly regulate the action of serine proteases through high-affinity binding. For instance, two serpins in the blood coagulation cascade, antithrombin III and plasminogen activator inhibitor (PAI-1), inhibit blood coagulation and clot dissolution, respectively, by keeping thrombin and plasminogen activators in check, thereby maintaining hemostatic balance. Similarly, α_1-antitrypsin principally inhibits neutrophil elastase, thereby controlling inflammatory cascades as found in pulmonary emphysema. Structure–function relationships suggest the possibility of mimicking the mechanism of action and molecular recognition of serpins in the design of novel protease inhibitors.[6] The inhibitory specificity of serpins is primarily conferred by a single amino acid at the reactive center of the molecule, the P1 residue in a specific peptide loop.[31] The inhibitor molecule captures the protease by presenting its reactive site and the surrounding amino acids as an ideal "no-turnover" substrate. The reactive centers can be identified by sequence homology: α_1-antitrypsin is a Met–serpin (an inhibitor that presents methionine); α_1-antichymotrypsin is a Leu–serpin (presents leucine); and antithrombin III is an Arg–serpin (presents arginine). The concept of inhibition by a reactive center acting as bait is supported by the crystal structure of α_1-antitrypsin,[32] which depicts a highly ordered structure with 40% β-sheet and 30% α-helix presumably designed to fix the reactive center in an exposed site. The structural features of serpins have been studied over the past decade by X-ray diffraction[33] and NMR[34] techniques, revealing in particular the conformation and constraints of the binding loop. The recognition epitope is contained within a *macrocyclic array* formed by the loop conformation, which is stabilized by an appropriately positioned disulfide bond. The inhibitor–enzyme complex is tight (K_a values up to 10^{13} M^{-1}) and dissociates slowly after cleavage to generate a stable deactivated form of the inhibitor.[6] Mechanistically, serpins differ from the smaller canonical serine protease inhibitors, such as bovine pancreatic trypsin inhibitor, which retain their inhibitory activity and may undergo resynthesis of the cleaved peptide bond.[6,32] This macrocyclic array concept has potential in the design of simplified macrocyclic inhibitors related to serpins.[35] As mentioned above, synthetic macrocyclic inhibitors of serine proteases are quite unusual.

Transition-state analogues (TSAs) are a subclass of reversible inhibitors that attain high affinity by mimicking the transition state of an enzyme–substrate complex. The premise of TSAs rests on the principle that an enzyme facilitates (catalyzes) a reaction predominantly by stabilizing the transition state relative to the ground state, the upshot being that interactions of the enzyme with the transition state are energetically very favorable.[36,37] Therefore, compounds that mimic the transition state should have a high affinity for the enzyme.[38,39] The known TSAs of serine proteases form negatively charged tetrahedral adducts by reacting with the hydroxyl group of Ser195 in the active-site cleft. These TSAs fall mainly into two classes: activated-carbonyl compounds and boronic acids (*vide infra*). Activated-carbonyl compounds react with Ser195 to form "oxyanions" of hemiacetals or hemiketals,[40,41] while the boronic acids form boronates.[42] A region of the active-site

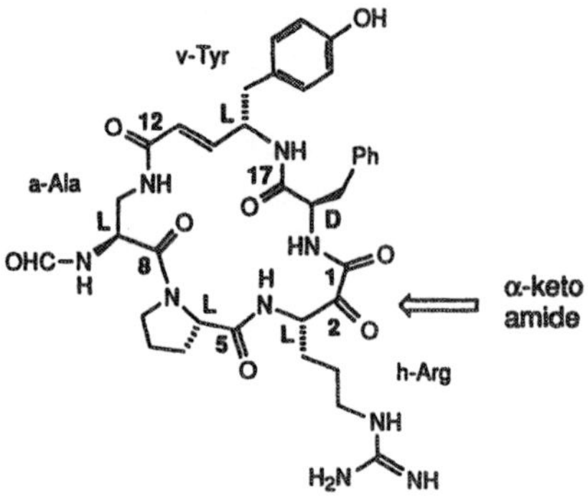

9 (Cyclotheonamide A; CtA)

cleft in serine proteases ("oxyanion hole") stabilizes the developed negative charge through a hydrogen-bonding network. Cyclotheonamide A (**9**, CtA), a compound isolated from the marine sponge *Theonella* sp., is a rare example of a naturally occurring TSA inhibitor. CtA inhibits thrombin, trypsin, and other serine proteases,[43-45] and is very intriguing because of its macrocyclic peptide nature, with a Pro-Arg S2-S1 recognition motif and an α-keto amide group, the latter of which forms a hemiketal adduct with Ser195.[45,46]

2.2. Trypsin and Thrombin

Trypsin is a prominent member of structurally related digestive enzymes and forms the functional principle, or prototype, of a large segment of the serine protease family.[47] It serves to activate itself (from trypsinogen) and other zymogens in pancreatic tissue.[48] There are many known trypsin inhibitors, and these have potential in the treatment of acute pancreatitis and hyperproteolytic conditions.[49,50] X-ray crystal structures of trypsin and inhibitor–trypsin complexes (*vide infra*) provide a three-dimensional understanding, which is important for a structural appreciation of all trypsin-like proteases and for the design of selective inhibitors. In general, the proteases in the blood coagulation cascade exhibit trypsin-like behavior, but with much greater specificity for substrates and inhibitors.[51]

α-Thrombin is a trypsin-like serine protease that is central to the bioregulation of hemostasis; its enzymatic actions can be procoagulant or anticoagulant, depending on conditions. Through cleavage of fibrinogen, thrombin is a primary promoter of coagulation: it exposes polymerization sites on fibrinogen by removal of four small fibrinopeptides and the resulting fibrin polymerizes to form a matrix for the nascent clot. Thrombosis is a pathological condition where inappropriate activity of the hemostatic mechanism interrupts normal blood flow, causing damage to tissues or organs. Since α-thrombin is widely accepted as the final key enzyme on all paths of thrombosis,[52,53] direct inhibition may offer advantages over currently used anticoagulants, such as heparin and warfarin. A potent, selective, long-acting, orally effective inhibitor of α-thrombin would be a significant advance in the prevention and/or treatment of acute and chronic thrombosis.

2.3. Thrombin Inhibitors

Several thrombin inhibitors have been under extensive study, and some have undergone clinical trials.[52,54] Argatroban (**10**) is marketed in Japan as a replacement for heparin;[54b,55] however, it is rapidly cleared and lacks useful oral activity.[54a] Recombinant hirudin, a naturally occurring 65-amino acid polypeptide from the salivary leech glands of the medicinal leech, and DuP-714 (**11a**), a boronic acid TSA, are important thrombin inhibitors that fall into the class of slow, tight-binding inhibitors. Native hirudin is the most potent and specific of all known thrombin inhibitors, having an incredible K_i value of 20 fM. Many synthetic thrombin inhibitors possess the D-Phe-Pro-Arg (fPR) recognition sequence, such as DuP-714 (**11a**), PPACK (**11b**), an irreversible inhibitor, and GYKI-14166 (**11c**, efegatran).

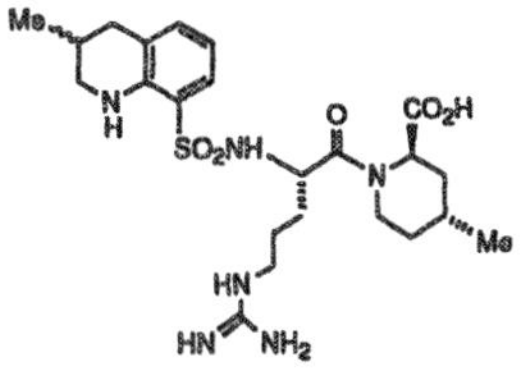

10

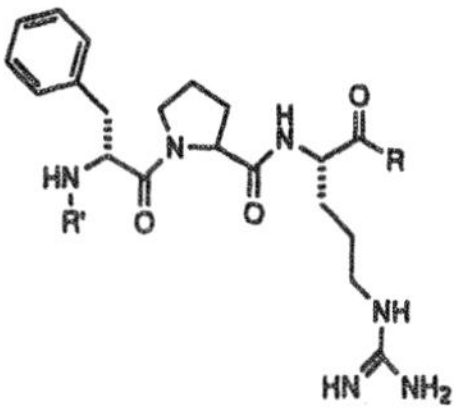

11

a R = $B(OH)_2$, R' = H (DuP-714)
b R = CH_2Cl, R' = H (PPACK)
c R = H, R' = Me (GYKI-14166)

The three-dimensional structure of trypsin, known from X-ray crystallography since the early 1970s,[47] has served for years as a prototype for the modeling of other trypsin-like serine proteases, such as α-thrombin.[56,57] The detailed molecular structure of α-thrombin was first established by X-ray crystallography in 1989 via the complex, PPACK–thrombin.[58] Subsequently, other crystal structures of inhibitor–thrombin complexes have been determined,[59,60] such as: *r*-hirudin–thrombin;[61] hirulog-1–thrombin, and hirugen–thrombin;[62] argatroban–thrombin and NAPAP–thrombin.[63,64] Structures have also been obtained for trypsin complexes of argatroban[65] and NAPAP.[66]

D-Phe-Pro-Arg-Pro-Gly-Gly-Gly-Gly-Asn-
Gly-Asp-Phe-Glu-Glu-Ile-Pro-Glu-Glu-Tyr-Leu

hirulog (**12**)

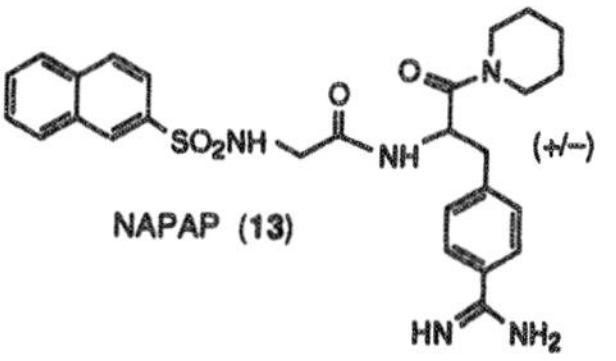

NAPAP (**13**) (+/−)

Comparison of these X-ray structures, relative to the active-site cleft of thrombin and trypsin, pinpoints important interactions for molecular recognition (depicted in Figure 2 for PPACK): (1) Formation of a salt bridge occurs in the specificity pocket (S1) between Asp189 and a basic moiety; for instance an amino, guanidino, or amidino group, on the inhibitor. (2) Interaction occurs with Ser195 in that the activated carbonyl or boronic acid group of TSAs usually react with Ser195 to form a negatively charged tetrahedral oxyanion or boronate that is stabilized through an intricate hydrogen-bonding network. (3) Formation of complementary hydrogen bonding to the Val213/Glu217 β-sheet of thrombin. In the case of PPACK, the D-Phe amino group forms a hydrogen bond with the carbonyl of Gly216 and the Arg α-NH group forms a hydrogen bond with the carbonyl of Ser214. (4) Nonpolar interaction in the hydrophobic S3 domain, especially for thrombin, in which there is also a pi-stacking interaction between an aromatic group of the inhibitor and Trp215 of thrombin.

The crystal structures of thrombin complexes reveal some interesting opportunities for additional positive interactions for exploitation in newly designed ligands. Specifically, Tyr60A, Trp60D, and Lys60F of the unique Tyr60A/Thr60I insertion loop are apparently available for new interactions. Although the fPR tripeptide sequence imparts high affinity to thrombin inhibitors, other motifs can offer high affinity, as well. The arylsulfonamides argatroban and NAPAP offer another molecular recognition paradigm to consider in the design of novel thrombin inhibitors.[63,64] Consideration of the arylsulfonamides is particularly relevant to serine protease selectivity because argatroban is recognized as one of the most selective thrombin inhibitors.[55c,d] By contrast, the tripeptide motif often does not confer selectivity for thrombin over trypsin, although it does confer selectivity relative to many other important serine proteases.

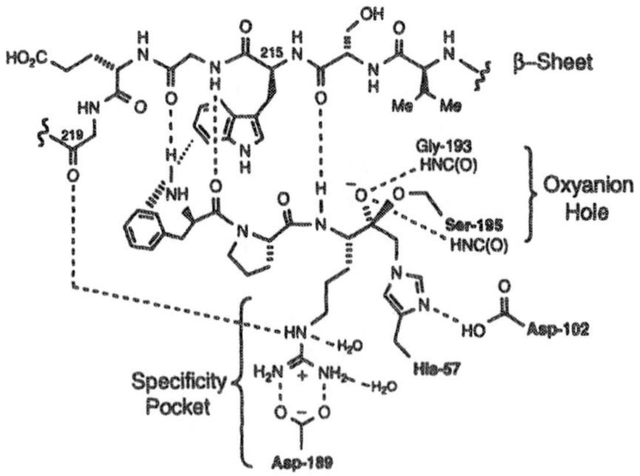

Figure 2. Schematic representation of the key interactions of PPACK–thrombin from an X-ray crystallographic study.

2.4. Cyclotheonamide Studies

Since CtA is a rare example of a serine protease inhibitor possessing a small cyclopeptide architecture, we focused attention on this unique natural product in the context of a drug design program aimed at discovering novel, small-molecule thrombin inhibitors. At the core of the strategy was our desire to understand fundamentals regarding the molecular basis for the interaction between CtA and thrombin.[45] Thus, our tactics involved: (1) determining the X-ray crystal structure of the bound complex of CtA–human α-thrombin; (2) obtaining a biochemical profile of CtA for the inhibition of other serine proteases; and (3) securing a total synthesis of CtA. An X-ray determination of CtA–human α-thrombin was expected to reveal intimate details regarding molecular recognition within the enzyme active site, while providing X-ray coordinates for computational docking of modified ligands. An inhibition profile of CtA in other serine proteases afforded further rationale for understanding molecular recognition within the enzyme site, especially when viewed in conjunction with X-ray structural details. A convergent total synthesis, amenable to the preparation of modified CtA structures, could be used to address structure-function issues via the preparation of analogues.

Structure of the Thrombin–CtA Complex

The molecular structure of a ternary complex between CtA, thrombin, and hirugen was determined by X-ray crystallography (2.3 Å).[44] Results of the X-ray structure, depicted in Figure 3, provide key insight into molecular recognition within the enzyme active site.

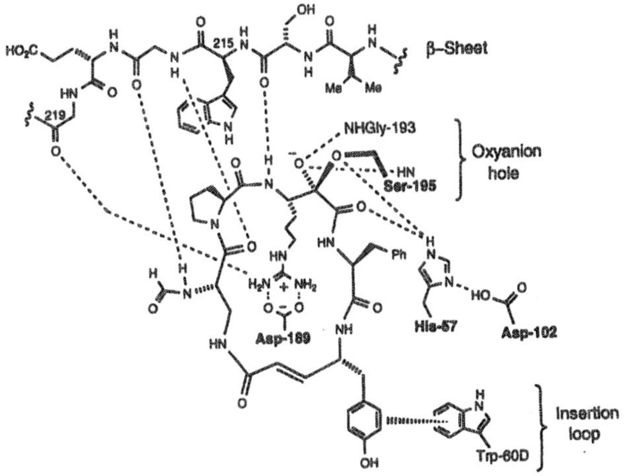

Figure 3. Schematic representation of the key interactions of CtA–thrombin from X-ray crystallography.

Clearly, the Pro-Arg unit of CtA interacts with the enzyme at the S2 apolar and the S1 specificity sites, respectively, similar to the Pro-Arg of PPACK.[58] The guanidinium group of Arg participates in a doubly hydrogen-bonded ion pair with Asp189 of thrombin and the –NH–C–C(O)-Pro-Arg– segment forms a hydrogen-bonded two-strand antiparallel β-sheet with Ser214/Trp215/Gly216. The two carbonyl groups of the α-keto amide are oriented at a dihedral angle of 109°, like the dicarbonyl conformation in FK-506, a macrocyclic- α-keto amide immunosuppressant, when it is complexed to the immunophilin FKBP.[67] However, the α-keto amide of CtA is involved in a complicated hydrogen-bonded, tetrahedral intermediate (hemiketal) that resembles a transition state for peptide hydrolysis. The Ser195 Oγ–C2 bond distance is 1.8 Å, which compares to a Ser195 Oγ–carbonyl bond distance in PPACK–thrombin of 1.6 Å[58] (C–O single bond distance in an aliphatic ether, ca. 1.45 Å). In the tetrahedral array, the γ-oxygen atom impinges orthogonally on the α-keto group from the *re* face (∠Oγ–C2–O2 = 86°, ∠C1–C2–Oγ = 93°, ∠C3–C2–Oγ = 97°); also, the keto oxygen (O2) makes a bifurcated hydrogen bond with thrombin.

As for other structural features, the CtA molecule adopts a relatively "open" conformation for the 19-membered ring with the proline ring orthogonal to the macrocyclic ring and one transannular hydrogen bond between O5 and N11 (3.1 Å). The aromatic groups of CtA engage in an aromatic stacking chain that encompasses Tyr60A and Trp60D of thrombin. This represents an unusual interaction for a thrombin inhibitor molecule because Trp60D of the insertion loop directly interacts with the ligand. However, the v-Tyr/Trp60D interaction is ostensibly weak in nature and turns out not to be significant for binding according to our analogue studies (*vide infra*). Another interaction exists between the D-Phe of CtA and Leu40/Leu41. The Tyr60A, Trp60D, and Leu99 residues also define a hydrophobic pocket that envelops the hydrocarbon portion of the proline of CtA. Significantly, the C13–C14 double bond of CtA is not involved in covalent attachment to the enzyme, but its proximity to Glu-192/Gly193 could be relevant to binding.

The three-dimensional solution structure of CtA was determined in an aqueous medium by employing NMR techniques.[68] The conformation of the D-Phe, h-Arg, and Pro segment of CtA is nearly identical in aqueous solution and the solid state; large differences occurred at the a-Ala and v-Tyr residues. Since the D-Phe, h-Arg, and Pro residues are more important for active-site binding, it can be inferred that CtA does not undergo a significant conformational change upon binding to the active site.

A crystallographic study of the binary complex formed between CtA and bovine β-trypsin revealed many comparable structural features for the ligand–enzyme interactions, although trypsin, being devoid of the insertion loop, lacks the loop-associated interactions.[46] In the CtA–trypsin complex, the aromatic rings of D-Phe, Tyr39 (instead of Glu39 in thrombin), and Phe41 (instead of Leu41 in thrombin) interact favorably, while the hydroxyphenyl of CtA is exposed to the solvent.

Enzyme Inhibition Studies

We investigated the action of CtA as an inhibitor of human α-thrombin and nine related serine proteases by enzyme kinetics.[44] The reversible, active-site-directed inhibitors GYKI-14766[69,70] and argatroban[54b,55] were also examined as reference standards. As such, our compilation affords a unique set of inhibition data for comparison. The data reveal that CtA is a much weaker competitive inhibitor of thrombin (K_i = 0.18 μM) than of trypsin (K_i = 0.023 μM) and streptokinase (K_i = 0.035 μM); its inhibition of thrombin is within the range of its inhibition of urokinase (K_i = 0.37 μM) and plasmin (K_i = 0.37 μM). GYKI-14766 has structural elements similar to those of CtA, including the possibility of forming a tetrahedral intermediate, although it is noncyclic and has a D-Phe group that affords an aromatic stacking interaction with Trp215. Our data indicate that GYKI-14766 is a potent, competitive inhibitor of thrombin (K_i = 0.016 μM), as well as trypsin (K_i = 0.032 μM); thus, it is not selective in this respect. By contrast, argatroban is an exceedingly selective inhibitor of thrombin (K_i = 0.0081 μM), in agreement with the literature.[55b]

Lewis et al.[45] found that CtA can exhibit slow, tight-binding inhibition of thrombin with a K_i value of 1.0 nM. The disagreement between our K_i value (at 37 °C) and that of Lewis et al. (at 25 °C) is related to the conditions employed for the enzyme kinetics. Neither the temperature (37 °C vs. 25 °C) nor pH differences turned out to be responsible for the disparity, rather the critical factors were the concentration of thrombin, the chromogenic substrate employed, and the time allowed for pre-equilibration. A comparison of results indicates that CtA can manifest standard competitive inhibition or slow-binding inhibition depending on the protocol of the experiment. Under conditions where thrombin was added to a mixture of substrate and inhibitor, the occurrence of slow-binding inhibition was dependent on the chromogenic substrate and on thrombin concentration, with slow binding more apparent at 0.1 nM than at 1.0 nM. However, under conditions where the substrate was added to an equilibrated mixture of CtA and thrombin, slow-binding inhibition occurred consistently, regardless of substrate or thrombin concentration. The K_i values were obtained by fitting the data to the equation $P = v_s t + (v_0 - v_s)(1 - e^{-kt})/k$, where P is the p-nitroaniline product, v_0, v_s, and k represent the initial velocity, steady-state velocity, and apparent first-order rate constant, respectively.[71,72] Under these conditions, our observation of slow-binding kinetics for thrombin inhibition by CtA (K_i = 4.1 ± 1.9 nM, N = 4) agrees well with the observation of Lewis et al.[45] We have since measured[73] the on-rate (k_{on}) for CtA with human α-thrombin and obtained a value of 3.3 ± 0.5 x 10^4 $M^{-1}s^{-1}$, which compares favorably with the value of 4.6 x 10^4 $M^{-1}s^{-1}$ reported by Lewis et al.[45] Because of the connection between the kinetic behavior and testing protocol, we determined K_i values under conditions that afford Michaelis–Menten and slow-binding kinetics in our structure–function studies of modified CtA analogues.

Total Synthesis of CtA and CtB

For the total synthesis of CtA, our intent was to devise a practical, convergent synthetic protocol amenable to the preparation of CtA and its analogues. Our overall approach centers around a [3 + 2] fragment condensation of segments **A** and **B** (Scheme 1).[44,74] In contrast to a linear sequence, this modular-type approach seemed more attractive for the assembly of analogues, especially for those requiring alteration of a single amino acid subunit. Hagihara and Schreiber[75] reported the first total synthesis of CtB via a linear assembly protocol, thereby reassigning the originally incorrect stereochemistry for the v-Tyr segment.[43] Other total syntheses of CtA or CtB have subsequently appeared.[76]

Synthesis of segment **A** commenced with the homologation of arginine derivative **14** (Scheme 2). The corresponding aldehyde derivative of **14** was converted to a diastereomeric mixture (3:2) of cyanohydrins **15**. Since the carbon bearing the stereoisomeric alcohols would ultimately exist as a carbonyl carbon in CtA, **15** was carried through the synthesis as a diastereomeric mixture. Application of the Pinner synthesis to **15** provided the intermediate α-hydroxy methylimidate, which was hydrolyzed to

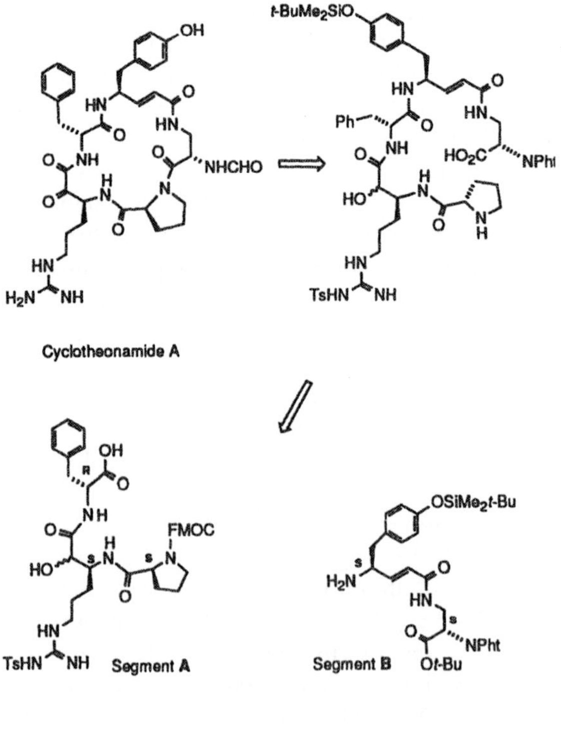

Scheme 1.

the α-hydroxy methyl ester in a one-pot procedure. Hydroxyl group protection followed by saponification afforded the key homologated arginine unit **16**. Routine peptide synthetic methods were employed to transform **16** into segment **A**. Of some concern to us was the concomitant loss of α-hydroxyl protection, since a free hydroxyl at this position has the potential to compete in an intramolecular lactonization process.

The key components of segment **B** are vinylogous tyrosine acid **17** and diaminopropionic acid derivative **18** (Scheme 3). The hydroxy and α-amino groups of methyl tyrosine were protected, and the corresponding tyrosine aldehyde derivative was formed from a DIBAL-H reduction of the ester functionality. Wittig olefination followed by selective cleavage of the *tert*-butyl ester afforded **17**. The amino groups of diaminopropionic acid were differentiated using carbobenzoxy and phthalimide protecting groups as in **19**. Hydrogenolysis of the β-amino CBZ group afforded **18**. It is noteworthy to point out that protection of the α-nitrogen as a phthalimide allows protection until a late stage in the synthesis, a strategy which enables efficient

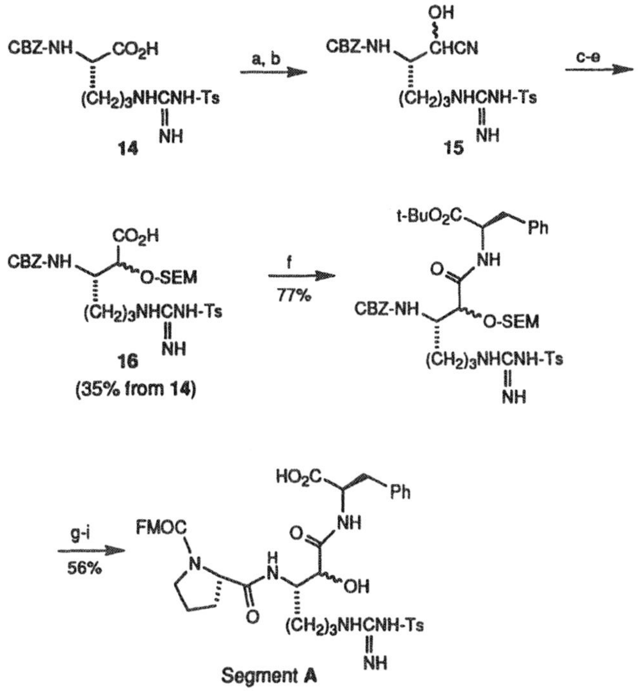

Scheme 2. **a.** (*i*) (1-imidazolyl)$_2$C=O, (*ii*) (i-Bu$_2$AlH; **b.** KCN; **c.** HCl, MeOH; **d.** SEM-Cl, 2,6-lutidine; **e.** LiOH; **f.** D-PheO-t-Bu, HOBt, dCC; **g.** Pd(OH)$_2$, H$_2$; **h.** FMOC-L-Pro, HOBt, DCC; **i.** CF$_3$CO$_2$H. i-Bu, Isobutyl; Me, methyl; SEM, trimethylsilylethoxymethyl; HOBt, 1-hydroxybenzotriazole; DCC, dicyclohexylcarbodiimide; FMOC, fluorenylmethoxycarbonyl; Ts, p-toluenesulfonyl.

construction of analogues in which this nitrogen bears substituents that might better satisfy the hydrophobic S3 binding pocket of thrombin (*vide infra*). Processing of **17** and **18** by using standard techniques completed the synthesis of segment **B**.

The final stage of the synthesis required coupling of segments **A** and **B**, macrocyclization, and incorporation of α-keto amide and N-formyl functionalities (Scheme 4). Segment condensation was successfully effected by employing the coupling agent BOP-Cl to afford **20**. Fortunately, intramolecular lactone formation involving the α-hydroxyl and carboxyl groups of segment **A** was not competitive in the condensation process since only a trace of this product could be detected in the product mixture. Following removal of FMOC and *t*-butyl protecting groups from **20**, the stage was set for macrocyclization. Again, we employed BOP-Cl and used high dilution conditions (0.001 M) to obtain the 19-membered macrolactam **21**. Subsequent optimiztion studies increased the yield to 41% when DCC/HOBt was used. We were gratified that macrocyclic lactonization to the 15-membered lactone was not competitive with lactamization, although a small amount of this material

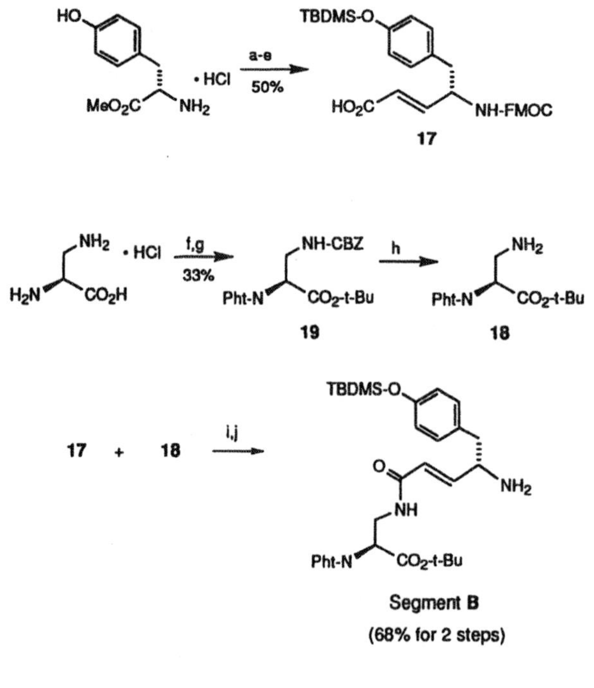

***Scheme* 3.** **a.** FMOC-Cl, K_2CO_3; **b.** TBDMS-Cl, imidazole; **c.** *i*-Bu_2AlH; **d.** $Ph_3P{=}CHCO_2$-*t*-Bu; **e.** CF_3CO_2H; **f.** (*i*) CBZ-Cl, (*ii*) isobutylene, H_2SO_4; **g.** *N*-carbethoxyphthalimide; **h.** $Pd(OH)_2$, H_2;. **i** HOBt, EDCI•HC1; **i.** Et_2NH. FMOC, fluorenylmethoxycarbonyl; TBDMS, *tert*-butyldimethylsilyl; Et, ethyl; *i*-Bu, isobutyl; Ph, phenyl; HOBt,1-hydroxybenzotriazole; EDCI,1-ethyl-3-(3'-dimethylaminopropyl) carbodiimide.

was detected by TLC. With **21** in hand, the final steps of the synthesis were carried out. Removal of the phthalimide afforded the corresponding free primary amine **22**, which was formylated with ethyl formate. This late-stage deprotection proved to be an efficient strategy, since N-acetylation of **22** (pentafluorophenyl acetate, 90%) gave the intermediate necessary for the synthesis of cyclotheonamide B (CtB, **23**). Dess–Martin periodinane oxidation followed by removal of silyl and tosyl protecting groups completed the synthesis of CtA. Similar processing of the N-acetylated intermediate afforded CtB, further demonstrating the versatility of our synthetic approach.

Biological Evaluation of CtA Analogues

Equipped with the tools acquired from the aforementioned studies, we set out to explore specific structure–function issues regarding inhibition of thrombin by

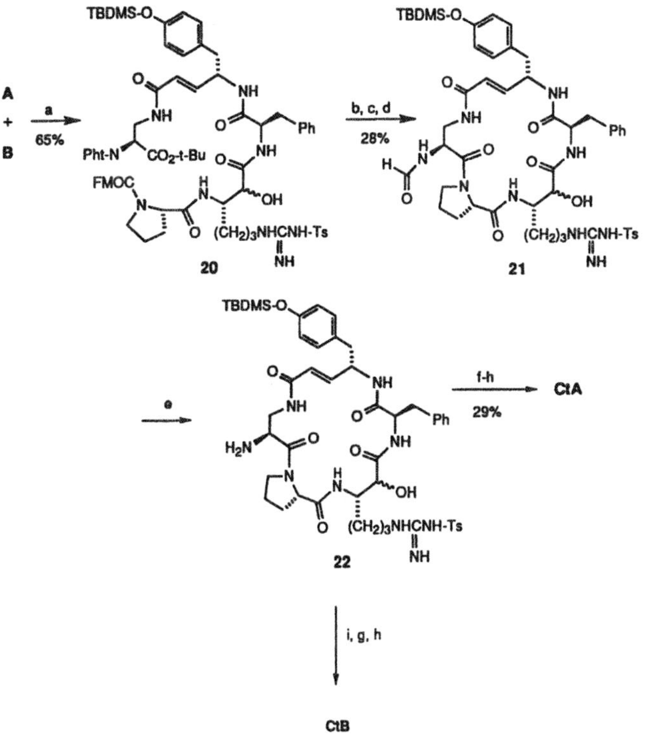

Scheme 4. a. BOP-Cl, Et_3N; **b**. Et_2NH; c. CF_3CO_2H; **d.** BOP-Cl, 4-dimethylaminopyridine; **e**. N_2H_4•H_2O, **f**. HCO_2Et; **g**. Dess–Martin reagent; **h**. HF; **i**. Pentafluorophenylacetate, DMF. Et, ethyl; BOP-Cl=bis(2-oxo-3-oxazolidinyl)phosphinic chloride; DMF, dimethylformamide.

CtA.[77] The formamide group of CtA is situated in the thrombin active site such that the hydrophobic S3 subsite is left vacant (Figure 4A).[44] In the PPACK–thrombin complex, the phenyl group of D-Phe nicely occupies this hydrophobic pocket, while pi-stacking with Trp215 (Figure 4B).[58a] It was our objective to modify the C9 formamide group in an attempt to take advantage of this hydrophobic space by incorporating the more hydrophobic phenylacetyl (**24**) and phenylethyl (**25**) substitu-

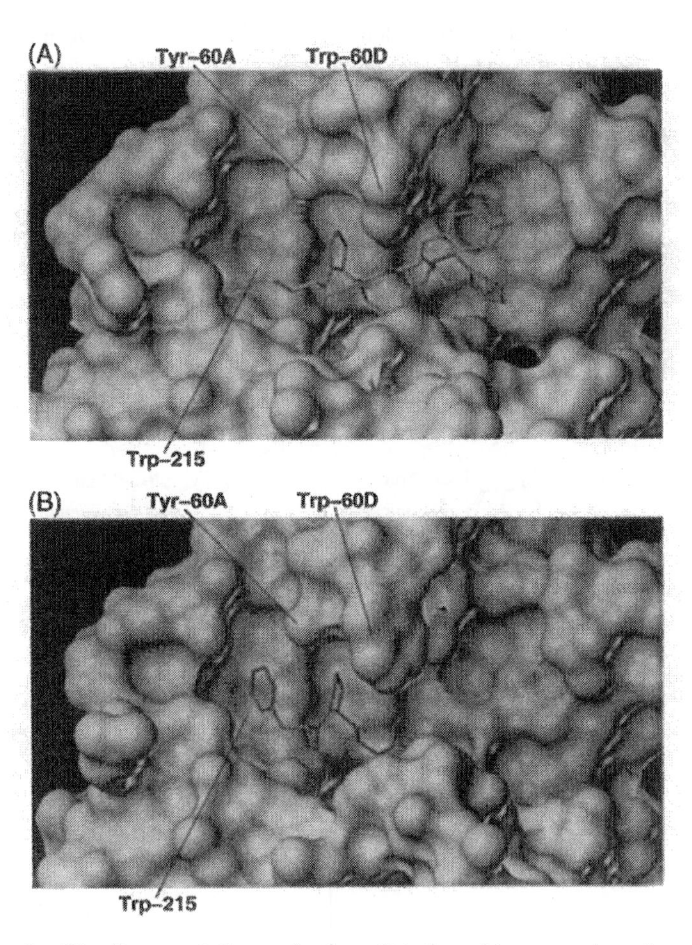

Figure 4. (**A**) Representation of the CtA–thrombin complex from X-ray crystallography[44] displaying the active-site region. Thrombin is shown with a Connolly surface; CtA is shown as a stick model of its non-hydrogen atoms. The S3 hydrophobic pocket, located between residues Tyr60A and Trp215, is vacant. (**B**) Representation of the PPACK-thrombin complex from X-ray crystallography[58] displaying the active-site region. Thrombin is shown with a Connolly surface; PPACK is shown as a stick model of its non-hydrogen atoms. The S3 pocket, located between residues Tyr60A and Trp215, is occupied by the phenyl ring of PPACK.

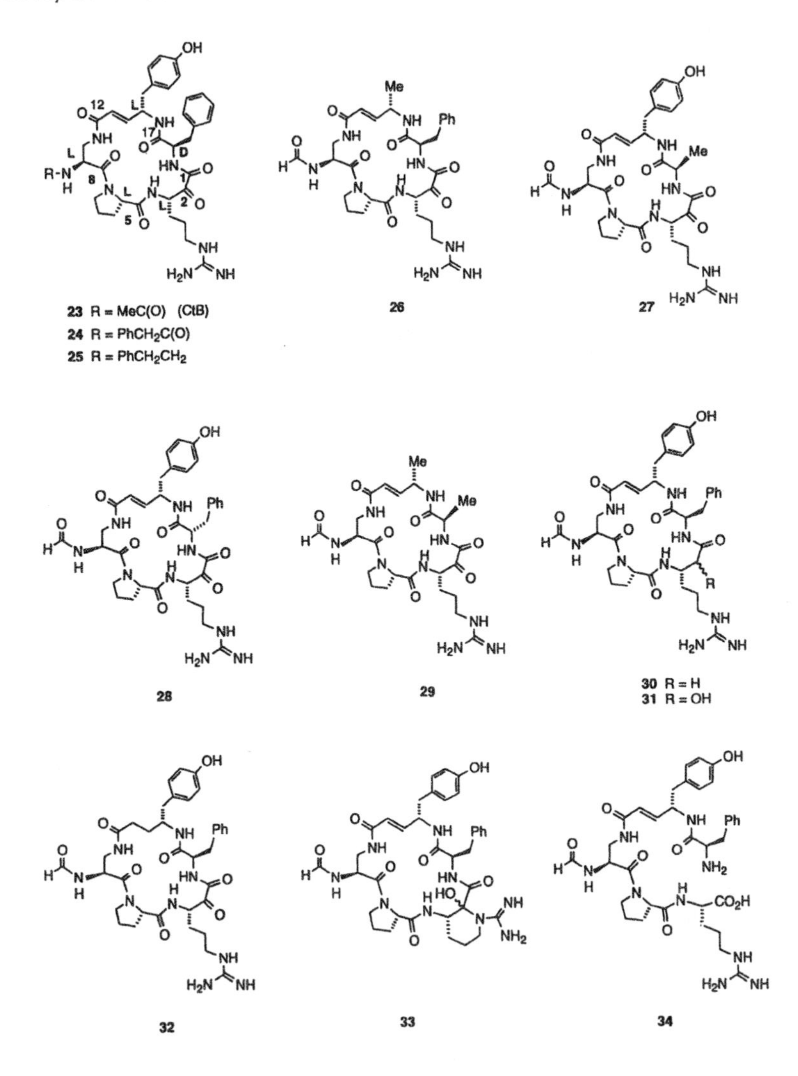

Figure 5. CtA analogues designed to probe specific active-site interactions.

ents. This took advantage of our synthetic route to cyclotheonamide derivatives that involves the late-stage, key primary amine intermediate **22**.

The structural features of the 19-membered ring and the Pro-Arg segment of CtA are virtually the same in its complexes with thrombin[44] and trypsin;[46] the main distinctions are subtle and relate to the orientations of the v-Tyr and D-Phe aromatic groups. Since thrombin has a more restrictive catalytic cleft, with the presence of a 60A–60I insertion loop that is proximate to the v-Tyr side chain, differences with respect to the aromatic side chains are not surprising. In comparing enzyme–inhibitor interactions and steric fit between the two X-ray structures,[44,46] there are only limited structural factors to explain the greater effectiveness of CtA for inhibition of trypsin relative to thrombin. Although the CtA–thrombin complex has an aromatic stacking interaction between v-Tyr of CtA and Trp60D,[44] which is absent from the CtA–trypsin complex, the CtA–trypsin complex has the D-Phe side chain ensconced in a hydrophobic area defined by two aromatic residues of the enzyme (Tyr39 and Phe41).[46] To assess the importance of the v-Tyr/Trp60D interaction, we decided to synthesize a v-Ala analogue that lacks the hydroxyphenyl group (viz. **26**) but retains the stereogenic center. We also addressed the role of the D-Phe residue at P1' by replacing it with D-Ala (**27**) and Phe (**28**).[78] Analogue **29**, in which the v-Tyr and D-Phe groups are simultaneously replaced with Ala was also prepared to further probe aromatic interactions.[78]

The ^{13}C NMR spectrum for CtA in D_2O (pH 4.5) indicates a predominance of the *gem*-diol form (covalent hydrate), as opposed to the α-keto amide form.[44,76] Methanol adducts of CtA have also been observed by mass spectrometry, presumably caused by addition of methanol solvent to C2.[44,76] It is unclear whether the tendency for CtA to exist as a covalent adduct at C2 has a bearing on its enzyme inhibitory properties, but the X-ray structures of the CtA–thrombin[44] and CtA–trypsin[46] complexes indicate the importance of covalent adduct formation with Ser195. In this connection, we synthesized the deoxy C2 analogue **30**,[78] and the keto-reduced (C2 hydroxy) analogue, as a diastereomeric mixture (viz. **31**), for biological evaluation.

We also synthesized an analogue in which the carbon–carbon double bond of CtA is saturated (viz. **32**). It was thought that this might exert an influence on the conformation of the 19-membered-ring macrocycle and thereby affect biological activity.

Our synthetic protocol served us well for preparing the desired analogues.[77] Analogues **24** and **25** were prepared from intermediate **22** by standard acylation and reductive alkylation procedures, respectively. Synthesis of **26** and **27** required prior assembly of the (v-Ala)-(a-Ala) and (v-Tyr)-(a-Ala) subunits, respectively, according to the [3 + 2] segment condensation approach, and **32** was prepared by introducing a late-stage diimide reduction of the v-Tyr double bond. Analogue **31** was prepared by simply omitting the Dess–Martin oxidation step, while **30**[78] was prepared via the deoxygenation product from the α-hydroxy methyl ester corresponding to **16** (thiocarbonyldiimidazole, followed by tributyltinhydride).

The phenylacetyl analogue **24** of CtA is approximately the same potency as CtA under the "Michaelis–Menten" and "slow-binding" conditions, which contradicts the idea about utilizing the S3 pocket to amplify the binding of CtA with thrombin (Table 1). Given this result, we considered the possibility that the amide linkage might not permit the hydrophobic group to properly access the S3 pocket due to conformational constraints. Thus, it seemed worthwhile to study the phenylethyl analogue **25**. Surprisingly, the K_i values for **25** are nearly the same as those for the parent compounds CtA and CtB (although the slow-binding K_i is twice as good). It is also interesting to note that there is little consequence in having a basic amine substituent **25** versus an amide substituent **24**. In the final analysis, our attempt to utilize the S3 pocket through analogues **24** and **25** failed to enhance inhibitor potency. Nevertheless, we feel that other hydrophobic substitution at P3 may serve to enhance binding within the active site (*vide infra*).

Another issue we wished to address concerned the potential stereochemical lability of the C3 stereogenic center of CtA.[78] Since it is adjacent to the α-keto amide group, we attempted to obtain the *epi*-C3 isomer for biological testing by treating CtA with mild base. However, upon treating CtA with triethylamine, we obtained a product whose ^1H NMR and mass spectral data were inconsistent with the expected C3 epimer. Instead, the product was assigned the cyclic hemiaminal structure **33** (4:1 mixture of diastereomers at C3) resulting from addition of the proximal guanidine nitrogen to the ketone carbonyl of the keto–amide. The relatively weak thrombin inhibitory activity displayed by **33** was consistent with our expectations; however, we sought more convincing structural information. Given the observed inhibition constant for **33**, we were able to probe its molecular structure through X-ray analysis of a complex with thrombin. Surprisingly, the thrombin-bound material proved to be the ring-opened pentapeptide **34**, resulting from enzymatic cleavage of the α-keto amide bond. This result gives rise to interesting speculation regarding molecular recognition and structural issues. The observed enzymatic cleavage of **33** implies that the putative hemiaminal structure is responsible for the changeover from competitive inhibitor to substrate. Furthermore, the macrocyclic framework might provide an entropic advantage for forming the hemiaminal structure, since acyclic D-Phe-Pro-Arg congeners failed to behave analogously on base treatment.[78] Further studies along these lines could shed new light on structure–function issues of related macrocyclic serine protease inhibitors.

Replacement of the hydroxyphenyl group, suggested to be involved in a stacking interaction with Trp60D and D-Phe of CtA, by a methyl group (**26**) did not provide the expected 10- to 50-fold loss of inhibition.[77] This result indicates that the stacking interaction alone is not critical for binding. However, replacement of the D-Phe residue with D-Ala reduced thrombin inhibition by about three-fold.[78] This result is consistent with the hydrophobic nature of the S1' subsite of thrombin. More notably, the simultaneous replacement of D-Phe and v-Tyr with Ala residues (**29**) caused a significant decrease in K_i (thr).[78] This result gives rise to the notion that D-Phe and v-Tyr can partially compensate for the absence of each other, since independent replacement of each residue affords compounds with similar potency (cf. **26** and **27**). The greater loss of activity in going from CtA to **27**, vs. CtA to **26**, indicates that the D-Phe residue is somewhat more important. Given the aromatic stacking chain involving Tyr60A, Trp60D, D-Phe, and v-Tyr, as viewed by X-ray crystallography,[44,46] a cooperative conformational effect between the two aromatic side chains of CtA is not unreasonable. Further insight into the interactions at S1' was garnered by changing the D-Phe residue to L-Phe (**28**).[78] The significant loss of thrombin inhibition in going from CtA to **28** can be attributed to an unfavorable steric interaction between the phenyl ring of L-Phe and the insertion loop of thrombin, an interaction for which we gained under-

Table 1. Kinetic Constants for Thrombin Inhibition by CtA, CtB, and Analogues

Compound	K_i *(nM)*[a,b]	K_i *(nM)*[a,c]
CtA (**1**)[d]	170 ± 80 (11)	4.1 ± 1.9 (4)
CtB (**23**)	200 ± 170 (3)	3.7 ± 2.1 (3)
24	320 ± 180 (3)	3.1 ± 0.8 (3)
25	290 ± 150 (6)	1.5 ± 0.2 (3)
26	200 ± 80 (3)	5.3 ± 0.5 (3)
27	380 ± 60 (3)	12 ± 3 (2)
28	10,800 ± 10 (2)	
29	230 ± 5 (6)	
30	inactive	
31	11,000 ± 7,000 (4)	100 ± 30 (3)
32	400 ± 230 (3)	6.3 ± 4.3 (3)
33	47,000 (1)	220 ± 90 (2)
argatroban (**10**)	10 ± 2 (6)	[e]
DuP-714 (**11a**)	0.92 ± 0.20 (4)	0.028 ± 0.014 (4)

Notes: [a] The number of experiments (*N*) is given in parentheses.
[b] Conditions biased to Michaelis-Menten kinetics.
[c] Conditions biased to slow-binding kinetics.
[d] The data reported here are for natural CtA.
[e] This inhibitor is not reported to show slow-binding behavior; however, we were able to detect a second steady-state slope that corresponds to K_i value of 0.84 ± 0.30 nM (3).

standing by using molecular modeling to analyze a similar situation for a related macrocycle (*vide infra*).

Analogue **30**, in which the electrophilic carbonyl has been removed, is completely devoid of activity, a result which underscores the importance of the covalent interaction between the Ser195 hydroxyl group and the keto–amide carbonyl of CtA. Hydroxy amide **31** is 25-fold less potent than CtA (slow binding). Lastly, saturation of the double bond of CtA (**32**) does not significantly affect activity, a result which is consistent with the X-ray crystal structure of the CtA thrombin complex.

The results of our structure–function studies are helpful in understanding the role that the various subunits of CtA play within the active site of thrombin. In summary, the following are important for CtA–thrombin interaction: (1) the C2 keto–amide carbonyl covalent interaction with the Ser195 hydroxyl group; (2) the phenyl ring of the D-Phe residue, which sufficiently fulfills an important hydrophobic interaction with the S1' subsite of thrombin; (3) stereochemistry at the P1' position; only D-Phe presents a phenyl ring in the proper orientation for hydrophobic interaction; and (4) a synergistic relationship between v-Tyr and D-Phe residues. Overall, our structure–function studies of CtA provide key insight into the molecular basis of the interaction between CtA and thrombin. We parlayed our experience to the design of novel thrombin inhibitors in the context of a drug discovery program.

3. MACROCYCLIC THROMBIN INHIBITORS: A HYBRID APPROACH

The considerable structural information available for thrombin and its interactions with ligands has contributed amply to the design of potent, active-site-directed thrombin inhibitors. For example, the mechanism of the interaction of thrombin with fibrinogen has been studied in detail, especially with respect to the cleavage of the Arg16/Gly17 bond in the Aα-chain of human fibrinogen.[80-89] The Aα (1-23) fragment (Figure 6) contains a fundamental structure element for thrombin binding. Kinetic studies indicate that the specific binding to thrombin lies within sequence Asp(7)-Arg(16);[81,82] no residues on the C-terminal side of Pro18 are required.[81] Further insight into the conformation of the Aα-chain was obtained through high-resolution NMR studies of the interaction of bovine thrombin with the Arg-Gly binding site on the Aa chain.[89] In particular, transfer NOE (TRNOE) and NMR distance geometry calculations were used to determine the thrombin-bound structure of the synthetic peptide, *N*-acetyl-Asp(7)-Arg(16).[89] The key fea-

Ala-Asp-Ser-Gly-Glu-Gly-Asp(7)-Phe(8)-Leu(9)-Ala-Glu(11)-Gly(12)-Gly-Gly-Val-Arg(16)-Gly(17)-Pro(18)-Val(20)-Val-Glu-Arg.

Figure 6. Amino acid sequence of the Aα (1-23) chain.

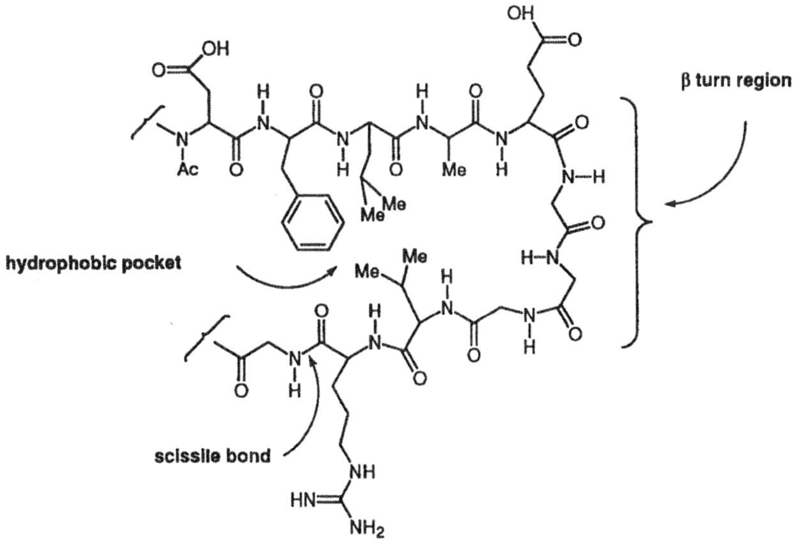

Figure 7. Secondary structure of *N*-acetyl-Asp(7)-Arg(16) bound to thrombin as determined by NMR.

tures of the thrombin-recognized structure that emerged from these studies, highlighted in Figure 7, are: (1) A complementary hydrophobic combining site, approximately 8 Å in diameter, formed by the nonpolar side chains of Phe8, Leu9, and Val15 within fibrinopeptide A, which is released on the conversion of fibrinogen to fibrin. (2) A chain reversal caused by a multiple turn conformation within residues Glu(11)-Val(15); Glu-11 and Gly12 accommodate a type II β-turn.[90]

This "hair-pin" geometry was corroborated by an X-ray crystal structure[91] of a complex of bovine thrombin and a peptide corresponding to residues 7-16 of the Aα -chain of fibrinogen. Conceptually, this defines a macrocycle-type spatial arrangement for the substrate within the enzyme complex. Fortuitously, CtA appears to confirm the viability of a cyclic inhibitor structure (Figure 8), in the form of a hybrid structure, by virtue of a bridge connecting the Pro-Arg recognition sequence and an α-keto amide group. Furthermore, as mentioned earlier, the various serpins function by use of a peptide recognition motif in a macrocyclic array. This formed a foundation for our interest in synthetic cyclic inhibitors of thrombin and other serine proteases. Despite the potential advantages for macrocyclic motifs, it is noteworty that small molecule cyclic inhibitors of serine proteases are unusual and underexplored.

3.1. General Approaches to Design

Conceptually, the simplest hybrid macrocyclic construct results from connecting an Arg at P1 to a D-Phe at P1' in the form of a cyclic keto amide (viz. **35**; syn-

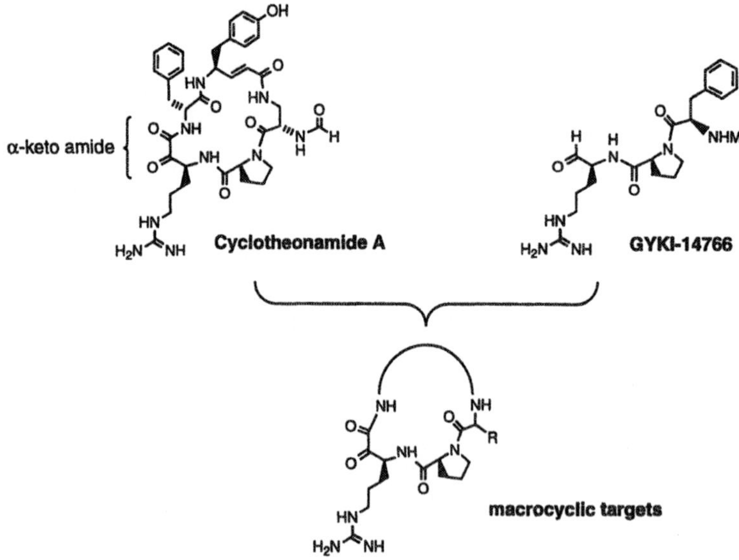

Figure 8. Design of macrocyclic inhibitors from CtA and a tripeptide thrombin inhibitor.

thetic route, *vide infra*). Such a structure contains the required P1 Arg residue for interaction with the Asp189 of the "specificity pocket", a D-Phe residue at P1' as in CtA, a hydrocarbon linker to form a macrocycle and thereby mimic the chain reversal of the Aα-chain of fibrinogen, and a keto–amide functionality for covalent, reversible interaction with the active-site Ser195 residue. As a measure of selectivity for thrombin versus other serine proteases, we routinely evaluated compounds for their ability to inhibit thrombin and trypsin. For pursuing thrombin inhibitors as potential therapeutic agents, we set a requirement that compounds of interest should have inhibition constants for thrombin, K_i (thr), that are less than ca. 1 μM. From this perspective, **35** proved to be ineffective as a thrombin inhibitor, and its relatively low K_i as a trypsin inhibitor, K_i (tryp), was an early omen that selectivity for thrombin versus trypsin would be an uphill battle. It is conceivable that the 14-membered macrocycle of **35** could be optimized with regard to thrombin inhibition; however, we chose instead to incorporate the customary proline residue at P2, as in **36**. In this case, a respectable K_i (thr) value was achieved with **36a**. Interestingly, variation in ring size is associated with a dramatic change in K_i values for thrombin inhibition, with a ca. 10-fold difference in K_i (thr) occurring between the 17-membered and 19-membered rings of **36a** and **36b**. Given the tight "proline pocket" available at the S2 position of thrombin, one might speculate that macrocycle ring size has a significant influence on the conformation that the Pro-Arg sequence can adopt within the active site of thrombin. Furthermore, since the fPR class of tripeptide inhibitors adopts an extended conformation on binding to thrombin,[56] one might further conclude that optimal macrocyclic ring sizes will permit an extended conformation of the Pro-Arg segment to be favored. With regard to trypsin, it is not surprising that **36a** and **36b** are approximately equipotent inhibitors. Since the 60A–60I insertion loop of thrombin is absent in trypsin, there is a less restrictive, more accommodating S2 subsite in the latter case.

The 60A–60I insertion loop of thrombin contains a lysine (Lys60F) residue, which presents an opportunity for a new binding interaction. According to docking experiments using thrombin/CtA coordinates,[92] the aminobutyl side chain of Lys60F is accessible by a suitably positioned carboxylic acid at the P1' position of a macrocycle, as in **37**. Importantly, stereochemistry is critical, as only the Glu stereoisomer, as opposed to D-Glu, has the proper orientation to interact with the amino group of the Lys60F side chain. Unfortunately, **37** was completely devoid of activity as a thrombin inhibitor. Although this result is somewhat perplexing, it served as an early warning that active-site ionic interactions, other than those occurring within the P1 specificity pocket, may be difficult to attain. Again, the indiscriminate nature of trypsin is underscored by the effectiveness of **37** as a trypsin inhibitor.

3.2. The Macrocyclic D-Phe-Pro-Arg Motif

With these preliminary results in hand, we pursued a more focused approach.[93] We concentrated our efforts on a study of a macrocyclic D-Phe-Pro-Arg motif with

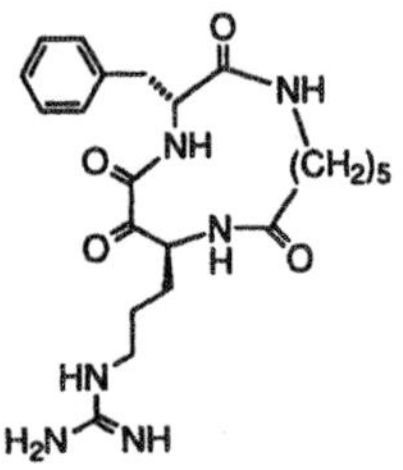

35

K_i (μM)
thr. tryp.
3.80 0.31

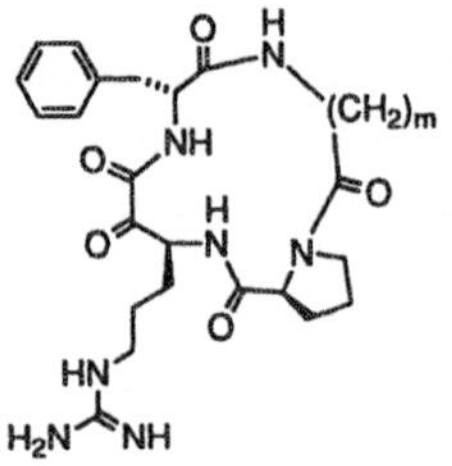

36

K_i (μM)
thr. tryp.
a) m = 5 0.15 0.026
b) m = 7 1.61 0.011

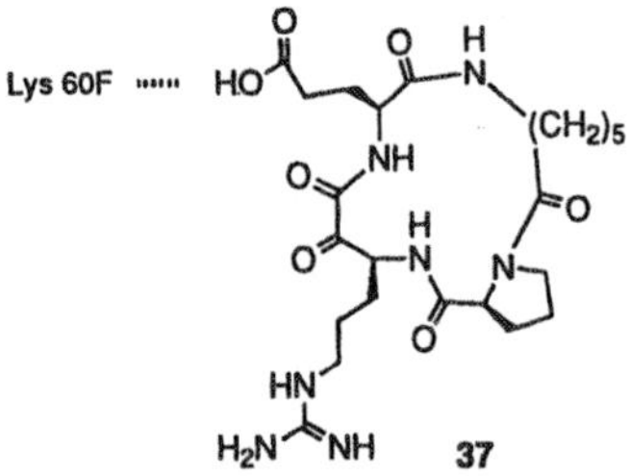

Lys 60F ······

37

K_i (μM)
thr. tryp.
85.90 0.62

the intent of evaluating effects of ring size and substitution at P3 and P1' (see Table 2). We parlayed our experience in the synthesis of CtA into a general synthetic route to the desired macrocycles. In Scheme 5, the route is exemplified for the synthesis of **38c**. For incorporation of P1' residues, the desired P1' amino acid *tert*-butyl ester is coupled to **16** prior to incorporation of **39** into the sequence.

Macrocyles **38a-j** (Table 2) provide general trends relating ring size and P3 substituents to activity. Inhibition data for CtA, efegatran, and the corresponding *N*-methyl keto-amide **40**[94] are included for comparison. Macrocyclic ring sizes of 19, 21, and 22 members appear to be in the optimal range for thrombin inhibition (cf. **38c-e** with **38a** and **38b**). As discussed previously, the relationship between ring size and thrombin inhibitory activity most likely centers around the ability of the fPR sequence to adopt an extended conformation. From this viewpoint, it could be concluded that the superior activity of the 19–21– and 22-membered ring analogues (i.e., **38c-e**) is related to their ability to present the fPR unit in an extended conformation. The reason for the seemingly anomalous behavior of **38b** as a thrombin inhibitor is unclear; however, it seems unlikely that ring size alone could be responsible for the high K_i (thr). Another implication with respect to ring size versus

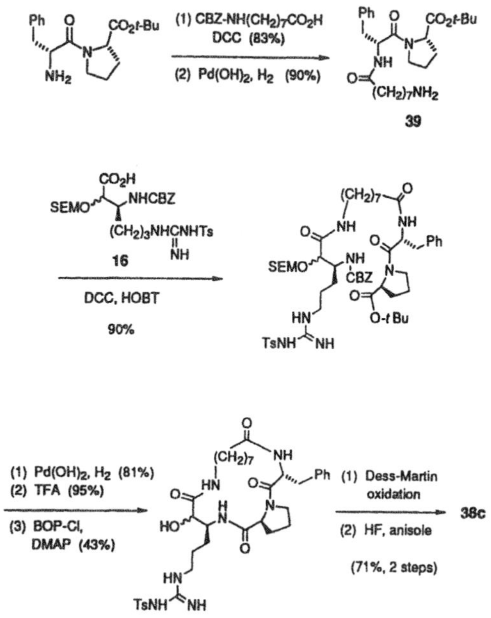

Scheme 5.

activity is that selectivity for thrombin versus trypsin does not seem to be greatly influenced by variation in macrocycle size. The trypsin inhibitory activity of compounds **38a-e** roughly correlates with thrombin inhibition, with the exception of **38b**. Thus, modification of other sites of the macrocycle seemed to be warranted.

The S2 apolar subsite of thrombin has a better-defined surface than the corresponding site of trypsin (*vide supra*), so we probed this site by altering the P2 proline subunit. Compound **38f**, in which a homoproline occupies P2, is ca. three-fold less potent as a thrombin inhibitor than **38c.** Compound **38g** positions a methoxy substituent into the S2 subsite; this resulted in a ca. six-fold drop in thrombin inhibitory activity relative to **38c**. These results indicate that the S2 subsite of thrombin is sensitive to steric bulk, and that even slight perturbations away from a P2 proline residue on the macrocycle are deleterious to thrombin inhibitory activity. Thus, as in acyclic fPR-based thrombin inhibitors, a proline residue appears to be optimal for P2 in the macrocyclic version. Interestingly, the trypsin S2 site seems to tolerate a methoxyproline (i.e., **38g**) well (cf. **38g** with **38c**); however, a corresponding homoproline residue (i.e., **38f**) causes a six-fold decrease in trypsin inhibitory activity. A possible explanation might be found by considering the X-ray structure of the trypsin-CtA complex.[46] The proline residue in the CtA/trypsin complex is partially exposed to solvent, making the relatively larger, hydrophobic homoproline residue less favorable than proline. However, similar positioning of the polar methoxy sub-

Table 2. Biological Data for Macrocyclic Thrombin Inhibitors

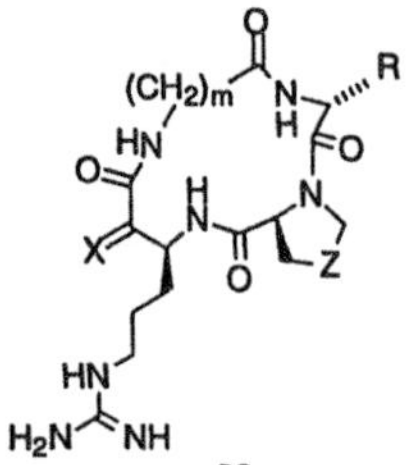

Compound	*m*	*X*	*Z*	*R*	K_i *(thr)*[a]	K_i *(tryp)*[a]
38a	5	O	CH_2	CH_2Ph	473 ± 23 (2)	710 ± 260 (3)
38b	6	O	CH_2	CH_2Ph	1800 ± 1000 (6)	280 ± 20 (6)
38c	7	O	CH_2	CH_2Ph	24 ± 4.2 (6)	10.0 ±1.0 (6)
38d	9	O	CH_2	CH_2Ph	9.9 ± 0.7 (7)	2.1 ± 0.5 (5)
38e	10	O	CH_2	CH_2Ph	15 ± 1.6 (6)	25 ± 12 (4)
38f	7	O	CH_2CH_2	CH_2Ph	86 ± 5 (3)	66 ± 15 (3)
38g	7	O	cis-CHOMe	CH_2Ph	159 ± 22 (9)	13 ± 2.4 (6)
38h	7	O	CH_2	Ph	163 ± 21(3)	11 ± 4 (2)
38i	7	O	CH_2	CH_2CH_2Ph	20 ± 2.8 (5)	14 ± 2.4 (5)
38j	7	H/OH	CH_2	CH_2CH_2Ph	IA[b]	
CtA (**9**)					4.1 ± 1.9(4)	1.0 ± 0.1(2)
efegatran (**11c**)					10.0 ± 5.0(6)	3.9 ± 1.2 (6)
Me-D-Phe-Pro-Arg-C(O)C(O)NHMe (**40**)					14 ± 5(5)	4.4 ± 1.5 (4)

Notes: [a] Data are reported in nM units. Standard error, given for N measurements, is indicated in parentheses.
[b] IA=inactive, as defined by the inability to inhibit thrombin at a concentration of 50 μM.

stituent does not result in a particularly unfavorable interaction, hence the similar K_i (tryp) for **38c** and **38g**.

To probe substitution at the P3 position, we prepared 19-membered ring analogues **38h** and **38i**, in which the positioning of the phenyl ring in the apolar S3 site of each enzyme is altered using methylene spacers. Interestingly, substitution of the D-Phe residue of **38c** with a phenylglycine residue, as in **38h**, resulted in a sevenfold decrease in thrombin inhibition. This result, in contrast to the acyclic Boc-D-Phe-Pro-Arg-H series in which a similar substitution resulted in a three-fold increase in thrombin inhibition,[95] suggests that the macrocycle confers a degree of constraint with respect to the positioning of the P3 substituent within the corresponding apolar S3 binding site of thrombin. Compound **38i**, which extends the phenyl substituent by a methylene relative to **38c**, is essentially equipotent to **38c**. These results are more or less in line with expectations with regard to the S3 site of both enzymes. The S3 subsite of thrombin and trypsin is overall hydrophobic; however, in trypsin the substitution of thrombin's Ile174/Arg175 by Gly174/Gln175 makes the S3 subsite more polar. This has been offered as a partial explanation for the trypsin selectivity of CtA since the S3 subsite of trypsin is better suited to accommodate the polar N-formyl residue.[46] From this perspective, macrocycles **38c**, **38e**, and **38i** represent improvements in thrombin selectivity brought about by the hybrid approach. In our CtA analogue studies, replacement of the N-formyl moiety with a hydrophobic residue at the P3 position (i.e., **24** and **25**, Figure 5; Table 1) did not result in the expected increase in thrombin inhibitory activity. To address this result in the context of a hybrid concept, we incorporated the *N*-formyl-(a-Ala)-Pro-(h-Arg) segment of CtA into the 19-membered macrocyclic keto amide **41**. As in the comparison between CtA and **24/25**, the biological profile of **41** versus **38c** (or **38i**) provides insight into ligand binding with respect to the S3 apolar binding site of trypsin and thrombin. In contrast to the CtA analogue studies, the nature of the P3 substituent has a significant effect on thrombin inhibitory activity, as evidenced by the ca. 10-fold drop in K_i (thr) in going from **38c** to **41**. It is noteworthy that the corresponding decrease in K_i (tryp) is consistent with the more polar, and thus more accommodating, S3 site of trypsin. These results leave open for speculation the inability of **24** or **25** to present the phenyl ring in a productive binding mode at the S3 subsite of thrombin. Perhaps the positioning of polar functionalities such as amide (in **24**) and amine (in **25**) within the S3 cavity offsets the advantage of the hydrophobic phenyl ring. Alternatively, it is reasonable to assume that the D-Phe-(v-Tyr) segment of CtA, in comparison to the hydrocarbon linker of **38**, imparts differences in the binding mode at the S3 subsite of thrombin.

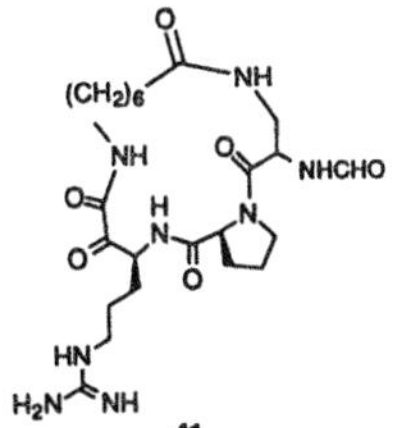

41

K_i (thr) = 260 ± 60 nM

K_i (tryp) = 3.2 ± 1.2 nM

As mentioned previously, the phenyl groups of CtA are involved in an aromatic-stacking interaction involving Tyr60A and Trp60D of thrombin. The D-Phe of CtA interacts with the hydrophobic residues Leu40 and Leu41. Therefore, it follows that positioning hydrophobic residues in the P1' position, as in **42** and **43**, should increase the potency for inhibition of thrombin (cf. **42** and **43** with **38c**). In the CtA/trypsin structure,[46] the D-Phe of CtA occupies an aromatic binding pocket consisting of Tyr39 and Phe41, which is consistent with the considerable potency of **42** and **43** as trypsin inhibitors. These results, together with our unsuccessful attempt to interact with the Lys60F residue, indicate that while good potency can be achieved through hydrophobic P1' substitution on a macrocyclic framework, obtaining selectivity for inhibition of thrombin versus trypsin is less than straightforward.[96] The relationship between stereochemistry and activity with respect to the S1' subsite becomes evident by comparing **43** and **44**. The ca. 300-fold drop in thrombin inhibitory potency that accompanies the change from D to L stereochemistry is most likely a result of an unfavorable steric interaction between the Phe side chain of **44** and the Trp60D residue of thrombin. We applied molecular modeling[92] to analyze this situation (Figure 9). Ligand **44** causes a spatial displacement of the side chain of Trp60D from its standard location, which is critical in defining the S2 binding pocket for the proline residue of fPR. Since the insertion loop is unique to thrombin, a similar situation cannot occur on the binding of **43** or **44** to trypsin, hence their comparable potency as trypsin inhibitors.

As in the case of CtA, the importance of the keto amide functionality is demonstrated by the loss of biological activity that occurs on changing the keto amide of **38i** to the hydroxy amide of **38j**. Unlike the hydroxyamide of CtA however, **38j** did not display slow-binding kinetics (Table 1).

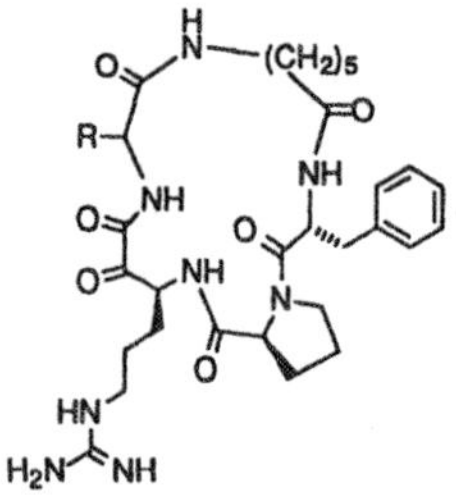

		K_i (nM)	
		thr.	tryp.
42	R = D-$CHPh_2$	5.3 ± 1.2	2.5 ± 0.4
43	R = D-CH_2Ph	3.1 ± 0.4	5.2 ± 1.2
44	R = L-CH_2Ph	1000 ± 54	17 ± 3.3

3.3. X-Ray Crystallographic Study of 38i

We chose **38i** as a representative macrocycle for obtaining an X-ray crystal structure complexed to thrombin. A ternary complex was formed by adding a 10-fold molar excess of **38i** to a complex of hirugen and human α-thrombin;[62] crystals of the complex were grown as described previously for CtA.[44,97] Overall, the D-homo-Phe-Pro-Arg segment of **38i** exhibits binding interactions typical of the fPR class of inhibitors (Figure 10). The D-homo-Phe residue occupies the corresponding apolar binding site of thrombin, while the Pro-Arg residues interact at the S2 "P pocket" and S1 specificity pocket, respectively. The keto amide carbonyl forms a

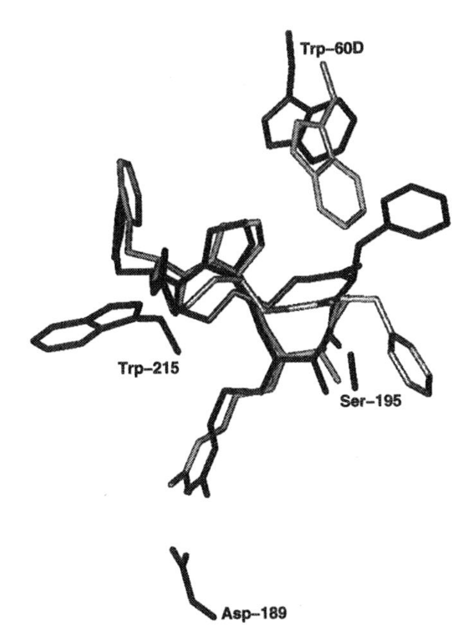

Figure 9. Simultaneous representation of docking experiments with **43** and **44** in the active site of thrombin. Macrocycle **43** and the corresponding Trp60D residue is lightly shaded; **44** and the corresponding Trp60D residue is shaded darkly.

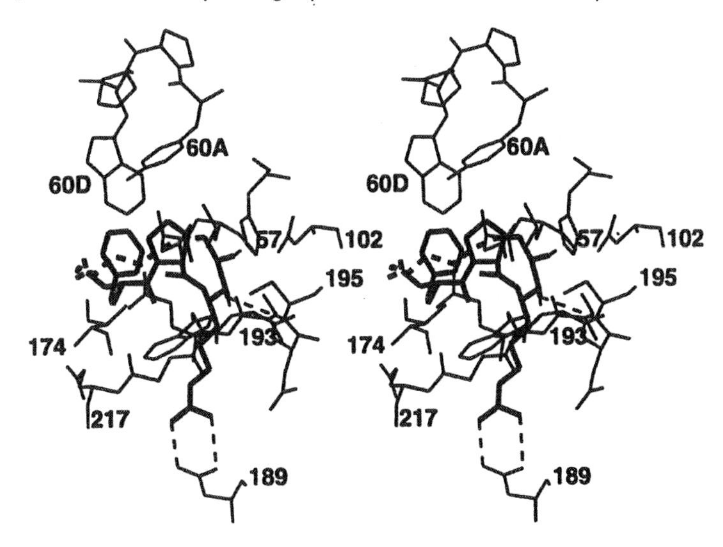

Figure 10. Stereoview of **38i** (*bold lines*) in the active site of thrombin (*thin lines*). The disordered methylene linker segment of **38i** is in bold broken lines. Hydrogen bonds are shown as broken thin lines.

tetrahedral hemiketal with the hydroxyl of Ser195 in a manner analogous to CtA–thrombin.[44] The carbonyl oxygen of the hemiketal is hydrogen-bonded to Gly193 and Ser195, while the keto oxygen of the α-keto amide forms a hydrogen bond with His57. The seven-methylene hydrocarbon segment, which is disordered in the crystal structure, does not influence the tripeptide array that adopts an extended conformation similar to that observed for acyclic fPR-based inhibitors. Thus, the rationale set forth above regarding the proposed extended conformation of the macrocyclic D-Phe-Pro-Arg motif is verified.

4. CONCLUSION

The established role of peptides in biological processes, combined with the burgeoning fields of genomics and combinatorial methods of synthesis,[98] promise to intensify the study of new peptide motifs and their interactions with bioreceptors. In a number of cases reviewed above, the study of macrocyclic peptide motifs has led to significant advances in understanding the molecular basis for the interactions of various peptide ligands with bioreceptors. Given the importance of serine proteases to a wide variety of critical biological processes, it is remarkable that macrocyclic peptide motifs are relatively underexplored for this class of enzymes.

Collectively, the studies involving CtA demonstrate how a protein-structure based approach can be applied to the design of macrocyclic serine protease inhibitors. A strategy which integrates X-ray crystal and solution structure information, computational methods, biochemistry, and synthetic chemistry was applied to the study of macrocyclic inhibitors of human α-thrombin. Features of the thrombin and trypsin-bound CtA structures were used to rationalize interactions between ligand and enzyme at specifc active-site positions, resulting in a class of novel macrocyclic thrombin inhibitors with good *in vitro* potency. From an overall perspective, these studies represent a starting point for drug design and provide a better appreciation for the potential of a macrocyclic motif. The search for novel, small-molecule thrombin inhibitors remains an area of active research. As the search for new inhibitory motifs intensifies for thrombin and other serine protease targets such as factor Xa, urokinase-type plasminogen activator,[99] tryptase,[100] etc., these studies serve as both a model protocol and fundamental knowledge base for the design of potential therapeutic agents.

5. APPENDIX

Diffraction intensity data were measured with an R-AXIS imaging plate detector by using CuKα. The X-rays were generated with a Rigaku fine-focus RU200 rotating-anode generator operating at 50 kV and 100 mA. The crystal scattered X-rays to about 1.8-Å resolution [48% observed with $I/\sigma(I) > 2.5$ in 2.0–1.8- Å resolution shell], and is isomorphous with crystals of hirugen–thrombin[44] and CtA–hirugen–thrombin (a = 71.08 Å, b = 72.14 Å, c = 72.82 Å, β = 100.8°, space group C2, 4 molecules per unit cell). A total of 47,859 reflections were measured and averaged to give 25,408 independent reflections (74.2% of theoretical) with an R merge = 4.5% ($I/\sigma(I) > 2.0$). The structure was solved using the coordinates of the hirugen–thrombin structure and it was refined by restrained least-squares methods with the program PROLSQ. The refinement converged at R = 16.6% including 135 water molecules in the calculations.

ACKNOWLEDGMENTS

We wish to thank Harold Almond for molecular modeling studies; Han-Cheng Zhang, Gene Powell, Len Hecker, Karen Glover, and H. Marlon Zhong for contributions to the preparation of CtA and its analogues; Prof. K. C. Nicolaou, Aijun Liu, and Peter Brungs for their collaboration on the first total synthesis of CtA; Prof. Alexander Tulinsky, X. Qiu, K. P. Padmanabhan, V. Ganesh, E. Zhang, and P. Vanderhoff-Hanaver for X-ray crystallographic work; P. Andrade-Gordon, Jack Kauffman, and Joan Lewis for biochemical studies; and Prof. N. Fusetani for generous samples of natural CtA.

REFERENCES AND NOTES

1. Milner-White, E. J. *Trends Pharmacol. Sci.* **1989**, *10*, 70.
2. Fairlie, D. P.; Abbenante, G.; March, D. *Curr. Med. Chem.* **1995**, *2*, 654.
3. Isolation: (a) du Vigneaud, V.; Ressler, C.; Trippett, S. *J. Biol. Chem.* **1953**, *205*, 949. (b) du Vignaud, V.; Ressler, C.; Swan, J. M.; Roberts, C. W.; Katsoyannis, P. G.; Gordon, S. *J. Am. Chem. Soc.* **1953**, *75*, 4879.
4. Synthesis: (a) du Vignaud, V.; Lawler, H. C.; Popenoe, E. A. *J. Am. Chem. Soc.* **1953**, *75*, 4880. (b) du Vignaud, V.; Bartlett, M. F.; Johl, A. *J. Am. Chem. Soc.* **1957**, *79*, 5572. (c) Light, A.; du Vignaud, V. *Proc. Nat. Exp. Biol. Med.* **1958**, *98*, 692.
5. Burgus, R.; Ling, N.; Butcher, M.; Guillemin, R. *Proc. Natl. Acad. Sci. USA* **1973**, *70*, 684.
6. Laskowski, M., Jr. *Adv. Exp. Med. Biol.* **1986**, *199*, 1.
7. Carrell, R. W.; Travis, J. *Trends Biochem. Sci.* **1985**, *10*, 20.
8. Schreiber, S. L. *Science* **1991**, *251*, 283.
9. Sigal, N. H.; Durmant, F. J. *Annu. Rev. Immunol.* **1992**, *10*, 519.
10. Fusetani, N.; Matsunaga, S. *Chem. Rev.* **1993**, *93*, 1793. For further examples of macrocyclic protease inhibitors, see: (b) Okino, T.; Murakami, M.; Haraguchi, R.; Munekata, H.; Matsuda, H. *Tetrahedron Lett.* **1993**, *34*, 8131; (c) Shin, H. J.; Murakami, M.; Matsuda, H.; Ishida, K.; Yamaguchi, K. *Ibid.* **1995**, 5235; (d) Murakami, M.; Kodani, S.; Ishida, K., Matsuda, H.; Yamaguchi, K. *Ibid.* **1997**, 3035.
11. Veber, D. F.; Holly, F.; Paleveda, W. J.; Nutt, R. F.; Bergstrand, S. J.; Torchiana, M.; Glitzer, M. S.; Saperstein, R.; Hirschmann, R. *Proc. Natl. Acad. Sci. USA* **1978**, *75*, 2636.
12. Veber, D. F.; Freidinger, R. M.; Perlow, D. S.; Paleveda, W. J.; Holly, F.; Strachan, R. G.; Nutt, R. F.; Arison, B. H.; Homnick, C.; Randall, W. C.; Glitzer, M. S.; Saperstein, R.; Hirschmann; R. *Nature* **1981**, *292*, 55.
13. Hirschmann, R.; Nicolaou, K. C.; Pietranico, S.; Salvino, J.; Leahy, E. M.; Sprengeler, P. A.; Furst, G.; Smith, A. B.; Strader, C. D.; Cascieri, M. A.; Candelore, M. R.; Donaldson, C.; Vale, W.; Maechler, L. *J. Am. Chem. Soc.* **1992**, *114*, 9217.

14. Nicolaou, K. C.; Salvino, J. M.; Raynor, K.; Pietranco, S.; Reisine, T.; Freidinger, R. M.; Hirschmann, R. *Peptides: Chemistry, Structure, and Biology*; Rivier, J. E.; Marshall, G. R. Eds.; Escom: Leiden, The Netherlands, 1990; p 881.
15. Laskowski, M., Jr.; Kato, I. *Annu. Rev. Biochem.* **1980**, *49*, 593.
16. Kraut, J. *Annu. Rev. Biochem.* **1977**, *46*, 331.
17. Tager, H. S.; Steiner, D. F. *Annu. Rev. Biochem.* **1974**, *43*, 509.
18. Davie, E. W.; Fugikawa, K.; Kurachi, K.; Kisiel, W. *Biochemistry* **1991**, *30*, 10363.
19. Landman, H. *Handbook of Experimental Pharmacology;* Markwardt, F., Ed.; Springer-Verlag, Berlin/New York, 1978, pp 3-48.
20. Erdos, E. G.; Steward, T. A. *Biological Function of Proteinases;* Holzer, H.; Tschesche, H., Eds.; Springer-Verlag, Berlin/New York, 1979, pp 260-268.
21. Morton, D. B. *Research Monographs in Cell and Tissue Physiology;* Barrett, A. J., Ed.; North-Holland Biomedical Press, Amsterdam/New York, 1978, Vol. 2, pp 445-500.
22. Beers, W. H.; Strickland, S.; Reich, E. *Cell* **1975**, *6*, 387.
23. Ossowski, L.; Buermann, C. W. *Cell* **1979**, *16*, 929.
24. Starkey, P. M.; Barrett, A. J.;Burleigh, M. C. *Biochim. Biophys. Acta* **1977**, *483*, 386.
25. Oronsky, A. L.; Buermann, C. W. *Annu. Rep. Med. Chem.* **1979**, *14*, 219.
26. Janoff, A. *Am. Rev. Respir. Dis.* **1985**, *132*, 417.
27. Travis, J.; Dublin, A.; Potempa, J.; Watorek, W.; Kurdowska, A. *Ann. N.Y. Acad. Sci.* **1991**, *624*, 81.
28. Duffy, M. J. *Fibrinolysis* **1993**, *7*, 295.
29. Freedman, M. D. *J. Clin. Pharmacol.* **1992**, *32* 196.
30. Fuster, V.; Badimon, L.; Chesebro, J. H. *New Engl. J. Med.* **1992**, *326*, 242.
31. Travis, J.; Salveson, G. S. *Annu. Rev. Biochem.* **1983**, *52*, 655.
32. Loberman, H.; Tokuoka, R.; Deisenhofer, J.; Huber, R. *J. Mol. Biol.* **1984**, *177*, 531.
33. Bode, W.; Huber, R. *Eur. J. Biochem.* **1992**, *204*, 433.
34. Pochapsky, T. C. (Brandeis University). Private communication, 1996.
35. For an example of cyclic peptide design based on the structure of an inhibitory loop of a serpin, see: Chen, S. T.; Yang, M. T.; Wu, C.Y.; Wang, K. T. *Bioorg. Med. Chem. Lett.* **1994**, *17*, 2123.
36. Wolfenden, R.; Kati, W. M. *Acc. Chem. Res.* **1991**, *24*, 209.
37. Shokat, K. M.; Ko, M.; Scanlan, T. S.; Kochersperger, L.; Yonkovich, S.; Thaisrivongs, S.; Schultz, P. G. *Angew. Chem., Int. Ed. Engl.* **1990**, *102*, 1339.
38. Mehdi, S. *Bioorg. Chem.* **1993**, *21*, 249.
39. Trainor, D. A. *Trends. Pharmacol. Sci.* **1987**, *8*, 303.
40. Takahashi, L. H.; Radhakrishnan, R.; Rosenfield, R. E., Jr.; Meyer, E. F., Jr.; Trainor, D. A. *J. Am. Chem. Soc.* **1989**, *111*, 3368.
41. Liang, T. C.; Abeles, R. H. *Biochemistry* **1987**, *26*, 7603.
42. (a) Bone, R.; Frank, D.; Kettner, C. A.; Agard, D. A. *Biochemistry* **1989**, *28*, 7600. (b) Bone, R.; Shenvi, A. B.; Kettner, C. A.; Agard, D. A. *Biochemistry* **1987**, *26*, 7609.
43. Fusetani, N.; Matsunaga, S.; Takebayashi, Y. *J. Am. Chem. Soc.* **1990**, *112*, 7053.
44. Maryanoff, B. E.; Qiu, X.; Padmanabhan, K. P.; Tulinsky, A.; Almond, H. R.; Andrade-Gordon, P.; Greco, M. N.; Kauffman, J. A.; Nicolaou, K. C.; Liu, A.; Brungs, P. H.; Fusetani, N. *Proc. Natl. Acad. Sci. USA* **1993**, *90*, 8048.
45. Lewis, S. D.; Ng, A. S.; Baldwin, J. J.; Fusetani, N.; Naylor, A. M.; Shafer, J. A. *Thromb. Res.* **1993**, *70*, 173.
46. Ganesh, V.; Lee, A. Y.; Clardy, J.; Tulinsky, A. *Protein Sci.* **1996**, *5*, 825.
47. Huber, R.; Bode, W. *Acc. Chem. Res.* **1978**, *11*, 114.
48. Rai, R.; Katzenellenbogen, J. A. *J. Med. Chem.* **1992**, *35*, 4150.
49. Markwardt, F.; Sturzebecher, J. *Design of Enzyme Inhibitors as Drugs;* Sandler, M.; Smith, H. J., Eds.; Oxford Univ. Press, Oxford, 1989, pp 619-649.
50. Anon. *Drugs Future* **1988**, *13*, 613.

51. Davie, E. W.; Fugikawa, K.; Kurachi, K.; Kisiel, W. *Adv. Enzymol.* **1978**, *48*, 277.
52. Talbot, M. D.; Butler, K. D. *Drug News Perspect.* **1990**, *3*, 357.
53. Maffrand, J. P. *Nouv. Rev. Fr. Hematol.* **1992**, *34*, 405.
54. (a) Wallis, R. B. *Drugs Today* **1989**, *25*, 597. (b) Brundish, D. E. *Curr. Op. Ther. Patents* **1992**, 1457. Kaiser, B.; Hauptmann, J. *Cardiovasc. Drug. Rev.* **1992**, *10*, 71. (c) Tapparelli, C.; Metternich, R.; Erhardt, C.; Cook, N. S. *Trends Pharmacol. Sci.* **1987**, *8*, 303. (d) Hauptman, J.; Markwardt, F. *Semin. Thromb. Haemostasis* **1992**, *18*, 200. (e) Banner, D. *Perspectives in Medicinal Chemistry;* Testa, B.; Kyburz, E.; Fuhrer, W.; Giger, R., Eds.; Springer-Verlag, Basel, 1993, pp 27-43.
55. (a) Okamoto, S.; Hijikata, A. *Biochem. Biophys. Res. Commun.* **1981**, *101*, 440. (b) Kikumoto, R.; Tamao, V.; Tezuka, T.; Tonomura, S.; Hara, H.; Ninomiya, K.; Hijikata, A.; Okamato, S. *Biochemistry* **1984**, *23*, 85. (c) Bush, L. R. *Cardiovasc. Drug. Rev.* **1991**, *9*, 247. (d) Hijikata-Okunomiya; Okamoto, S. *Semin. Thromb. Haemostasis* **1992**, *18*, 200. (e) Anon. *Drugs Future* **1990**, *15*, 1115.
56. Matsuzaki, T.; Sasaki, C.; Umeyama, H. *J. Biochem.* **1989**, *21*, 249.
57. Chow, M. M.; Meyer, E. F.; Bode, W.; Kam, C. M.; Radhakrishnan, R.; Vijayalakshmi, J.; Powers, P. C. *J. Am. Chem. Soc.* **1990**, *112*, 7783.
58. (a) Bode, W.; Mayr, I.; Bauman, U.; Huber, R.; Stone, S.; Hofsteenge, J. *EMBO J.* **1989**, *8*, 3467. (b) Bode, W.; Turk, D.; Karshikov, A. *Protein Sci.* **1992**, *1*, 426.
59. Tulinsky, A.; Qiu, X. *Blood Coagul. Fibrinol.* **1993**, *4*, 305.
60. Stubbs, M. T.; Bode, W. *Thromb. Res.* **1993**, *69*, 1.
61. Rydel, T. J.; Ravichandran, K. G.; Tulinsky, A., Bode, W.; Huber, R.; Roitsch, C.; Fenton, J. W. *Science* **1990**, *249*, 277.
62. Skrzypczak-Jankun, E.; Carperos, V.; Ravichandran, K. G.; Tulinsky, A.; Weatbrook, M.; Maraganore, J. M. *J. Mol. Biol.* **1991**, *221*, 1379.
63. Banner, D. W.; Hadvary, P. *J. Biol. Chem.* **1991**, *266*, 20085.
64. Brandstetter, H.; Turk, D.; Hoeffken, H. W.; Grosse, D.; Stuerzebecher, J.; Martin, P. D.; Edwards, B. F.; Bode, W. *J. Mol. Biol.* **1992**, *226*, 1085.
65. Matsuzaki, T.; Sasaki, C.; Okumura, C.; Umeyama, H. *J. Biochem.* **1988**, *103*, 537.
66. Bode, W.; Huber, R. *Eur. J. Biochem.* **1992**, *204*, 433.
67. Van Duyne, G. D.; Standaert, R. F.; Karplus, P. A.; Schreiber, S. L.; Clardy, J. *Science* **1991**, *252*, 839.
68. McDonnell, P. A.; Caldwell, G. W.; Leo, G. C.; Podlogar, B. L.; Maryanoff, B. E. *Biopolymers* **1997**, *41*, 349.
69. Bajusz, S.; Szell, E.; Bagdy, D.; Barabas, E.; Horvath, G.; Dioszegi, M.; Fittler, Z.; Szabo, G.; Juhasz, A.; Tomori, E.; Szilagyi, G. *J. Med. Chem.* **1990**, *33*, 1729.
70. Jackson, C. V.; Crowe, V. G.; Frank, J. D.; Wilson, H. C.; Coffman, W. J.; Utterback, B. G.; Jakubowski, J. A; Smith, G. F. *J. Pharmacol. Exp. Ther.* **1992**, *261*, 546.
71. Williams, J. W.; Morrison, J. F. *Methods Enzymol.* **1979**, *63*, 437. Also, see: Morrison, J. F. *Trends Biochem. Sci.* **1982**, *7*, 102.
72. Cha, S. *Biochem. Pharmacol.* **1975**, *24*, 2177.
73. Andrade-Gordon, P. Unpublished results.
74. Maryanoff, B. E.; Greco, M. N.; Zhang, H.-C.; Andrade-Gordon, P.; Kauffman, J. A.; Nicolaou, K. C.; Liu, A.; Brungs, P. H. *J. Am. Chem. Soc.* **1995**, *117*, 1225.
75. Hagihara, M; Schreiber, S. L. *J. Am. Chem. Soc.* **1992**, *114*, 6570.
76. (a) Wipf, P.; Kim, H. *J. Org. Chem.* **1993**, *58*, 5592. (b) Bastianns, H.; van der Baan, J. L.; Ottenheijm, C. J. *Tetrahedron Lett.* **1995**, *30*, 5963. (c) Deng, J.; Hamada, Y; Shiori, T. Matsunaga, S.; Fusetani, N. *Angew. Chem. Int. Ed. Engl.* **1994**, *33*, 1729. (d) Deng, J.; Hamada, Y; Shiori, T. *Tetrahedron Lett.* **1996**, *37*, 2261.
77. Maryanoff, B. E.; Zhang, H.-C.; Greco, M. N.; Glover, K. A.; Kauffman, J. A.; Andrade-Gordon, P. *Bioorg. Med. Chem.* **1995**, *3*, 1025.

78. Unpublished results from our laboratory.
79. Maryanoff. B. E.; Zhang, H.; Greco, M. N.; Zhang, E.; Vanderhoff-Hanaver, P.; Tulinsky, A. *Tetrahedron Lett.* **1996**, *37*, 3667.
80. Van Nispen, J. W.; Hageman, T. C.; Scheraga, H. A. *Arch. Biochem. Biophys.* **1977**, *182*, 227.
81. Meinwald, Y. C.; Martinelli, R. A.; van Nispen, J. W.; Scheraga, H. A. *Biochemistry* **1980**, *19*, 3820.
82. Scheraga, H. A. *Ann. NY Acad. Sci.* **1983**, *408*, 330.
83. Scheraga, H. A. *Ann. NY Acad. Sci.* **1986**, *485*, 124.
84. Marsh, H. C.; Meinwald, Y. C.; Lee, S.; Scheraga, H. A. *Biochemistry* **1982**, *21*, 6167.
85. Marsh, H. C.; Meinwald, Y. C.; Thannhauser, T. W.; Scheraga, H. A. *Biochemistry* **1983**, *22*, 4170.
86. Chang, J. *Eur. J. Biochem.* **1985**, *151*, 217.
87. Ni, F.; Scheraga, H. A.; Lord, S. T. *Biochemistry* **1988**, *27*, 4481.
88. Ni, F.; Konishi, Y.; Frazier, R. B.; Scheraga, H. A. *Biochemistry* **1989**, *28*, 3082.
89. Ni, F.; Meinwald, Y. C.; Vasquez, M.; Scheraga, H. A. *Biochemistry* **1989**, *28*, 3094.
90. For thrombin inhibitors involving β-turn mimetics of fibrinogen, see: Wu, T. P.; Yee, V.; Tulinsky, A.; Chrusciel, R. A.; Nakanishi, H.; Shen, R.; Priebe, C.; Kahn, M. *Protein Eng.* **1993**, *6*, 471.
91. Martin, P. D.; Robertson, W.; Turk, D.; Huber, R.; Bode, W.; Edwards, B. F. P. *J. Biol. Chem.* **1992**, *267*, 7911.
92. The X-ray structure of CtAthrombin was used to model the candidate ligand in a hydration sphere conformationally reduced by simulated annealing using AMBER 4.0.
93. Greco, M. N.; Powell, E. T.; Hecker, L. R.; Andrade-Gordon, P.; Kauffman, J. A.; Lewis, J. M.; Ganesh, V.; Tulinsky, A.; Maryanoff, B. E. *Bioorg. Med. Chem. Lett.* **1996**, *6*, 2947.
94. Compound **40** was disclosed in Australian patent (no. 86 52881).
95. Shuman, R. T.; Rothenberge, R. B.; Campbell, C. S.; Smith, G. F.; Gifford-Moore, D. S.; Gesellchen, P. D. *J. Med. Chem.* **1993**, *36*, 314.
96. For an example of thrombin inhibitors that probe the S1' subsite, see: Costanzo, M. J.; Maryanoff, B. E.; Hecker, L. R.; Schott, M. R.; Yabut, S. C.; Zhang, H.; Andrade-Gordon, P.; Kauffman, J. A.; Lewis, J. M.; Krishnan, R.; Tulinsky, A. *J. Med. Chem.* **1996**, *39*, 3039.
97. Details regarding the X-ray determination can be found in the Appendix of this chapter.
98. Caflisch, A.; Karplus, M. *Persp. Drug Discovery Des.* **1995**, *3*, 51.
99. Billstrom, A.; H.; Hartley-Asp, B.; Lecander, I.; Batra, S.; Astedt, B. *Int. J. Cancer* **1995**, *61*, 542.
100. Clark, J. N.; Moore, W. R.; Tanaka, R. D. *Drugs Future* **1996**, *21*, 811.

MIMICKING EXTENDED CONFORMATIONS OF PROTEASE SUBSTRATES:

DESIGNING CYCLIC PEPTIDOMIMETICS TO INHIBIT HIV-1 PROTEASE

Robert C. Reid and David P. Fairlie

Advances in Amino Acid Mimetics and Peptidomimetics
Volume 1, pages 77-107

ISBN: 0-7623-0200-3

ABSTRACT

There are many examples of structurally distinct cyclic peptides that potently regulate biological processes (e.g. *Curr. Med. Chem.* **1995**, 2, 654-686), but the method of altering their structures to obtain optimal bioactivity is still largely trial and error and thus is painstakingly slow. In this article we describe a more rational approach to inhibitor design that focuses on directly mimicking the conformations of peptidic substrate analogues that bind to the protease of Human Immunodeficiency Virus type 1 (HIVPR). This application of inhibitor design to the specific example of HIVPR is illustrative of a general approach to the structural mimicry of receptor-binding conformations of substrates for other proteases. The macrocycles are easily synthesized, conformationally rigid, proteolytically stable, water and lipid soluble, and can be readily varied by altering the size or substituents of a cycle. Such molecules promise to be important mechanistic probes of biological processes and potential drug leads.

A novel hypothesis that extended (strand) substrate conformations are selectively recognized by proteolytic enzymes in general is demonstrated here for HIV-1 protease and supported by our evidence elsewhere for this and other proteases. Evidence includes (1) a consensus extended conformation in X-ray crystal structures of all substrate-mimicking peptidic inhibitors bound to HIV-1 protease, (2) slower processing for turn- or helix-favoring peptide substrates, (3) lack of turn/helical/sheet structure in peptide and also protein substrates that are most rapidly degraded by HIV-1 protease, and (4) factors that promote stretching of the amide bond for substrate processing.

Based upon this apparent conformational selection by HIV-1 protease of peptides and analogues in extended strand conformations, we have designed macrocycles that constrain segments of peptide sequences to an extended conformation. These include cyclic replacements for N-terminal tri- and tetrapeptides, C-terminal tri- and tetra- peptides, and bicyclic hexapeptide analogues which replace both ends. X-ray crystal structures show that these are all structural mimics of protease-binding conformations of substrate analogues. Molecular modeling and NMR spectroscopic studies identify the extent of conformational restriction afforded by the macrocycles. This structural mimicry leads to functional mimicry, since the cyclic compounds selectively inhibit HIV-1 protease at nanomolar concentrations when attached to appropriate peptidic and non-peptidic appendages. Unlike acyclic peptide analogues which do not inhibit HIV-1 replication in cells, these macrocycles are pre-organized for receptor-binding, less conformationally flexible than acyclic components of peptidomimetic inhibitors, more resistant to proteolysis than their acyclic analogues, and lipophilic enough to penetrate cell membranes for antiviral activity. This *regioselective* fixing of the conformation of components of inhibitors for HIV-1 protease may also be useful in designing structural and functional peptidomimetics for substrates of other proteolytic enzymes.

The design, synthesis, structure and activities of inhibitors of HIV-1 protease are now described, focusing on some of our work towards incorporating cycles into peptidomimetics.

1. HIV AND HIV-1 PROTEASE

The Human Immunodeficiency Virus type 1 (HIV-1) [1] is now established as the etiological agent for acquired immunodeficiency syndrome (AIDS) and infects over 20 million people worldwide.[2] It is transmitted through sexual, intravenous, and parenteral routes and infects by binding to immune cells (e.g. T-cells, macrophages) bearing the surface receptor protein CD4. After fusion of the virus with a target cell, the protective envelope is removed by proteolysis leaving single-stranded viral RNA which is copied to single- and then double-stranded DNA by reverse transcriptase (Figure 1). The RNA is degraded by ribonucleaseH while the viral DNA is integrated into the nucleus of the host cell and transcribed, mRNA is exported to the cytoplasm followed by translation and protein synthesis, myristolation, glycosylation, viral assembly, and maturation involving budding and protease activity (Figure 1). Many viral proteins that are necessary for viral replication are targets for drug development,[3-5] but the most topical at present is HIV-1 protease [6-11] due to the efficacy of several inhibitors now licensed for the treatment of HIV infections.

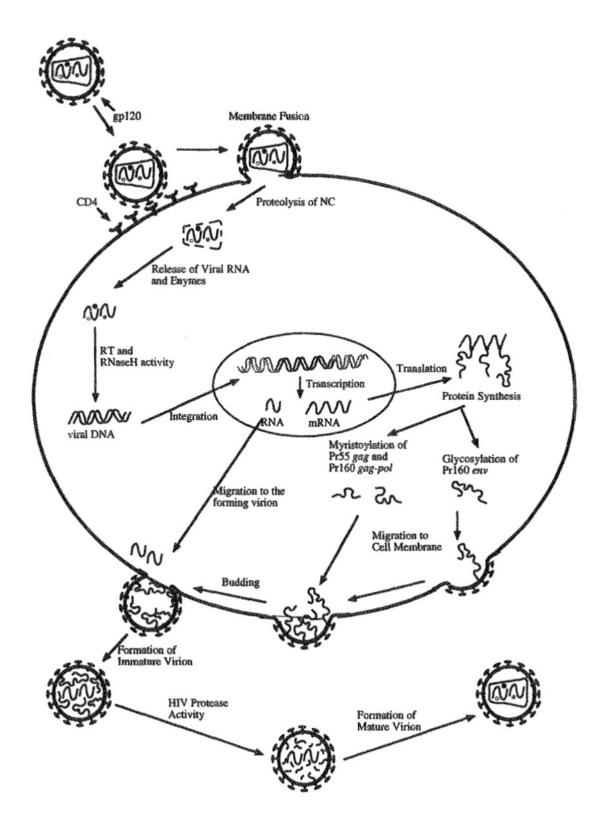

Figure 1. Simplified schematic of the replicative cycle of HIV.

HIV-1 protease (HIVPR) and its inhibitors have been comprehensively reviewed elsewhere.[6-11] In summary, the enzyme is a structurally unique aspartic protease[12] that is essential for processing of *gag* and *gag-pol* polyproteins to form the individual structural and functional viral proteins needed for host recognition, host infection, and replication by the virus in host cells.[13,14] Inactivation of this enzyme through mutation or inhibition prevents assembly of viral proteins and leads to immature non-infective virus.[15,16] The protease is a homodimer with 99 amino acids and one catalytic residue (Asp25) per subunit, high (C_2) symmetry, a conserved Asp-(Thr/Ser)-Gly sequence and a hydrophobic substrate-binding active site (Figure 2a).[7] Each of the monomers provides a mobile flap and both flaps form the "roof" of the substrate-binding groove (Figure 2a). HIVPR is inhibited weakly by the classic aspartic protease inhibitor, pepstatin, and by substrates in which the scissile amide bond is replaced by a non-cleavable unit.

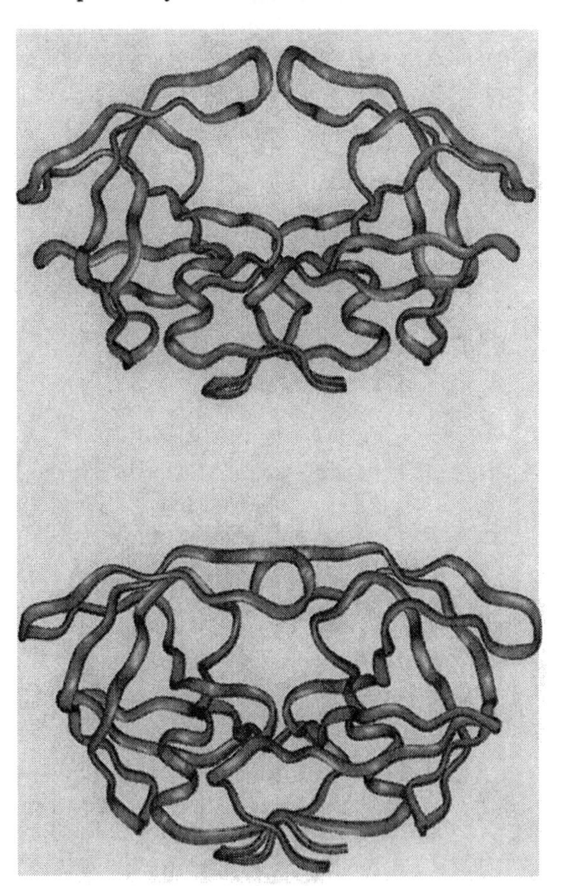

Figure 2. (**a**, Top) Uncomplexed HIV-1 protease. (**b**, Bottom): Inhibitor-bound form of HIV-1 protease.

Upon inhibitor binding (Figure 2b), asymmetry is induced in the unusually symmetrical protease dimer.[7] Most of the conformational change in the enzyme occurs in the flaps region, with the volume of the substrate-binding active site decreasing by about 30%. One of the great difficulties in designing protease inhibitors is the unpredictability of subtle conformational changes in the substrate-binding groove of a protease in response to changes in the inhibitor, the so-called "induced fit", or mutations in the enzyme which can alter inhibitor affinity leading to drug resistance even when mutation is remote from the active site.

The mechanism of substrate processing by HIVPR is not precisely known but is commonly thought to involve general acid/base catalysis, where one of the two catalytic aspartates is protonated.[17,18] The unprotonated aspartate acts as a base in abstracting a proton from the catalytic water molecule while the protonated aspartate is acidic and donates a proton to the carbonyl oxygen (Figure 3). In contrast to serine and cysteine proteases, there is no covalent interaction between the protease and substrate to form an acyl intermediate. Noncovalent activation of the water nucleophile is consequently dependent upon substrate interactions between the protease and several amino acid residues in the substrate flanking each side of the scissile amide bond.

Although *substrates* have not been co-crystallized with proteolytic enzymes, numerous complexes of HIVPR with substrate analogues that have no cleavable amide bond have been crystallized and their three-dimensional structures have been determined by X-ray crystallography.[7,11] For example a heptapeptide inhibitor known as JG365 (*vide infra*) has the same sequence of amino acids as in a polypeptide substrate, but the cleavable amide bond (–CONH–) is replaced by a hydroxyethylamine "transition state isostere" (–CH(OH)CH_2NH–)[19]. Figure 4a is a view of the X-ray crystal structure of HIVPR-bonded JG365 showing the fitting of the inhibitor into the substrate-binding groove of HIVPR. In particular it illustrates how the inhibitor side chains position in the active site indentations formed by residues that line the groove. Figure 4b shows the hydrogen bonding (dashed lines) between this inhibitor and the enzyme. As in most such HIVPR–peptide X-ray structures, a water molecule is observed to bridge between the enzyme flaps (Ile50,

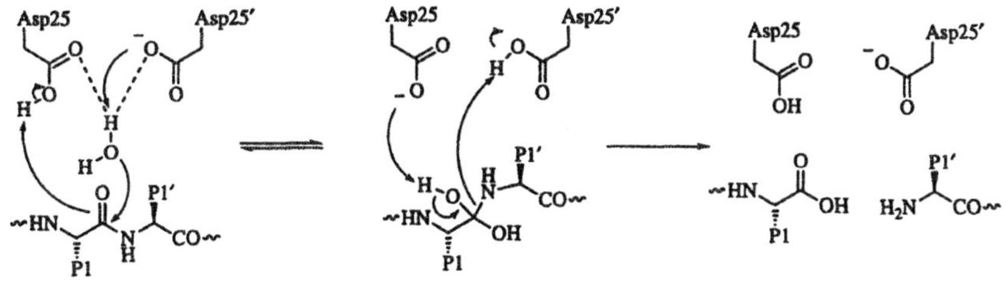

Figure 3. Proposed mechanism of peptide cleavage by HIVPR and aspartyl proteases.

Ile50') and the carbonyl oxygens of the inhibitor (Figure 4b). This hydrogen-bonded water, which probably stabilizes the transition state by assisting to stretch the amide bond towards the tetrahedral transition state, appears to be incorrectly positioned and orientated to be the putative catalytic water molecule of Figure 3.

Interestingly, the numerous enzyme–inhibitor X-ray crystal structures all show a common receptor-bound inhibitor conformation (Figure 5).[20] Instead of helices,

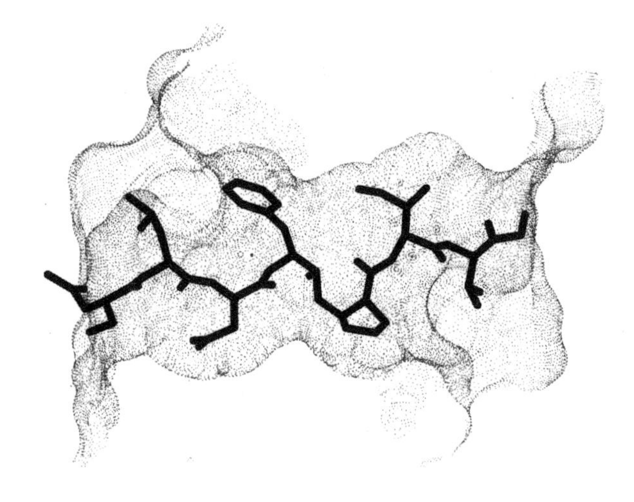

Figure 4. (Top) Conformation of JG365 bound in the active site of HIV-1 Protease.[19] (Bottom). Schematic for hydrogen bonds (dashed lines) between inhibitor JG365 and HIV-1 Protease based on the X-ray crystal structure.[19]

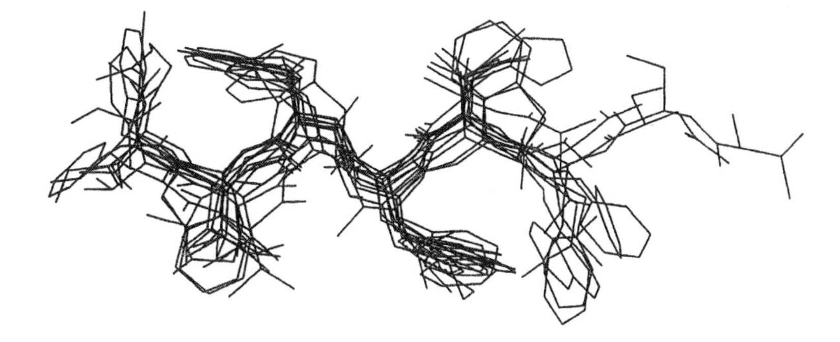

Figure 5. Superimposition of the X-ray structures of 12 enzyme-bound inhibitors of HIV-1 protease showing a common extended (β strand) conformation.[20]

turns, or β-sheets, all these substrate analogues are bound in the active site in identical extended (β-strand) or elongated conformations. This observation suggests that conformational selection is important for HIV-1 protease and, if typical for proteases generally as appears to be the case,[11,21] has important and profound implications for design of protease inhibitors as potential drugs.

2. SUBSTRATE CONFORMATIONS

To complete virus maturation, viral *gag* and *gag-pol* polypeptides need to be cleaved by HIVPR at eight specific sites between P1 and P1' positions (Figure 6). Substrate optimization by varying amino acids on either side of these cleavage sites to optimize enzyme kinetics has brought mixed results.[6-11] Independent optimization of component amino acids is difficult because of unpredictable cooperativities between residues well separated from each other and/or the scissile peptide bond, causing conformational changes in the enzyme in response to fitting different substrate side chains. Nevertheless there are clear advantages for substrates that have hydrophobic/hydrophobic or hydrophobic/proline cleavage sites and clear preferences for specific types of residues in each position. It is well known that protease activity in general is sequence-specific but an important question concerning substrate processing, with implications for drug design, is whether *conformational* features of the substrate are also crucial for processing.

Polypeptides are capable of folding into a variety of different conformations that are distinguished by different intramolecular hydrogen-bonding patterns (e.g. helices, sheets, turns, strands), but short peptides are most commonly observed in random conformations.[23] In reality, many discrete conformations are probably present in equilibrium, but is the highest populated conformer necessarily the one recognized by a protease? There is now strong evidence that protease substrates are only recognized and processed in a single conformation, the extended or β-strand form, even when this may not be abundant or even de-

P5	*P4*	*P3*	*P2*	*P1*	*P1'*	*P2'*	*P3'*	*P4'*	*P5'*
Val	Ser	Gln	Asn	**Tyr**	**Pro**	Ile	Val	Gln	Asn
Lys	Ala	Arg	Val	**Leu**	**Ala**	Glu	Ala	Met	Ser
Thr	Ala	Thr	Ile	**Met**	**Met**	Gln	Arg	Gly	Asn
Arg	Pro	Gly	Asn	**Phe**	**Leu**	Gln	Ser	Arg	Pro
Val	Ser	Phe	Asn	**Phe**	**Pro**	Gln	Ile	Thr	Leu
Cys	Thr	Leu	Asn	**Phe**	**Pro**	Ile	Ser	Pro	Ile
Ile	Arg	Lys	Ile	**Leu**	**Phe**	Leu	Asp	Gly	Ile
Gly	Ala	Glu	Thr	**Phe**	**Tyr**	Val	Asp	Gly	Lys

Figure 6. Substrate sequences near the eight cleavage sites (between bolded residues) for HIVPR. [22]

tectable in solution.[20,21] Such *conformational selection* may be an important feature of protease activity in general.

We have found[21] that 30 residue sequences spanning the cleavage sites of the eight natural polypeptide substrates for HIVPR have different preferred peptide conformations (Table 1). For example, three shorter derivatives of these substrates adopt random coil (Figure 7a), α-helical (Figure 7c) and β-turn (Figure 7b) confor-

Table 1. Predicted[25] Substrate Conformations (italics)[a] Around HIVPR Cleavage Sites (bold)

AAAGT	GNSSQ	VSQN**Y**	**P**IVQN	LQGQM	VHQAI
rrrrr	*rrrrr*	*rrrrs*	*hhhhh*	*hhhhh*	*hhhhh*
ACQGV	GGPGH	KARV**L**	**A**EAMS	QVTNP	ANIMM
hhhhh	*rrttt*	*hhhhh*	*hhhhh*	*hhsrt*	*trrrr*
LAEAM	SQVTN	PANI**M**	**M**QRGN	FRNQR	KTVKC
hhhhh	*hhhsr*	*tthhh*	*hhhrr*	*rrrrt*	*tssss*
LGKIW	PSYKG	RPGN**F**	LQSRP	EPTAP	PEESF
rrrrr	*ttttt*	*rtttt*	*rrrrr*	*rrrrr*	*rrrrr*
SEAGA	DRQGT	VSFN**F**	PQITL	WQRPL	VTIRI
rrrrr	*rttts*	*sssss*	*tssss*	*sssts*	*tssss*
IGRNL	LTQIG	CTLN**F**	**P**ISPI	ETVPV	KLKPG
hhhhh	*sssss*	*sssss*	*ssstt*	*tssss*	*ssstt*
WYQLE	KEPIV	GAET**F**	**Y**VDGA	ANRET	KLGKA
rsssr	*rrrsss*	*ssss*	*ssttt*	*rrrrr*	*rrrrr*
EQVDK	LVSAG	IRKV**L**	**F**LNGI	DKAQE	QHEKY
hhhhh	*hhhhh*	*hhhhh*	*hhhrrr*	*rrrrr*	*hhhhh*

Notes: [a] *h*(helix); *s*(strand/extended); *t*(turn); *r*(random).

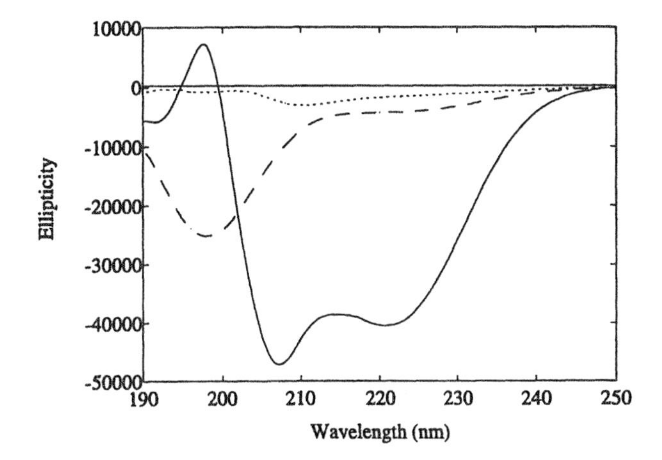

Figure 7. Circular Dichroism at 22 ° C of Ac-KARVLAEAMSQUTNP-NH_2 (**a**,- - -) and Ac-PSYKGRPGNFLQSRP-NH_2 (**b**,.......) in 0.1M phosphate buffer at pH 7, and Ac-LVSAGIRKALFLNGI-NH_2 in 30% TFE(**c**,—).[21]

formations in solution, [21] yet all the polypeptide substrates are processed by HIV-1 protease albeit at different rates.[24] We have demonstrated that the protease only recognizes the extended conformation of peptidic substrate derivatives (Figure 5);[20,21] neither the helix nor turn conformations can fit into the substrate-binding groove. We conclude that HIV-1 protease only processes peptide substrates in an extended conformation and that the position of the conformational equilibria (Figure 8) must influence the rate of processing, since non-extended conformations need to rearrange to the extended form prior to peptide cleavage.

Elsewhere we demonstrated that this extended peptide conformation is also selected by cysteine, serine, metallo, and other aspartic proteases,[21] and we also present structural data for truncated peptide substrates of HIVPR, an argument related to stretching of the amide bond towards the transition state, and evidence of processing of protein substrates that all support the notion of conformational selection by HIVPR.[21] HIVPR is also known to cleave a wide range of viral and non-viral pro-

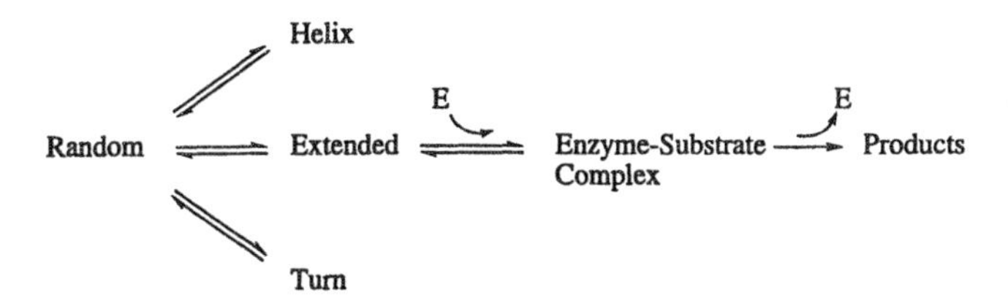

Figure 8. Proposed conformational equilibria for substrates of HIVPR.[21]

teins with extensive variation in cleavage sequences.[26] There is evidence that such cleavages occur primarily in protein regions between folded domains,[26] where there is sufficient flexibility to adopt an extended conformation. There is additional support in the form of observations of the greater effectiveness of HIVPR-induced degradation of some proteins after loss of structural integrity due to unfolding by pH denaturation, acid-rich domains, or removal of structure-stabilizing ligands.[26]

It seems reasonable then to argue that the substrate/inhibitor main chain is required to be in an extended conformation in which the amide bonds contribute substantial binding energy to the interaction with the protease. In this conformation, maximum torque is provided by the side chains to stretch the amide bond towards the transition state. A logical consequence of this conformational selection hypothesis, if it proves to be universal, is that protease–inhibitor design in general might reasonably be based only upon knowledge of the substrate sequence alone, the goal being to structurally mimic the unique (extended) substrate conformation that is recognized by a proteolytic enzyme. We now describe an approach to the general mimicry of protease–substrates in extended conformations using HIVPR–substrates as an example.

3. CYCLIC PEPTIDOMIMETIC INHIBITORS

Problems that normally make peptides unsuitable as drugs include proteolytic instability to peptidases in the gut, blood, and cells, and poor pharmacokinetic profiles due to low absorption through membranes and rapid clearance from circulation. Thus a rational approach to developing inhibitors of proteolytic enzymes such as HIVPR to overcome these problems is the systematic modification of their peptide substrates to proteolytically stable, low molecular weight, and pharmacologically more acceptable non-peptidic inhibitors. Peptidomimetics could ideally reproduce the precise interactions made by an optimal peptide substrate with the proteolytic enzyme. There are two key problems in designing such mimetics.

First, a major difficulty lies in finding an appropriate replacement for each information-rich peptide (containing a conformationally constraining planar amide bond, a proton-accepting CO, a proton-donating NH, and amino acid side chains that flank either side of the amide bond). There have simply been no chemical units available to date to accurately reproduce the geometry of, and bonds made by, a peptide. Second, the sequential optimization of amino acids or non-peptidic units in an acyclic inhibitor of HIVPR has proven to be very difficult because the changes are not independent of one another. Substitution at one position in an inhibitor sequence can influence the conformation of both inhibitor and enzyme, thus altering the requirements for enzyme complementarity further along the inhibitor sequence. Such unpredictable cooperative effects of variations to multiple side chains seriously compromise the effectiveness of analogue- and receptor-based drug design.

One approach to these problems is a minimalist solution involving retention of the peptide units in macrocycles. This approach utilizes the fact that protease substrates in extended conformations have neighboring side chains that are close enough to be covalently linked to form macrocycles without disturbing their enzyme-binding interactions (see ahead). The macrocycle can potentially maintain amino acid components in the same positions and orientations as in the receptor-bound peptide, while increasing peptide stability to proteolytic cleavage. Research with renin inhibitors and other peptides had indicated enhanced proteolytic stability for cyclic versus acyclic peptides.[27,28]

We now summarize our development of hydrolytically stable protease inhibitors using macrocyclic components for independent structural (hence functional) mimicry of peptide segments in protease substrates.[20,29-31] The precise *copying* of three-dimensional positions of side- and main-chain atoms in amino acids of a peptidic substrate/inhibitor, using hydrolytically stable isostructural macrocyclic replacements to fix the conformations of amino acid components, allows regional structural mimicry and permits localized structure–activity optimization in inhibitors of HIV-1 protease. Such regioselective optimization of amino acid sequences in peptidic inhibitors of HIVPR has not previously been successful. This new approach, which reduces conformational rearrangement of either inhibitor or enzyme by mimicking a specific known enzyme-binding conformation of the inhibitor, might be useful for quickly generating potent and selective inhibitors of other enzymes.

3.1. N-Terminal Macrocycles

Tripeptide Mimetic of Leu-Asn-Phe

A common method of rapidly obtaining potent inhibitors of proteolytic enzymes is to replace the scissile amide bond (–CO–NH–) of a proteolytic substrate with one of the many non-cleavable "transition state isosteres", for example the hydroxyethylamine (–CHOH–CH_2–NH–) isostere. In the case of HIV-1 protease, one such substrate-derived peptidic inhibitor is JG365, Ac-Ser-Leu-Asn-Phe-{(*S*)-CHOH–CH_2}-Pro-Ile-Val-OMe (**1**).[32] While a potent inhibitor of HIV-1 protease (*K*i 1 nM), **1** shows no anti-HIV activity in cell culture even at 100 μM concentration. We attribute this either to its low lipid solubility (log $P_{o/w} < 0$) preventing cellular uptake or to cleavage by intracellular peptidases. Because of the early availability of X-ray crystal structural coordinates for **1** bound in the active site of HIV-1 protease,[19] we chose this peptide for structural mimicry.

The X-ray crystal structure of this inhibitor–protease complex showed that neighboring amino acid side chains of JG365 were close enough (Figure 4a) to be covalently linked in a small macrocyclic ring. We expected that amide bonds in such a small constrained macrocycle would be resistant to cleavage by cellular proteases. At the same time it was anticipated that the cycle would increase enzyme affinity through an entropy advantage associated with conformationally restricting otherwise flexible amino acids to an enzyme-binding conformation. Initially we designed[29] and incorporated a cyclic mimetic of the tripeptide Leu-Asn-Phe into the slightly shorter peptidic inhibitor of HIV-1 protease, Ac-Leu-Asn-Phe-CHOH–CH_2-Pro-Ile-Val-NH_2 (**2**, K_i = 3 nM).

The resulting molecule (**3**, Table 2) was found to be a potent inhibitor of HIV-1 protease. Inhibitor potencies (Table 2) were used in conjunction with computer modeling of inhibitors in the JG365-bound conformation of HIVPR to guide structural and functional mimicry.[29,31] Later with the aid of X-ray crystallography we

Table 2.

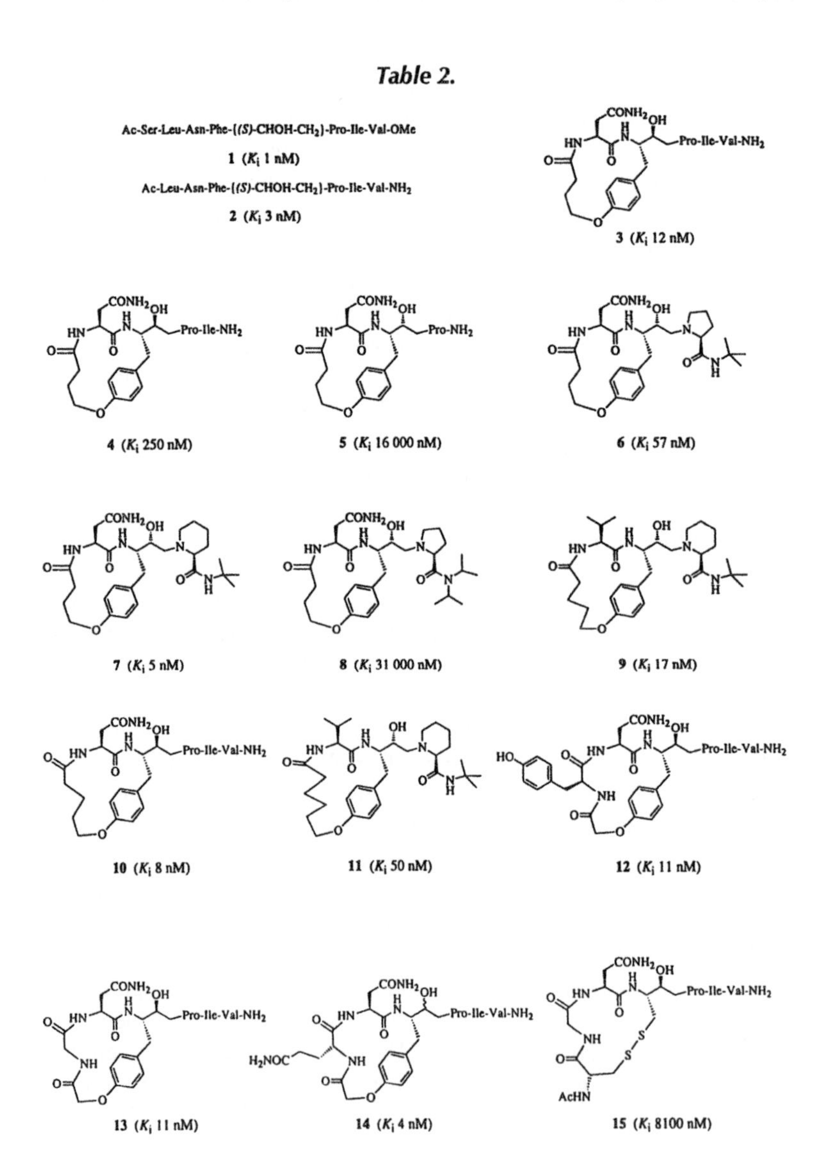

were able to establish structural mimicry for inhibitor **3** co-crystallized with HIV-1 protease. For example, the HIVPR-bound conformations of cyclic and acyclic inhibitors **3** and **1** were superimposable (Figure 9) and both inhibitors formed very similar interactions with the enzyme (cf. Figure 10 vs. Figure 4).[29,31] Key interactions include hydrogen bonds between the hydroxyl group of the chiral alcohol and the catalytic aspartates (25, 125) of HIVPR, as well as between inhibitor main-chain atoms and enzyme flap residues or flap water molecule.

This structural mimicry translated into functional mimicry as demonstrated by comparable inhibition of the protease by acyclic inhibitor **2** and its cyclic mimic **3**

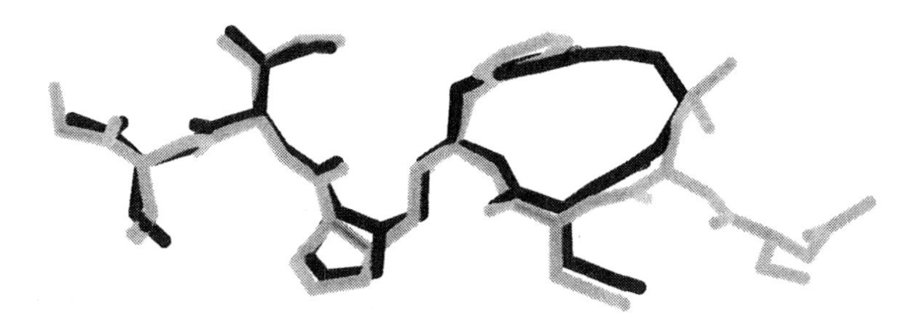

Figure 9. Superimposed HIVPR-bound X-ray structures of the (*S*)-diastereomers of JG365 (light) and the macrocyclic inhibitor **3** (black).

Figure 10. Schematic for hydrogen bonds (dashed lines) between inhibitor **3** and HIV-1 protease based on the X-ray crystal structure.[41]

(Table 2). (*S*)-Stereochemistry of the alcohol diastereomers was generally preferred by HIVPR, the (*R*)-diastereomers being 10–50-fold less potent.[29,31] Thus by fixing the position and orientation of the amino acid side chains at P1, P2, and P3, the cycle can theoretically be useful as a template to independently vary the P1', P2', and P3' subsites of the inhibitor C-terminus without affecting interactions between the macrocyclic N-terminus and enzyme. We now report preliminary observations of such regioselective structure–activity optimization of cyclic inhibitors of HIVPR.

Tetra- to Hexa-Peptidomimetics

The next stage in the development of these inhibitors was to decrease their size and peptidic nature by removing the hydrolyzable acyclic C-terminal tripeptide segment.[29,31] Progressive truncation of Pro-Ile-Val in **3** led to compounds **4** and **5** with significantly reduced inhibitor potency (Table 2). Deletion of Val (**4**) decreased activity 20-fold; further loss of Ile (**5**) reduced activity 100-fold more. The activity was essentially restored when Pro-Ile-Val was substituted by the less hydrolyzable and more lipophilic *t*-butyl amides of L-Pro (**6**) and L-pipecolinic acid (**7**), but not by the sterically bulkier diisopropyl amide of Pro (**8**). The 10-fold preference for L-pipecolinic acid over L-Pro is due to more efficient fitting of this bulkier residue in the S1' subsite of the protease. For all three bulky P2' substituents, the stereochemical preference was reversed with the (*R*)-diastereomeric alcohols of **6-8** being more potent inhibitors of HIVPR than the (*S*)-isomers.[31] This is consistent with other findings where an (*S*) to (*R*) inversion in isomer preference by HIVPR accompanies bulky P1' or P2' substituents, such as the decahydroisoquinoline replacement for Pro in Ro-318959.[33,34]

Variation in the size of the macrocycle from a 15-membered ring (**3–7**) to 16-membered (**9** and **10**) or 17-membered (**11**) rings had only a small effect on inhibitor potency, with 16-membered rings being better. Replacing part of the alkyl linker in the cycle with a third amino acid, which was expected to further constrain the cycle, gave the 16-membered cycles (**12–14**).[31] Additional interactions were anticipated between the inserted amino acid and the enzyme at Arg8 and either Asp29 and/or Asp30. However only inhibitor **14** proved to be more active than **9**, the (*S*)-diastereomer being predicted to hydrogen bond through the D-Gln side chain to Asp30 and Asp29 of HIVPR. In contrast, **15**, with a 14-membered macrocycle and disulfide linker, was a much less potent inhibitor of HIVPR.[31]

We next consider elaboration of the P1' and P2' substituents in analogues **16–26** (Table 3).[31] Replacing Pro in **3** with an N-alkylated glycine derivative (**16**) considerably reduced activity, which was further decreased by additional replacement of the *N*-isobutyl substituent with the bulkier benzyl substituent (**17**). These results are attributed to poor orientation or positioning of the carbonyl oxygen for hydrogen bonding to the flap-connected water "301" of the enzyme. The similar inhibitor SC52151 which lacks the methylene unit is more active.[7-9,11] On the other hand, incorporation of *N*-isoamyl benzenesulfonamide, a component of inhibitior VX-

478,[35] into P1'-P2' positions gave inhibitors **18** (Asn in P2) and **19** (Val in P2) that are at least as potent as the parent acyclic inhibitor **2**.[31] The activity of **19** was expected to be slightly less than for **18** because the Asn side chain at P2 is known[19] to make two hydrogen bonds with the enzyme (Figure 4b).

Compounds **19–21** compare the effect of increasing macrocycle size in this sulfonamide series. Clearly there is not a great effect, with the 16-membered cycle again proving to be slightly better than the others. Substitution of the aromatic ring of the sulfonamide is tolerated but without improvement in inhibitor potency in the examples shown (**22–24**). Compounds **25** and **26**, with the geometrically more constrained urea substituent in P2', were less effective inhibitors.

Conformational Restrictions

The N-terminal macrocycle in **3** possesses conformational rigidity due to the presence of three planar units (one aromatic ring and two *trans* amides). Some insight[36] into the extent of conformational restriction is provided by Figure 11, which

Table 3.

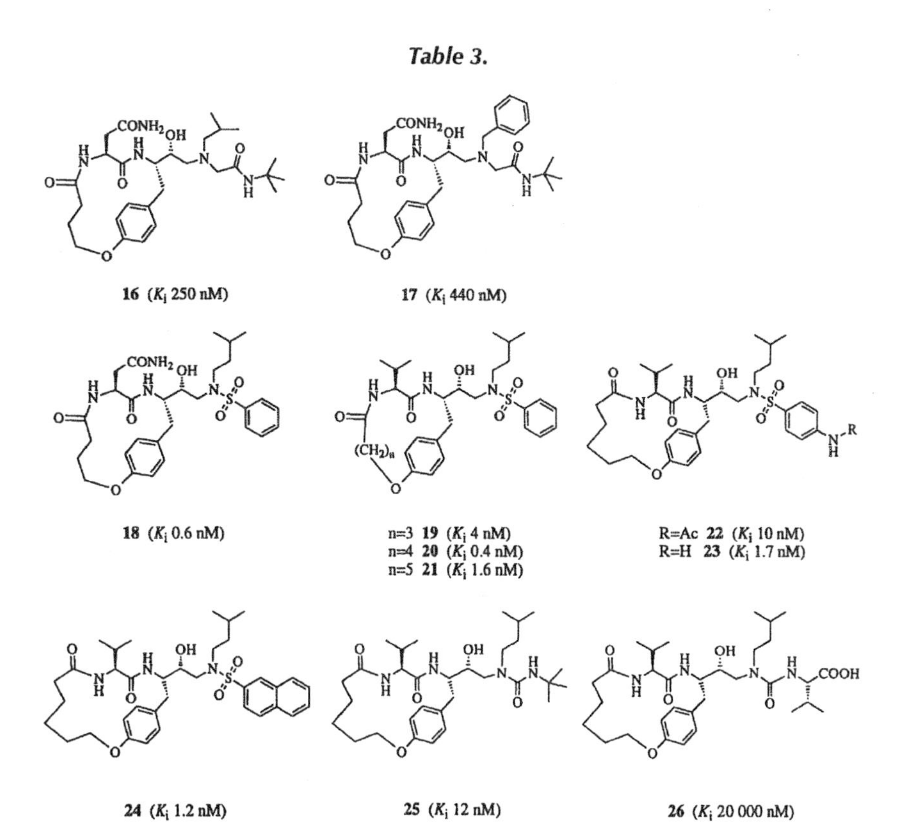

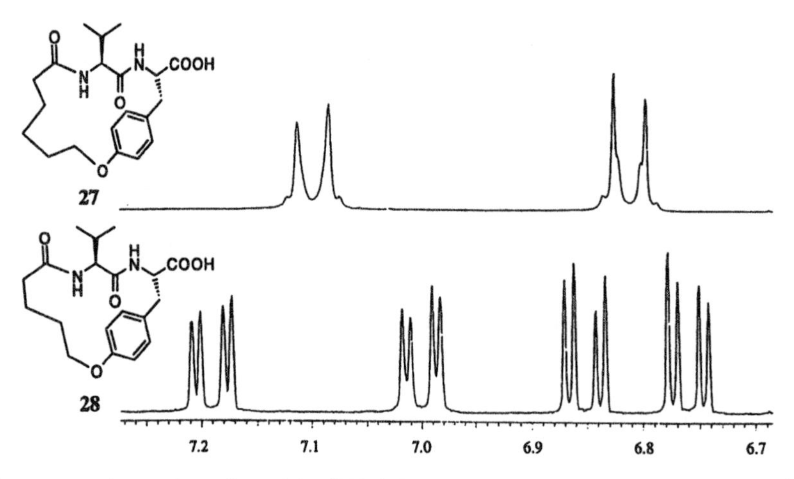

Figure 11. Aromatic region of the ^{1}H NMR spectra (300 MHz, CD_3OD, 298 K) of **27** and **28**.

shows a comparison of the aromatic regions of the proton NMR spectra for the different size cycles **27** and **28**. Although both cycles contain three planar constraints, the spectra reveal a conformational distinction between the cycles at ambient temperature. The upper trace in Figure 11 is typical of an AA'XX' spin system for **27**, while the lower trace shows a doublet of doublets for each of the for aromatic protons in **28**. Thus for the smaller macrocycle **28**, rotation of the aromatic ring is more restricted. Variable temperature studies (not shown) demonstrate that the barriers to rotation of the aromatic rings in **27** and **28** are respectively ~220 K and ~320 K.

A further measure of the entropic advantage that the cyclic component brings to the inhibitors by virtue of its conformational restrictions comes from a comparison of the linear or open precursor **29** (K_i = 2100 nM) with the closed cycle **6** (K_i=57 nM).[31] The ~37-fold increase in potency for the cycle is likely to be an underestimate of the enhancement, since the linear compound can potentially make additional interactions with the enzyme via hydrogen bonding of its ester group and hydrophobic space-filling by its Boc substituent.

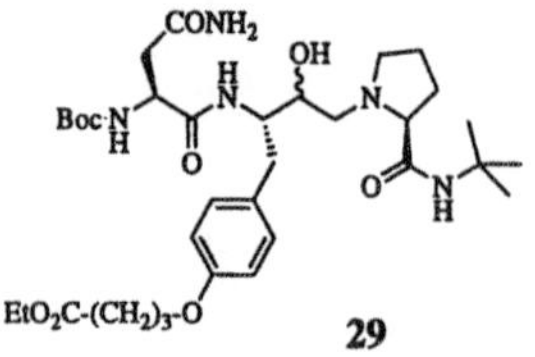

Additional support for restricted conformational flexibility in the N-terminal macrocycles comes from unrestrained molecular dynamics studies. [20] We reported molecular dynamics (200 ps) performed at 300 K with Discover using the cvff forcefield and a dielectric medium of 1. The non-hydrogen atoms of the N-terminal cycle of **28**, from all 200 dynamics frames saved at 1ps intervals and energy minimised, superimposed to rmsd's < 0.3 Å. Thus the cycles are significantly more restrained than the acyclic tripeptide components which they structurally mimic.

Synthetic Routes To Macrocycles

The solid-phase synthesis of the cyclic inhibitor **3** is outlined in Scheme 1. Boc-tyrosine was O-alkylated and then converted in two steps to a bromomethyl ketone (51%). The C-terminus was then coupled to the tripeptide PIV (assembled on MBHA resin using HBTU), the ketone was reduced with $NaBH_4$, and the N-terminus was extended with Asn, before cleavage from the resin (HF). De-esterification with NaOH gave a mixture of diastereomeric alcohols (54%), which were cyclized with BOP in dilute (mM) solutions to **3** and the diastereomers were separated by HPLC.

The cyclic inhibitors of HIVPR reported in Table 2 were synthesized via two similar procedures outlined in Schemes 1 and 2. Cycles attached to the C-terminal fragment Pro-Ile-Val-NH_2 were prepared via solid-phase synthesis (Scheme 1), [29] while other macrocyclic analogues were synthesized in solution via the strategy illustrated for **7** (Scheme 2). Each procedure requires synthesis of O-alkylated tyro-

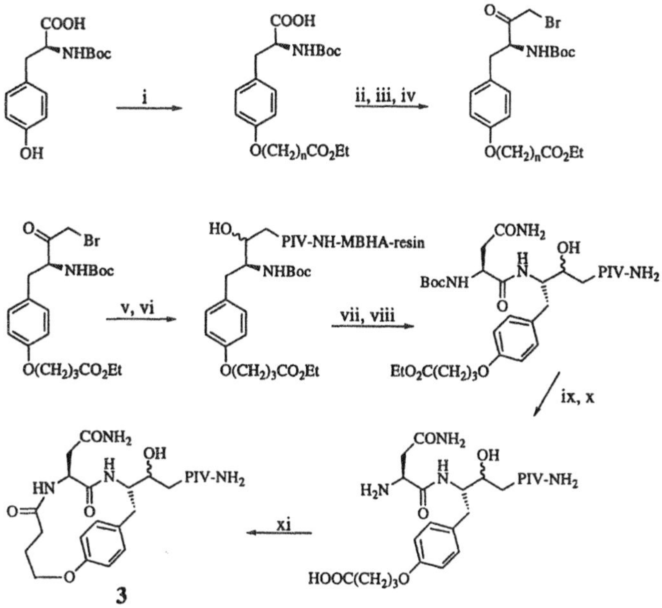

Scheme 1. Reagents : **i** X-$(CH_2)_n CO_2Et$ (5 equiv), NaH (7 equiv); **ii** NMM; EtOCOCl; **iii** CH_2N_2; **iv** HBr (EtOAc solution); **v** H-Pro-Ile-Val-NH-MBHA resin; **vi** $NaBH_4$; **vii** TFA; **viii** Boc-Asn-OH, HBTU, DIPEA; **ix** HF; **x** NaOH; **xi** BOP, DIPEA.

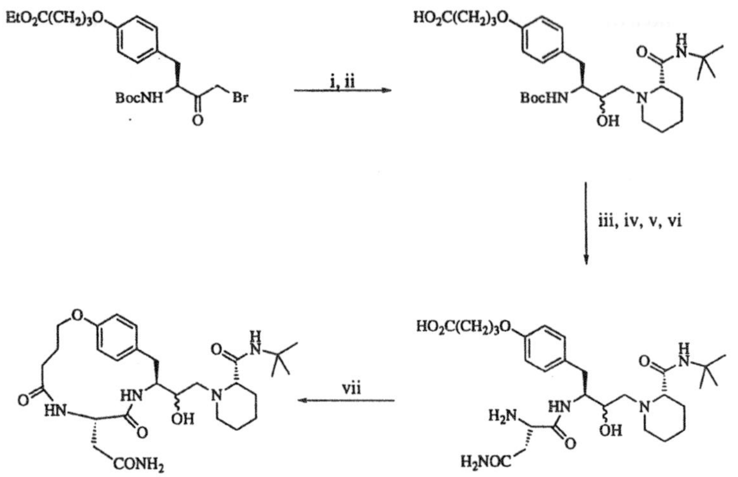

Scheme 2. Reagents: **i** Pip-NH^{t}Bu; **ii** $NaBH_4$; **iii** TFA, **iv** Boc-Asn-OH, HBTU, DIPEA; **v** TFA; **vi** NaOH; **vii** BOP, DIPEA.

sine ketobromide derivatives which were obtained (yield 70–90%) by direct alkylation of Boc-tyrosine with ethyl 4-bromobutyrate (–$(CH_2)_3$– derivatives)[29] or ethyl 5-bromopentanoate (–$(CH_2)_4$– derivatives) or ethyl chloroacetate (–CH_2– derivatives)[31] in the presence of NaH (Scheme 1). These modified amino acids were converted to α-bromoketones via reported procedures.[29, 33]

The α-bromoketones reacted with various nitrogen nucleophiles to form amino methyl ketones which were reduced with $NaBH_4$ to give a diastereomeric mixture (typically 1:1) of hydroxyethylamine derivatives. These intermediates were elaborated to linear precursors (Scheme 2) and cyclized in dilute solution (10^{-4} M, DMF) using BOP/DIPEA to form macrocycles (yields ~20–60 %). Diastereomers were separated and purified by rp-HPLC.

The synthesis of N-terminal macrocyclic inhibitors including **19–26** (Table 3) and the bicyclic inhibitor **41** (Scheme 6 ahead) was most conveniently achieved by the use of a common macrocyclic epoxide intermediate. Ring opening by a nucleophilic amine and subsequent acylation by either a sulphonyl chloride or isocyanate (Scheme 3) quickly gave access to inhibitors with widely varying substituents intended to occupy the P1' and P2' pockets. In each case, a single diastereomer was obtained which facilitated purification of the products.

3.2. C-Terminal Macrocycles

Tripeptide Mimetic of Phe-Ile-Val

The strategy of replacing flexible segments of peptide substrates with conformationally constrained macrocyclic structural mimics can also be applied to the C-

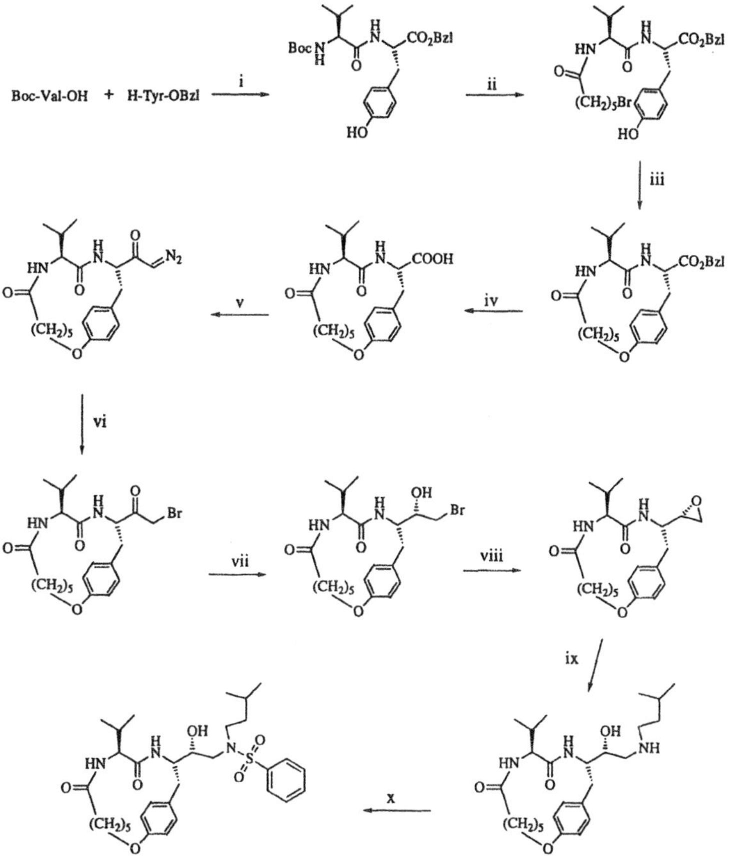

Scheme 3. Reagents : **i** DCC, HOBT, Et_3N; **ii** TFA then $Br(CH_2)_4COCl$ and $KHCO_3$; **iii** NaI in acetone, then K_2CO_3 in DMF; **iv** H_2 and 10%Pd on C in MeOH; **v** NMO then $^{i}BuOCOCl$ in THF at -15 °C, then CH_2N_2; **vi** HBr in dry EtOAc; **vii** $NaBH_4$ in EtOH at -10 °C for 2 min then separation of diastereomers; **viii** NaOMe in MeOH; **ix** isoamylamine, DMF, 70 °C; **x** $PhSO_2Cl$

terminus.[30] We chose to demonstrate this by incorporating similar 15- to 17-membered rings in a derivative of **2**, Ac-Leu-Val-Phe-CHOHCH$_2$-{Phe-Ile-Val}-NH$_2$ (**30**, K_i=1.5 nM), this time imitating the reverse tripeptide sequence Phe-Ile-Val. Compound **31**, containing a 15-membered macrocycle, proved to be an inhibitor of HIVPR with slightly improved potency over acyclic **30** (Table 4). Molecular modeling studies suggested that the new macrocycle in **31** would superimpose accurately upon the acyclic peptidic inhibitor **30** (Figure 12) indicative of structural mimicry, and this seemed to be supported by functional mimicry as there was comparable inhibition of the protease by cyclic and acyclic compounds (Table 4).

Table 4.

Ac-Leu-Val-Phe-[(*S*)-CHOH-CH_2]-Phe-Ile-Val-NH_2

30 (K_i 1.5 nM)

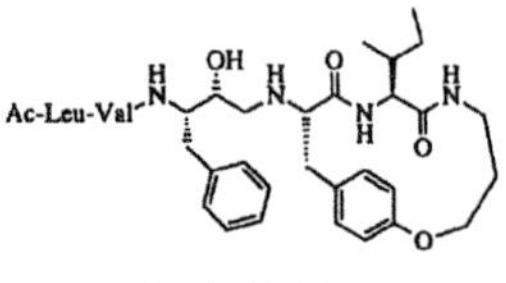

31 (K_i 0.6 nM)

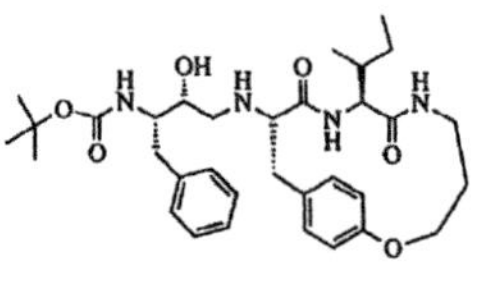

32 (K_i 4 nM)

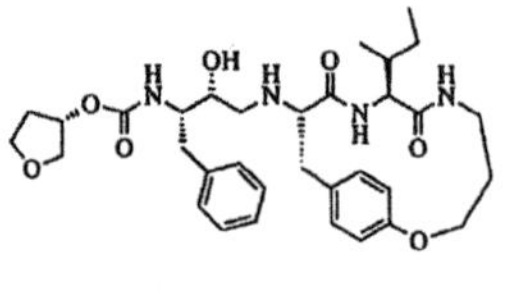

33 (K_i 4 nM)

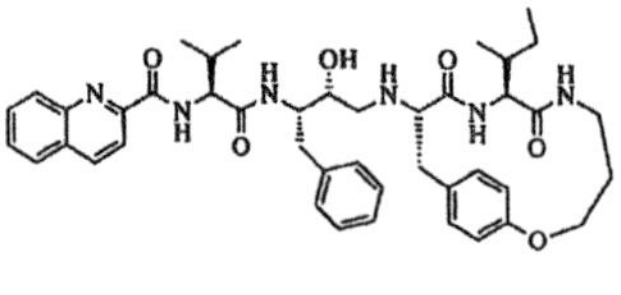

34 (K_i 0.3 nM)

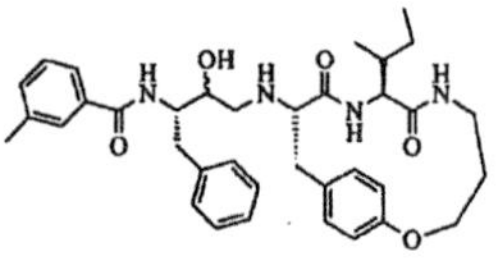

35 (K_i 15 nM)

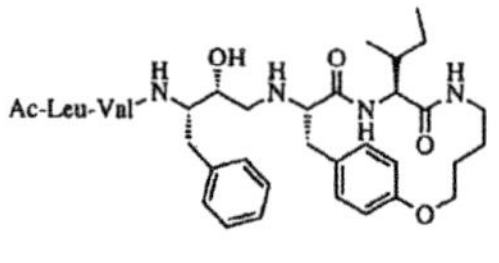

36 (K_i 1.5 nM)

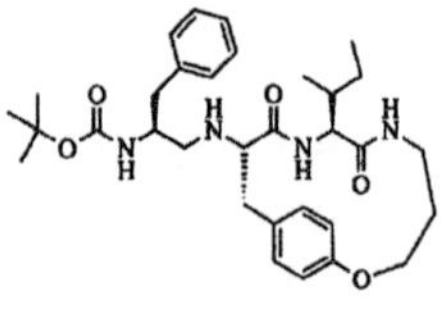

37 (K_i 44 nM)

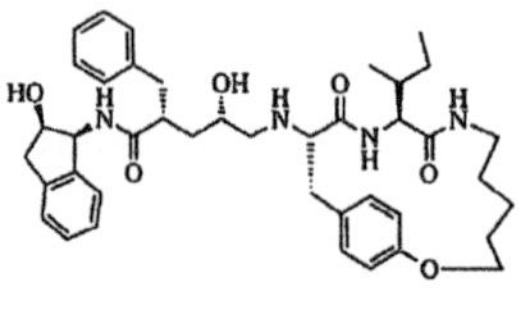

38 (K_i 0.6 nM)

However, we did not obtain an X-ray crystal structure of **30** or **31** bound to HIVPR to prove this structural mimicry. Instead we "leap frogged" to the smaller compound **32**. When the modeled receptor-bound structure of cyclic inhibitor **32** (Figure 13, black image) was compared with that for the acyclic inhibitor **30** (not shown) assuming that the hydroxyethylamine nitrogen of **32** is not protonated (uncharged), [30] the tyrosine, isoleucine, and trimethylene units of **32** all superimpose well on the Phe-Ile-Val residues of **30**, both side- and main- chain atoms matching corresponding atoms of the protease-bound acyclic peptide.

On the other hand, the hydroxyethylamine nitrogen would be expected to be protonated at physiological pH (pK_a ~10 for a secondary amine). When **32** is modeled

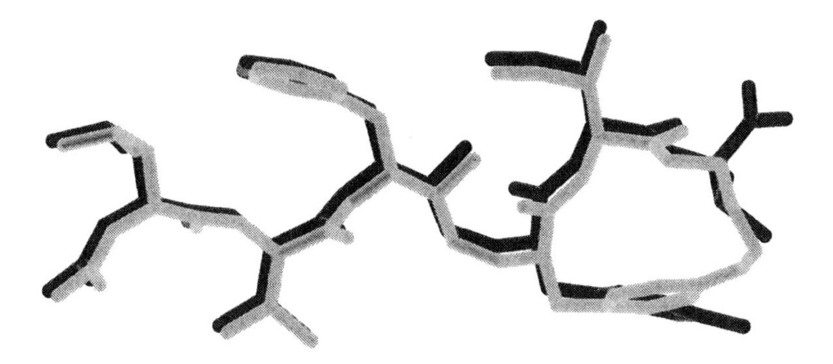

Figure 12. Superimposed HIVPR-bound energy-minimized modeled structures of the (*R*)-diastereomers of **30** (dark) and **31** (light).

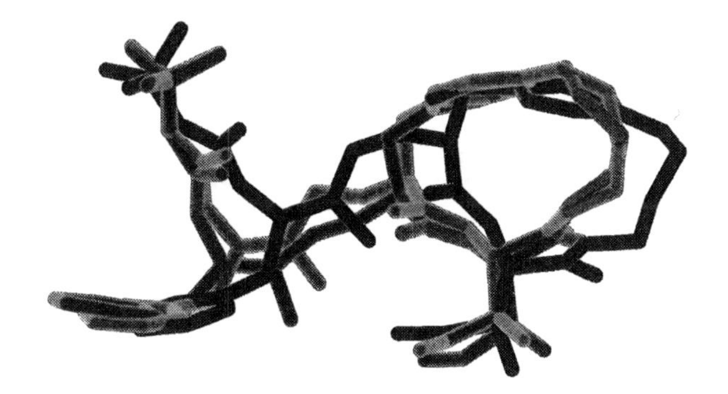

Figure 13. Comparison of HIVPR-bound energy-minimized structures of **32** modeled with unprotonated amine (black) and protonated amine (light). X-ray structure of **32** bound to HIVPR is also shown as a second light image.[41]

as the protonated form, a different enzyme-binding mode is observed (Figure 13, light image), the transition state isostere being translated further along the substrate-binding groove. Subsequently we were able to obtain an X-ray structure of **32** crystallized with HIVPR (Figure 13, second light image, rmsd = 0.4 Å) at pH 5.5 and this more closely matches the modeling for the protonated amine than for the unprotonated form (black image, rmsd = 2.8 Å). Further, when the X-ray structure was energy-minimized with its amine nitrogen uncharged, the structure converted to the conformation in black (Figure13). These results [30] are consistent with the nitrogen of the hydroxyethylamine isostere in inhibitor **32** being protonated when bound in the hydrophobic active site of the enzyme. The charge clearly plays an important role in positioning the inhibitor in the active site of HIVPR. This was an interesting result because acidity can dramatically change in nonaqueous environments and it was not clear if amines would exist in a protonated form in such cavities, protected from the solvating effects of water.

By building the cycle one atom from the hydroxyethylamine nitrogen rather than incorporating the nitrogen in the cycle, the nitrogen is sterically less restricted and can interact with the catalytic Asp residues in the enzyme. The X-ray structure of **32**, together with modeling for **30–38**, suggests that the protonated nitrogen is within hydrogen-bonding distance from both catalytic Asp carboxylates (Figure 14).[30] The hydroxyl substituent remains close enough to hydrogen bond to one Asp in this unsymmetrical arrangement. This shift in the transition state isostere thus causes a disturbance in P1 and P1' residues of **32** (Figure 15, black), these occupying very different positions from those of Phe and Pro in JG365 (Figure 15, light).[19]

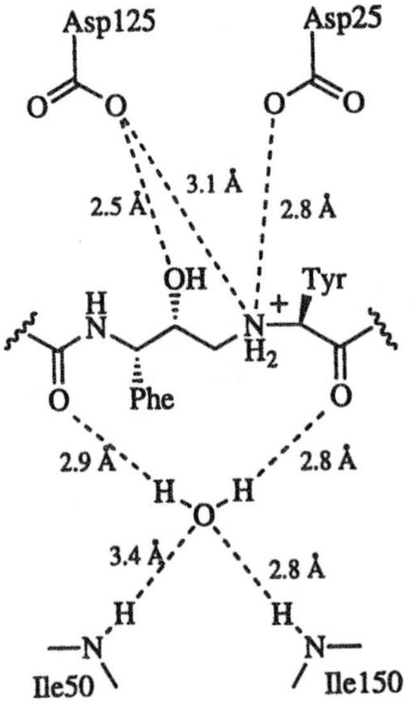

Figure 14. Schematic of hydrogen bonding (dashed lines) between **32** and HIVPR in the active site, taken from the X-ray crystal structure.[41]

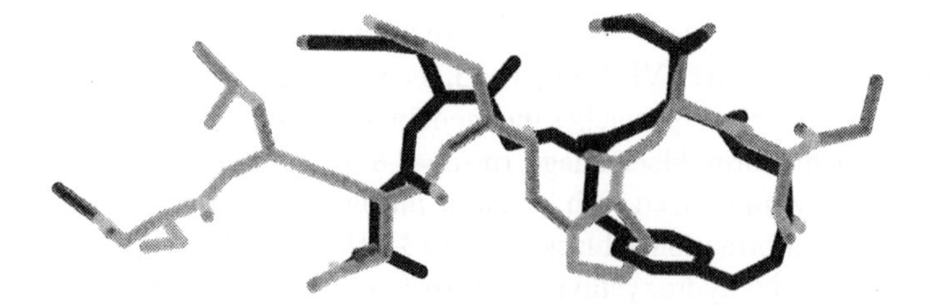

Figure 15. Comparison of X-ray structures of **32** (black) and JG365 (light) bound to HIVPR.

At the time this was a unique inhibitor-binding mode for HIVPR, the hydroxyethylamine isostere being able to bind to HIVPR through *both* its hydroxyl and *protonated* nitrogen to the anionic Asp 25 catalytic residues (Figure 14). There was no effect however upon the positions of more remote side-chain residues (P2, P2', P3') which occupy similar positions to those of JG365 (Figure 15).

Structure–Activity Relationships

The C-terminal cycles were thus used as templates to produce a series of potent inhibitors of synthetic HIVPR as demonstrated in Table 4.[30,36] Under assay conditions[37,38] in which JG365[19] and DM323[39] potently inhibit cleavage of a synthetic fluorogenic substrate,[37] the larger 16-membered ring in **36** was no more effective than the 15-membered ring of **31**, and neither ring on its own was a significant inhibitor (IC_{50} ~ 1–10 μM). Compounds **31–38** combine different N-terminal ends with the C-terminal cycle. Truncating the N-terminal tripeptide LVF to Boc-Phe reduced the potency by an order of magnitude (**31** vs. **32**), and removing the important hydroxyl substituent of the transition state isostere further reduced potency by a second order of magnitude (**37**), despite the presence of a likely protonated amine in the active site.

Compounds **33**, **34**, and **38** couple the C-terminal cycle to different N-terminal groups of known inhibitors.[6-11] The tetrahydrofuran terminus in **33** did not improve potency over the Boc group in **32**; in both cases the cycle most likely shifts the furan oxygen slightly out of hydrogen-bonding contact distance. By contrast the longer quinoline-Val-Phe of **34**, which can make more interactions with HIVPR, increased potency significantly. Compound **35** has a benzamide substituent that appears to be comparable to the other N-termini since even the diastereomeric mixture had IC_{50} ~50 nM. The *cis*-1-amino-2-hydroxy-indane substitutent at P2 in **38** was as effective as the quinoline-Val- substituents at P3 and P2 in **34**.

Synthetic Routes to Macrocycles

Compounds **31–38** in Table 4 were synthesized in a straightforward manner as outlined in Scheme 4 for compounds **32** and **34** and in Scheme 5 for compound **36**. Compound **37** was synthesized by reductive amination of an equimolar mixture of the cycle **40** [29,30] and Boc-Phe-CHO.[40] Inhibitors **32** and **34** were made by solution phase coupling of the known epoxide from Boc-Phe **39** [33] with the C-terminal cycle **40**, followed by removal of the Boc group and capping with one of the N-terminal groups depicted in Table 4. The use of the (2*S*, 3*S*)-Boc-Phe-epoxide **39** ensured a diastereoselective synthesis of the chiral alcohol. Compounds **31, 33, 35,** and **36** were made as diastereomeric mixtures by coupling the cycle to the N-terminal bromomethyl ketone derivatives, followed by reduction with $NaBH_4$.[30] The (*R*)-diastereomer was the major product for **31**, **33** and **36**, and these compounds were easily purified by reverse phase HPLC. Compound **35** was obtained as an equal

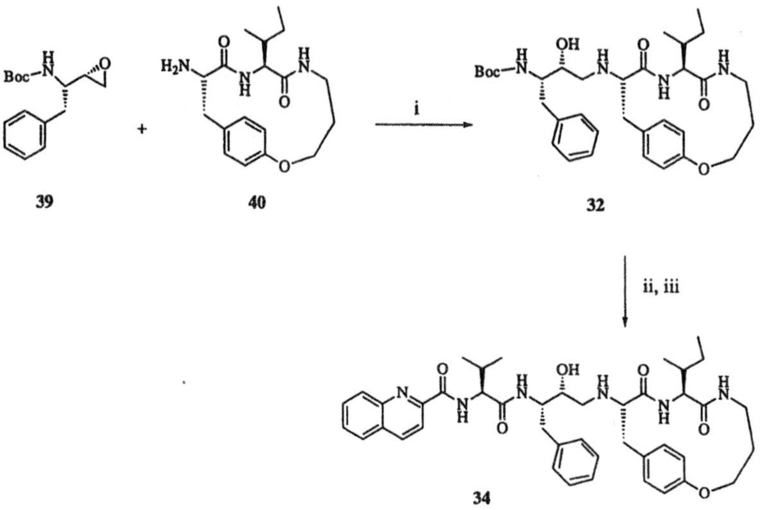

Scheme 4. Reagents : **i** DMF, 60 °C, 6 h; **ii** TFA; **iii** Quin-Val-OH, BOP, DIPEA.

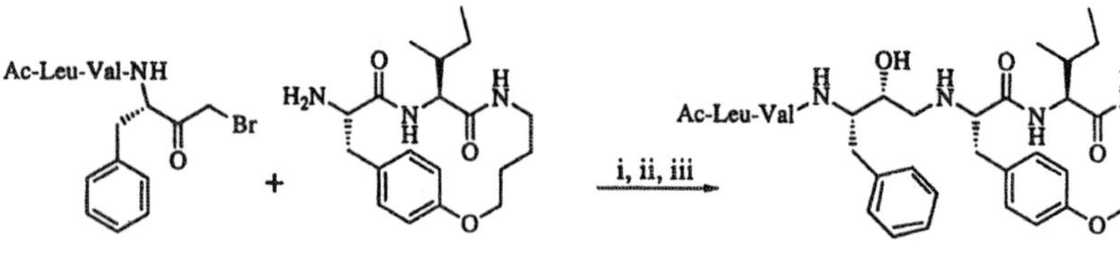

Scheme 5. Reagents : **i** DMF; **ii** $NaBH_4$; **iii** Separate diastereomers

mixture of diastereomeric alcohols that did not separate well under a range of HPLC conditions.

3.3. Bicyclic Inhibitors

Design and Synthesis

A logical progression from the previous sections is to combine N-terminal cycles with C-terminal cycles to obtain bicyclic peptidomimetic inhibitors[20] such as **41**. This consists of two cycles connected by a hydroxyethylamine transition state isostere. Each of the macrocyclic components, independently formed through side chain to backbone condensation, contains two proteolytically resistant amide bonds and either isoleucine or valine linked via a short aliphatic spacer to tyrosine. The aromatic ring and two *trans* amide bonds serve to constrain each macrocycle to a conformation that is pre-organized for receptor binding.[20]

Although we did not initially have an X-ray crystal structure of **41** bound to HIVPR, we had determined X-ray crystal structures for compounds **3** and **32**, containing N- or C-terminal cycles, bound to HIV-1 protease.[29,30] For those solid-state structures, the N-terminal cycle of **3** and the C-terminal cycle of **32** precisely superimposed upon the protease-bound conformations of acyclic peptides. This strongly supported our idea that each cycle in **41** would also be conformationally pre-organized for receptor binding.

Modeling studies of **41** in the active site of the enzyme suggested that this compound would structurally mimic Ac-Leu-Val-Phe-{$CHOHCH_2N$}-Phe-Ile-Val-$-NH_2$ **30**, the two cycles being mimics for the tri-peptides LVF and FIV, and indeed **30** and **41** had identical inhibitor potencies (IC_{50} = 5 nM) against HIVPR.[20] The (*S*)-diastereomeric alcohol of **41** is a ~20-fold less potent inhibitor[36] than the (*R*)-diastereomer, a finding supported by modeling predictions for the two diastereomers.

Neither of the component cycles **28** and **40** of the bicyclic **41** was a potent inhibitor of HIV-1 protease (IC_{50} ~15 μM and ~8 μM, respectively). Thus it is apparent that additional groups need to be appended to the cycles to provide sufficient interactions with HIVPR for potency and selectivity at nM concentrations of inhibitor. Constraining two flexible tripeptides to the receptor-binding cyclic conformations in **41** was expected to bring a significant entropy advantage for inhibitor binding over the acyclic inhibitor Ac-Leu-Val-Phe-{$CHOHCH_2N$}-Phe-Ile-Val-NH_2 **30**. However the latter makes H-bonds between its acetyl, terminal Leu-C$\underline{O}$ and Val-C$\underline{O}\underline{N}\underline{H}_2$ and the enzyme which cannot be made by the cycles of **41**, as shown in seven X-ray crystal structures of the individual cycles in our inhibitors complexed to HIVPR,[41] due to twisting of the amide out of the extended conformation recognized by the protease. Despite this loss of possible hydrogen bonding, as well as some rotational freedom about the transition state isostere,[20] the bicyclic inhibitor **41** must gain some entropy advantage over the conformationally more flexible acyclic inhibitor since it is equipotent as an inhibitor of HIV-1 protease.

A feature of the design process was a comparison of a modeled structure of **41** bound to HIVPR with the HIVPR-bound structures of other inhibitors. The three-dimensional structures[19,42-51] of 12 substrate-based protease-bound peptidic inhibitors of HIV-1 protease are shown overlayed in Figure 5.[20] Inhibitors include MVT101 and JG365 (**1**), which are not antiviral in cell culture at 100 μM, and SKF108738, U75875, A77003, and SB206343 which are antiviral *in vitro* but did not advance to the clinic for other reasons. We found [20] that the bicyclic compound **41** was a consensus mimetic, accurately superimposing on the other 12 structures (Figure 16).

The synthesis of **41** was straightforward because five of the six chiral centers originate from L-amino acids and the sixth is readily controlled via a diastereoselective[33] synthesis involving the chiral (2*S*)-epoxide intermediate[36] **42**. Compound **41**

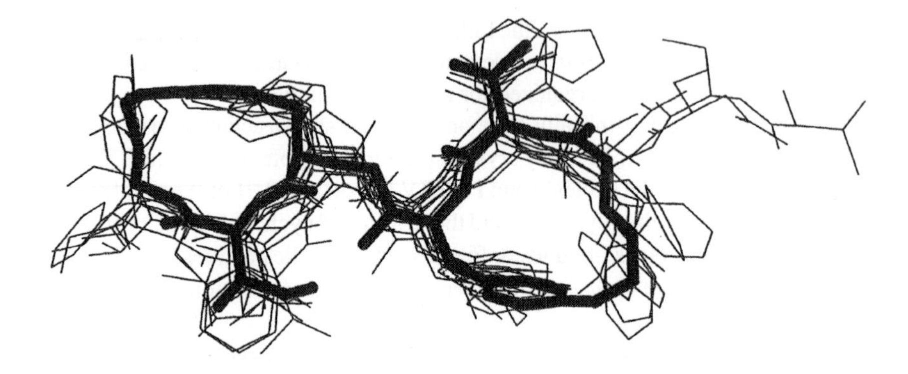

Figure 16. Energy-minimized modeled structure of **41** (black), in a putative HIVPR-binding conformation, overlayed upon the receptor-bound conformations of 12 peptidomimetic inhibitors of HIV-1 protease (X-ray crystal structures).[20]

was prepared in solution by coupling its component halves **42** and **40** (Scheme 6), and was characterized by 2D-NMR (^{1}H, ^{13}C) spectroscopy and mass spectrometry.[20]

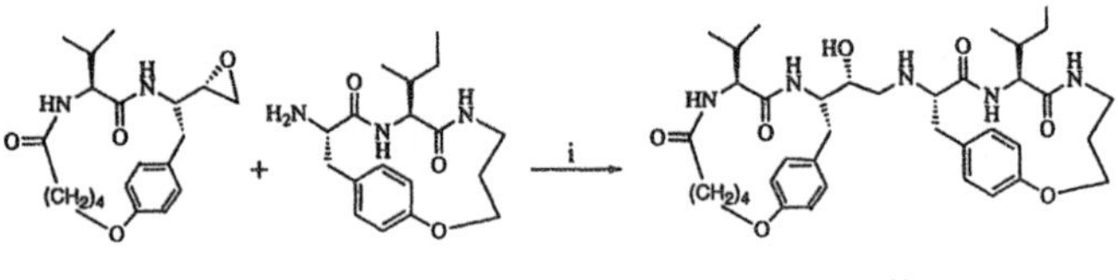

42 **40** **41**

Scheme 6. Conditions : **i** DMF, 80 °C, 24 h

Conformational Restrictions

It was not possible to determine the three-dimensional solution structure of **41** by 2D NMR spectroscopy because of the lack of long-range NOE data. However 1D ^{1}H-NMR spectra in CD_3OH for each of the component cycles of **41** incorporated into other compounds (Tables 2–4)[29,30] show four distinct aromatic proton resonances for each cycle, indicative of restricted rotation of the aromatic ring. Also bicycle **41** in CD_3OH at 300 K shows distinct ^{1}H NMR resonances (1D and COSY) for each of the eight aromatic protons belonging to the tyrosine-derived aromatic rings, consistent with restricted rotation for both aromatic rings. At this temperature signals for the larger N-terminal cycle were clearly closer to coalescence than those for the smaller C-terminal cycle. Variable temperature ^{1}H-NMR experiments in d_6-DMSO (not shown) led to coalescence into two signals at ~330 K of the four resonances attributed to the larger N-terminal macrocycle, suggesting that free rotation of its aromatic ring is occurring at the higher temperature. On the otherhand the four remaining aromatic proton resonances for the smaller C-terminal cycle of **41** persisted even at 400 K (upper limit for experiments), implying highly restricted rotation. These observations are consistent with limited conformational freedom for both macrocyclic components of **41** at physiologically relevant temperatures.

Further support for limited conformational flexibility in each cycle comes from molecular modeling studies of **41**. A Dreiding model of **41** reveals that the short aliphatic spacer in each macrocyclic component only just connects the planar aromatic ring to the rigid region containing the two amide planes. Using computer modelling, unrestrained molecular dynamics on the bicycle supports the constrained nature of each monocycle.[20] All carbon, nitrogen, and oxygen atoms of the N-terminal cycle superimposed to rmsd's < 0.3 Å, while these same atoms in the C-terminal cycle superimpose to rmsd's < 0.7 Å.[20]

Although each cyclic component of **41** is conformationally rigid, the bicycle possesses some conformational freedom as a result of each cycle being able to rotate relative to the other about the linking transition state isostere. The bicycle might be further conformationally constrained if additional bulk were incorporated into this isostere (e.g. an NEt substituent). Nevertheless examination of all available HIV-1 protease structures reveals that only one conformation (the extended conformation) is recognized by the enzyme.[21] In summary, the cyclic components of the bicycle impart significant conformational rigidity to > 90% (47 of the 51 heavy atoms) of the molecule, making it significantly more conformationally constrained than acyclic hexapeptides.

4. ADVANTAGES OF THE MACROCYCLES

Using macrocyclic replacements for short amino acid sequences offers a potentially valuable solution to the problem of precisely mimicking the interactions made by linear bioactive peptides with their receptors. This minimalist approach of retaining the peptide components in the macrocycle helps to ensure that the same interactions will be made with a receptor.

Second, the conformational restrictions placed upon the side chains and backbone amides within a small cycle helps to "fix" the receptor-binding conformation for this portion of the inhibitor. By locking otherwise flexible amino acids into a protein-binding conformation, independent structure–activity changes can be made elsewhere in the inhibitor without affecting the conformation of the cycle. This localized regioselective approach to structure–activity optimization is a major advantage for mimicking a peptide, minimizing the unpredictable cooperative effects that normally compromise structure–activity optimization involving multiple variations to a peptide sequence. This control of conformation also minimizes the potential for non-specific binding to other receptors, a major reason for toxicity of linear peptides.

Other cyclic inhibitors of aspartic proteases, namely HIV-1 protease[52,53] and renin,[28] have been synthesized with even larger rings than those reported here but their protease-bound conformations have not been determined. One of these compounds[53] does, however, exemplify this concept of regioselective optimization since it consists of a macrocycle attached to a difluoroketone transition state isostere. The use of conformationally constrained cycles to fix portions of peptide sequences in their receptor-binding conformations enables a powerful template approach to inhibitor design and development. Appendages can be rationally designed and added to the cycles to rapidly obtain structurally optimized bioactive peptidomimetics.

Peptidic inhibitors of proteases usually suffer from degradation by peptidases in the gut, bloodstream, or in cells. Constricting the tripeptide sequences into macrocycles[20,29-31] was expected to make the amide bonds less recognizable to such proteolytic enzymes. This expectation of greater resistance to cleavage was realized, with all of the macrocyclic compounds described herein being stable to hydrolysis and proteolysis. For example,[54] cycles **28** and **40** : (a) remain entirely intact in aqueous 3M HCl for at least a week at 80 °C; (b) survive pepsin A (pH 3), trypsin (pH 8), chymotrypsin (pH 8), gastricsin, and cathepsin D, and rat gastric juice (pH 2, 4 and 6) for at least 24 hours; and (c) withstand exposure to human serum and lysates from human lymphocytes and polymorphonuclear neutrophils (37 °C, 24 hours). Under these same conditions control acyclic peptides (e.g. **30** and YSFKDMQLGR) were degraded within 1 hour.

This stability was confirmed by the detection of antiviral activity for numerous cyclic inhibitors of the general classes described above.[55] Most acyclic peptides (e.g. JG365, MVT101) are inactive as antiviral agents *in vitro* even at 100 μM concentrations. While some of these are highly degradable by intracellular peptidases, another major stumbling block is their high water solubility but low lipophilicity. In order to penetrate cellular membranes, their log $P_{o/w}$ values should be ~2–4. Lipophilicity can be regulated in the proteolytically more stable macrocycles either by using apolar side chains in/on the cycle or by attaching hydrophobic appendages. We find that lipophilic compounds containing macrocycles (e.g. **22** and **38**)[55] are antiviral at nanomolar concentrations when exposed for 4 days to cultured cells infected with HIV. There is evidence of intracellular metabolism by unknown mechanisms over several days. In rats, dogs and cats compounds from these inhibitor classes persist in blood for several hours at up to 10–20 μM concentrations.[55]

In summary, we have described a general strategy for rationally developing proteolytically stable structural and functional mimics of short peptidic enzyme inhibitors. The use of macrocycles permits predictable regioselective structure–activity optimization to facilitate the generation of potent and selective protease inhibitors. Their ease of synthesis, variability of side chains, hydrolytic stability, ease of incorporating chiral centers, and water and lipid solubility make this an attractive approach for developing protease inhibitors. This method of mimicking bioactive peptides may have wider application to protease inhibitors, peptidomimetic design, and drug development.

ACKNOWLEDGMENTS

Support from an R&D syndicate, the National Health and Medical Research Council, and the Commonwealth AIDS Research Grant Scheme are gratefully acknowledged. We would also like to acknowledge the contributions made to this research at the University of Queensland by Darren March, John Abbenante, Doug Bergman, Jenny Martin, Belinda Todd, Peter Hunt, Ross Brinkworth, Michael Dooley, Bronwyn Garnham, Robert Dancer, Ian James, Wasantha Wickramasinghe, Di Alewood, Angela Finch, Margaret Passmore, and Trudy Bond.

ABBREVIATIONS

DIPEA = diisopropylethylamine;
MBHA = *p*-Methylbenzhydrylamine resin; HCl, 0.79 meq/g; DMF = dimethylformamide.
BOP = (Benzotriazol-1-yl-oxy-tris(dimethylamino) phosphonium)hexafluorophosphate.
HBTU = *O*-Benzotriazole N',N',N',N'-tetramethyluronium hexafluorophosphate.
TFA = Trifluoroacetic acid.
DMF = Dimethylformamide.
NMM = *N*-Methylmorpholine.

REFERENCES

1. Ratner, L. *Perspect.Drug Disc. Design* **1993**, *1*, 3.
2. World Health Organization, Global Programme on AIDS [web page] **1996**, http://www.unaids.org/
3. Mitsuya, H.; Yarchoan, R.; Broder, S. *Science* **1990**, *249*, 1533.
4. Johnston, M. I.; Hoth, D. F. *Science* **1993**, *260*, 1286.
5. DeClerq, E. *J. Med. Chem.* **1995**, *38*, 2491.
6. Tomasselli, A. G.; Howe, W. J.; Sawyer, T. K.; Wlodawer, A.; Heinrikson, R. L. *Chim. Oggi* **1991**, *9*, 6.
7. Wlodawer, A.; Erickson, J. W. *Annu. Rev. Biochem.* **1993**, *62*, 543.
8. Darke, P. L.; Huff, J. R. *Adv. Pharmacol.* **1994**, *25*, 399.
9. West, M. L.; Fairlie, D. P. *Trends Pharmacol. Sci.* **1995**, *16*, 67.
10. Kempf, D. J.; Sham, H. L. *Curr. Pharma. Design* **1996**, *2*, 225.
11. March, D. R.; Fairlie, D. P. In *HIV-1 Protease and its Inhibitors*; Wise, R., Ed.; R. G. Landes, Austin, Texas, 1996, pp 1-88.
12. Fitzgerald, P. M.; Springer, J. P. *Ann. Rev. Biophys. Chem.* **1991**, *20*, 299.
13. Kohl, N. E.; Emini, E. A.; Schleif, W. A.; Davis, L. J.; Meimbach, J. C.; Dixon, R. A. F.; Scolnick, E. M.; Sigal, I. S. *Proc. Natl. Acad. Sci. USA* **1988**, *85*, 4686.
14. McQuade, T. J.; Tomasselli, A. G.; Liu, L.; Karacostas, V.; Moss, B.; Sawyer, T. K.; Heinrikson, R. L.; Tarpley, W. G. *Science* **1990**, *247*: 454.
15. Lambert, D.M.; Petteway, S.R. Jr.; McDanal, C. E.; Hart, T.K.; Leary, J.J.; Dreyer, G.B.; Meek, T.D.; Bugelski, P.J.; Bolognesi, D.P.; Metcalf, B.W.; Matthews, T. J. *Antimicrob. Agents Chemother.* **1992**, *36*, 982.
16. Ashorn, P.; McQuade, T. J.; Thaisrivongs, S.; Tomaselli, A. G.; Tarpley, W. G.; Moss, B. *Proc. Natl. Acad. Sci. USA* **1990**, *87*, 7472.
17. Chatfield D. C.; Brooks, B. R. *J. Am. Chem. Soc.* **1995**, *117*, 5561.

18. Smith, R.; Brereton, I. M.; Chai, R. Y.; Kent, S. B. H. *Nature Struct. Biol.* **1996**, *3*, 946 and references therein.
19. Swain, A.; Miller, M. M.; Green, J.; Rich, D. H.; Schneider, J.; Kent, S. B. H.; Wlodawer, A. *Proc. Natl. Acad. Sci. USA.* **1990**, *87*, 8805.
20. Reid, R. C.; March, D. R.; Dooley, M. J.; Bergman, D. A.; Abbenante, G.; Fairlie, D. P. *J. Am. Chem. Soc.* **1996**, *118*, 8511.
21. Wong, A.; March, D. R.; Tyndall, J.; Burkett, B.; Abbenante, G.; Chai, C.; Fairlie, D. P. Submitted for publication.
22. Ratner, L.; Haseltine, W.; Patarca, R.; Livak, K. J.; Starcich, B.; Josephs, S. F.; Doran, E. R.; Rafalski, J. A.; Whitehorn, E. A.; Baumeister, K.; Ivanoff, L.; Petteway, S. R., Jr.; Pearson, M. L.; Lautenberger, J. A.; Papas, T. S.; Ghrayeb, J.; Chang, N. T.; Gallo, R. C.; Wong-Staal, F. *Nature* **1985**, *313*, 277.
23. Milner-White, E. J. *Trends Pharmacol. Sci.* **1989**, *10*, 70.
24. Dunn, B. M.; Gustchina, A.; Wlodawer, A.; Kay, J. *Meth. Enzymol.* **1994**, *241*, 255.
25. Ptitsyn, O. B.; Finklestein, A. V. *Biopolymers* **1983**, *22*, 15.
26. Tomaselli, A. G.; Heinrikson, R. L. *Meth. Enzymol.* **1994**, *241*, 279.
27. Weber, A. E.; Halgren, T. A.; Doyle, J. J.; Lynch, R.J.; Siegl, P. K. S.; Parsons, W.H.; Greenlee, W. J.; Patchett, A. A. *J. Med. Chem.* **1991**, *34*, 2692.
28. Fairlie, D. P.; Abbenante, G.; March, D. R. *Curr. Med. Chem.* **1995**, *2*, 654 and references therein.
29. Abbenante, G.; March, D. R.; Bergman, D. A.; Hunt, P. A.; Garnham, B.; Dancer, R. J.; Martin, J. L.; Fairlie, D. P. *J. Am. Chem. Soc.* **1995**, *117*, 10220.
30. March, D. R.; Abbenante, G.; Bergman, D. A.; Brinkworth, R. I.; Wickramasinghe, W.; Begun, J.; Martin, J. L.; Fairlie, D. P. *J. Am. Chem. Soc.* **1996**, *118*, 3375.
31. Abbenante, G.; Bergman, D. A.; Brinkworth, R. I.; March, D. R.; Reid, R. C.; Hunt, P. A.; James, I. W.; Dancer, R. J.; Garnham, B.; Stoermer, M. J.; Fairlie, D. P. *Bioorg. Med. Chem. Lett.* **1996**, *6*, 2531.
32. Rich, D. H.; Green, J.; Toth, M. W.; Marshall, G. R.; Kent, S. B. H. *J. Med. Chem.* **1991**, *33*, 1285.
33. Rich, D. H.; Sun, C. Q.; Prasad, J. V. N. V.; Pathiasseril, A.; Toth, M. V.; Marshall, G. R.; Clare, M.; Mueller, R. A.; Houseman, K. *J. Med. Chem.* **1991**, *34*, 1222.
34. Roberts, N. A.; Martin, J. A.; Kinchington, D.; Broadhurst, A. V.; Craig, J. C.; Duncan, I. B.; Galpin, S. A.; Handa, B. K.; Kay, J.; Krohn, A.; Lambert, R. W.; Merrett, J. H.; Mills, J. S.; Parkes, K. E. B.; Redshaw, S.; Ritchie, A. J.; Taylor, D. L.; Thomas, G. L.; Machin, P. J. *Science* **1990**, *248*, 358.
35. Kim, E. E.; Baker, C. T.; Dwyer, M. D.; Murcko, M. A.; Rao, B. G.; Tung, R. D.; Navia, M. A. *J. Am. Chem. Soc.* **1995**, *117*, 1181.
36. Reid, R. C.; Bergman, D. A.; Abbenante, G.; March, D. R.; Fairlie, D. P. Submitted for publication.
37. Toth, M. V.; Marshall, G. R. *Int. J. Pept. Protein Res.* **1990**, *36*, 544.
38. Bergman, D. A.; Alewood, D.; Alewood, P. F.; Andrews, J. L.; Brinkworth, R. I.; Engelbretsen, D. R.; Kent, S. B. H. *Lett. Peptide Sci.* **1995**, *2*, 99.
39. Lam, P. Y. S.; Jadhav, P. K.; Eyermann, C. J.; Hodge, C. N.; Ru, Y.; Bacheler, L. T.; Meek, J. L.; Otto, M. J.; Rayner, M. M.; Wong, Y. N.; Chang, C. H.; Weber, P. C.; Jackson, D. A.; Sharpe, T. R.; Ericksonviitanen, S. *Science* **1994**, *263*, 380.
40. Fehrentz, J.-A.; Castro, B. *Synthesis* **1983**, 676.
41. Martin, J. L.; manuscript in preparation.
42. Dreyer, G. B.; Lambert, D. M.; Meek, T. D.; Carr, T. J.; Tomaszek, T. A., Jr.; Fernandez, A. V.; Bartus, H.; Cacciavillani, E.; Hassell, A. M.; Minnich, M.; Petteway, S. R., Jr.; Metcalf, B. W. *Biochemistry* **1992**, *31*, 6646.
43. Krishno, H. M.; Winborne, E. L.; Minnich, M. D.; Culp, J. S.; Debouck, C. *J. Biol. Chem.* **1992**, *267*, 22770.
44. Thonki, N.; Rao, J. K. M.; Foundling, S. I.; Howe, W. J.; Tomaselli, A. G.; Heinrikson, R. L.; Thaisrivongs, S.; Wlodawer, A. Unpublished work.

45. Hosur, M. V.; Bhat, T. N.; Kempf, D. J.; Baldwin, E. T.; Liu, B. S.; Gulnik, S.; Wideburg, N. E.; Norbeck, D. W.; Appelt, K.; Erickson, J. W. *J. Am. Chem. Soc.* **1994**, *116*, 847.
46. Miller, M.; Schneider, J.; Sathyanarayana, B. K.; Toth, M. V.; Marshall, G. R.; Clawson, L.; Selk, L.; Kent, S. B. H.; Wlodawer, A. *Science* **1989**, *246*, 1149.
47. Jaskolski, M.; Tomaselli, A. G.; Sawyer, T. K.; Staples, D. G.; Heinrikson, R. L.; Schneider, J.; Kent, S. B. H.; Wlodawer, A. *Biochemistry* **1991**, *30*, 1600.
48. Erickson, J.; Neidhart, D. J.; Van Drie, J.; Kempf, D. J.; Wang, X. C.; Norbeck, D. W.; Plattner, J. J.; Rittenhouse, J. W.; Turon, M.; Wideburg, N.; Kohlbrenner, W. E.; Simer, R.; Welfrich, R.; Paul, D. A.; Knigge, M. *Science* **1990**, *249*, 527.
49. Thompson, S. K.; Murthy, K. H. M.; Zhao, B.; Winborne, E.; Green, D. W.; Fisher, S.; DesJarlais, R. L.; Tomasek, T. A.; Meek, T. D.; Gleeson, J. G.; Abdel-Meguid, S. S. Unpublished work.
50. Bone, R.; Vacca, J. P.; Anderson, P. S.; Holloway, M. K. *J. Am. Chem. Soc.* **1991**, *113*, 9382.
51. Abdet Meguid, S. S.; Metcalf, B. W.; Carr, S. T. J.; Demarsh, P.; DesJarlais, R. L.; Fisher, S.; Green, D. A.; Ivonoff, L.; Lambert, D. M.; Murthy, K. H. M.; Petteway, S. R.; Pitts, W. J.; Tomasek, T. A.; Winborne, E.; Zhao, B.; Dreyer, G. B.; Meek, T. D. Unpublished work.
52. Smith, R. A.; Coles, P. J.; Chen, J. J.; Robinson, V. J.; Macdonald, I. D.; Carriere, J.; Krantz, A. *Bioorg. Med.Chem.Lett.* **1994**, *4*, 2217.
53. Podlogar, B. L.; Farr, R. A.; Friedrich, D.; Tarnus, C.; Huber, E. W.; Cregge, R. J.; Schirlin, D. *J. Med. Chem.* **1994**, *37*, 3684.
54. March, D. R. Ph.D. Dissertation, University of Queensland, 1996.
55. Reid, R. C.; Todd, B.; March, D. R.; Passmore, M.; Fairlie, D. P. Submitted for publication.

[illegible]

PEPTIDOMIMETIC SYNTHETIC COMBINATORIAL LIBRARIES

Barbara Dörner, John M. Ostresh, Sylvie E. Blondelle,
Colette T. Dooley, and Richard A. Houghten

ABSTRACT

Combinatorial chemistry is a rapidly expanding field focusing on the simultaneous synthesis of hundreds to millions of compounds for the discovery of pharmaceutical

Advances in Amino Acid Mimetics and Peptidomimetics
Volume 1, pages 109-125

ISBN: 0-7623-0200-3

leads. The contributions from our laboratory to the combinatorial chemistry field include the development of methods for the solid-phase syntheses and screening in solution of a wide range of peptide, peptidomimetic, and small organic molecule combinatorial libraries. This review concentrates on the development and use of peptidomimetic combinatorial libraries. The peptidomimetic libraries discussed were prepared either through chemical modification of existing resin-bound peptide libraries by applying the "libraries from libraries" concept developed in our laboratory, or through stepwise synthesis on a solid support. Amide alkylation and reduction reactions on the solid phase were used to generate a wide variety of peptidomimetic combinatorial libraries. The combinatorial libraries that are described in this chapter include: permethylated hexapeptides, heptamines, peralkylated pentamines, acylated triamines, and dipeptidomimetics. Examples of the screening of these libraries to identify functional, individual peptidomimetics having opiate receptor binding and antimicrobial properties are described.

1. INTRODUCTION

The utility of synthetic combinatorial libraries (SCLs) for basic research and drug discovery has been extensively reported (see reviews[1-3]). All of the seminal concepts and early practical examples of SCLs were developed using peptides of varying size and composition. The poor oral bioavailability and lack of biostability of L-amino acid peptides make them of less general utility as pharmaceutical leads as compared to non-peptidic compounds. Therefore, research in combinatorial chemistry is currently focused on the generation of peptidomimetic and small organic molecule combinatorial libraries (see reviews[3-6]).

Merrifield's solid phase synthesis approach,[7] originally developed for the synthesis of peptides, is used almost exclusively in the synthesis of SCLs. In this approach, the first building block is covalently bound to a polymeric support and further components are successively added from a defined group of building blocks. This method allows excess reagents to be removed by simple washing and filtration steps, and avoids the laborious isolation steps associated with typical solution-phase synthesis. Following cleavage from the solid support, compound mixtures or individual compounds are screened in solution in a wide variety of biological systems.

Two methods for the determination of active sequences or structures from the assay data are used in this laboratory. The first is the iterative deconvolution method,[8] involving a stepwise process of synthesis and screening, in which all positions of the active sequences are successively defined to identify individual compounds. The second is the positional scanning (PS) deconvolution method.[9] A PS-SCL is synthesized as sublibraries such that each sublibrary illustrates the structural details of specific positions of diversity. This, in turn, allows the identification of individual active compounds in a single screening, and provides information on structure-activity relationships of each variable position in the pharmacophore being studied.

Two synthetic approaches are used in our laboratory to synthesize mixture positions in SCLs and PS-SCLs. The "divide, couple, and recombine" (DCR) approach,[8] also known as "split synthesis"[10] or "mixing and portioning",[11] was originally used with the iterative deconvolution format. With the DCR method, a mixture position is obtained by physically mixing resin portions, which are individually reacted with one reagent prior to mixing. The second approach to the synthesis of mixture positions involves the use of reagent mixtures. The relative reaction rates for each reagent have to be predetermined under controlled reaction conditions.[12,13] The main advantage of this method is that both defined and mixture positions can be incorporated into a molecule at any given position, which is a requirement for the synthesis of PS-SCLs. Alternatively, if the chemical ratios are not available for a particular group, the more labor-intensive DCR method is used for the synthesis of PS-SCLs.

Taking advantage of the large diversity afforded by peptide libraries (typically tens of millions of compounds), one focus of research in our laboratory has been on the development of methods to chemically transform existing resin-bound peptide libraries. Termed the "libraries-from-libraries"concept,[14] this approach enables the preparation of combinatorial libraries of non-peptidic compounds having very different physical, chemical, and biological properties from the peptide libraries used as starting materials.

2. GENERATION OF PEPTIDOMIMETIC LIBRARIES

Peptidomimetic libraries have been generated using the "libraries-from-libraries" concept.[14-16] Chemical reactions, such as alkylation, reduction, etc. can be applied to an existing resin-bound or free peptide library. Examples to illustrate the breadth of the approaches described in this chapter are: a permethylated hexapeptide library; a heptamine library generated by exhaustively reducing a hexapeptide library; peralkylated tri- and tetrapeptide libraries prepared using different alkyl halides such as methyl iodide, allyl bromide or benzyl bromide; and combinations of reduced and peralkylated tetrapeptide libraries. Acylated triamines, which were produced by permethylation of resin-bound dipeptides, subsequent reduction of the dipeptides, and finally acylation of the N-terminal amino groups, are also described. Finally, dipeptidomimetic combinatorial libraries were prepared directly by a stepwise synthesis. These can also be further modified through chemical transformation of the resin-bound compounds.

Resin-bound parent peptide libraries were prepared using standard *t*-butyloxycarbonyl synthesis protocols in conjunction with simultaneous multiple peptide synthesis[17] using methylbenzhydrylamine (MBHA) polystyrene resin as solid support.[18] In step wise peptidomimetic synthesis, standard 9-fluorenylmethyloxycarbonyl (Fmoc) protocols were applied. Following their synthesis, the resin-bound peptides or library mixtures were further chemically modified. Initial development of the methods for chemical transformation of resin-bound compounds involved the synthesis of individual peptidomimetic model sequences, their cleavage from their solid support, and their characterization (routine analysis included MALDI-MS, RP-HPLC and when necessary NMR). These or similar sequences were then always included as controls during the library syntheses.

The use of peptidomimetic SCLs in biological assay systems is illustrated through some examples of screening profiles and individual compounds showing activity in an antimicrobial or a radio receptor binding assay.

2.1. Mimetics Derived from Hexapeptides

The initial synthesis of peptidomimetic libraries was performed by chemically modifying hexapeptide libraries composed of approximately 38 million compounds (Table 1). The parent resin-bound hexapeptide libraries were prepared in a positional scanning format using 20 different L-amino acids in the defined positions (O-position), and 18 different L-amino acids in the mixture positions (X-position) having either free N-terminal amino groups or N-terminal acetyl groups. The hexapeptide PS-SCL was composed of six separate positional sublibraries (O_1XXXXX, XO_2XXXX, XXO_3XXX, $XXXO_4XX$, $XXXXO_5X$, $XXXXXO_6$). The 38 million compounds result from the 20 amino acids making up the defined position times the 18 amino acids making up the five mixture positions (i.e. 20×18^5). A hexapeptide PS-SCL having a free N-terminal amino group was permethylated, while the acetylated form of such a hexapeptide library was reduced.

Table 1. Peptidomimetic SCLs Generated from Resin-Bound Hexapeptide Libraries through Permethylation or Reduction

	Library Format[a]	*Chemical Transformation*	*Amino Acids in O/X*	*Total # of Compounds*
1	OXXXXX	permethylation	20/18	each PS-SCL:
	XOXXXX			37,791,360
	XXOXXX			(20×18^5)
	XXXOXX			
	XXXXOX			
	XXXXXO			
2	Ac-OXXXXX	reduction	20/18	each PS-SCL:
	Ac-XOXXXX			37,791,360
	Ac-XXOXXX			(20×18^5)
	Ac-XXXOXX			
	Ac-XXXXOX			
	Ac-XXXXXO			

Notes: Ac, acetyl; O, defined position; X, mixture position.

Permethylated Hexapeptide PS-SCL

The existing resin-bound hexapeptide PS-SCL described above was exhaustively permethylated to yield the peptidomimetic library **1** (Table 1).[14] Among various methods reported for the permethylation of peptides in solution,[19,20] the permethylation, using sodium hydride for the amide anion formation and further reaction with methyl iodide, was used for the solid phase transformation.[21] The reaction scheme is shown in Figure 1. The resin-bound peptides, having a free N-terminal amino group and all of the amino acid side chain functionalities protected, were reacted for 16 h at 25 °C with a solution of sodium hydride (10-fold excess) in dimethyl sulfoxide. Methyl iodide was then added to the reaction mixture in a 30-fold excess, and the reaction allowed to proceed for 15 min. Solid phase synthesis allows the use of such high reagent excess to drive the reaction to completion. Following appropriate washing steps to readily remove excess reagent, permethylated compounds were cleaved from the solid support using hydrogen fluoride.[22] Individual control compounds were prepared in parallel during the library synthesis. These were cleaved from the solid support to allow for RP-HPLC and MS analysis.

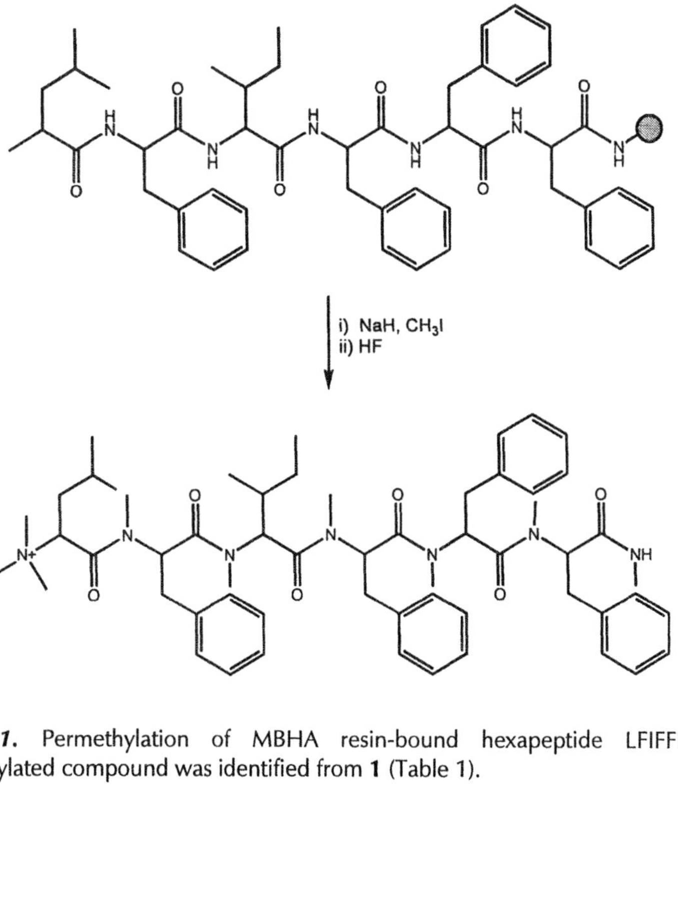

i) NaH, CH_3I
ii) HF

Figure 1. Permethylation of MBHA resin-bound hexapeptide LFIFFF; the permethylated compound was identified from **1** (Table 1).

During the permethylation, the amide nitrogens were methylated and the trimethyl quaternary ammonium salt of the N-terminal amino group was formed. A number of the protected amino acid side-chain functionalities were also modified during the permethylation reaction.[14] Representative examples are: monomethylation of the e-amine of lysine, dimethylation of the amide of asparagine and glutamine, N-methylation of the indole in tryptophan, methylation of the ether in tyrosine, and trimethylation of the guanidine of arginine. The amino acid side chains of serine or threonine (in their protected form during the permethylation) were not modified.

Racemization was studied as a possible side reaction during the strong base treatment. RP-HPLC analysis of permethylated model pentapeptides showed the extend of racemization to be less than 1%.[14,23]

This permethylated peptidomimetic PS-SCL was screened for inhibition of antimicrobial activity against *Staphylococcus aureus*.[14,21,24] From the results of the screening of this library, 144 permethylated sequences were synthesized. The four permethylated (pm) sequences—pm[LFIFFF-NH_2] (structure shown in Figure 1), pm[FFIFFF-NH_2], pm[FFFFFF-NH_2], and pm[LFFFFF-NH_2]—were found to be the most active peptidomimetic compounds, with MIC (minimum inhibitory concentration) values of 11–15, 11–15, 11–15, and 21–31 μg/mL respectively. The corresponding peptide sequences were not active.

Reduced Hexapeptide PS-SCL

The exhaustive reduction of an acetylated hexapeptide PS-SCL was first performed in solution[25,26] to yield a heptamine library **2** (Table 1). More recently, the reaction was modified for use on the solid phase. The solution-phase reaction scheme is outlined for the most active sequence found in an opioid receptor assay in Figure 2. In the solution phase reduction, the peptides were treated with a solution of borane, trimethylborate, and boric acid in tetrahydrofuran. Following reduction, the boron complexes formed were hydrolyzed with 2 N HCl methanol at 65 °C for 19 h. Repetitive evaporation yielded the hydrochloride salts of the reduced compounds and removed all of the volatile by-products. A number of amino acid side chains underwent modification during the reduction: the acid functionalities of aspartic and glutamic acid were reduced to the alcohols; the amide functionalities of asparagine and glutamine were reduced to the corresponding amines; the 2-chlorobenzyloxycarbonyl protected ε-amine of lysine was reduced to the methylamine; and the formyl protected tryptophan was reduced to the *N*-methyl indole. The borane complexed arginine was eliminated during HF cleavage to give ornithine, while borane–sulfur complexed methionine yielded the dehydro analogue of homoalanine.

This library was screened in a radioreceptor assay selective for the μ-opioid receptor for its ability to inhibit [^{3}H]-DAMGO binding to crude rat brain homogenate.[26-28] Screening each of the six positional SCLs making up the PS-SCL permits

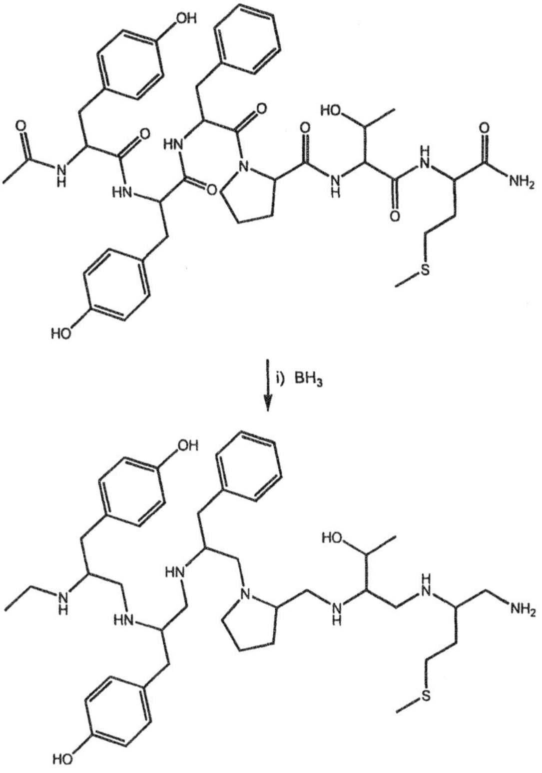

Figure 2. Exhaustive reduction of MBHA resin-bound hexapeptide Ac-YYFPTM; the reduced compound was identified from **2** (Table 1).

the determination of the most effective amino acids at each position of a hexapeptide. The most active polyamines had IC_{50} (concentration which inhibits 50% binding) values in the range of 14–20 nM, and the general formula of red[Ac-YYFPTO-NH_2], with O = F, M, S, P, or Y where "red[ ]" specifies the reduced polypeptide. Truncation analogues (see Table 2) of these heptamines indi-

Table 2. IC_{50} Values for Individual μ-Opioid Receptor Ligands

Sequence[a]	IC_{50} *(nM)*
red[Ac-YYFPTM-NH_2]	14
red[Ac-YYFP-NH_2]	16
red[Ac-YYF-NH_2]	44
red[Ac-YY]	334

Notes: [a] Ac, acetyl; red, reduced.

cated that the pentamine, red[Ac-YYFP-NH_2], retained activity. A decrease in activity was found for shorter sequences such as red[Ac-YYF-NH_2].

2.2. Mimetics Derived from Tri- and Tetrapeptides

The molecular diversity inherent in the parent SCL was transformed by alkylating the amide groups of resin-bound peptide libraries with reagents such as methyl iodide, ethyl iodide, allyl bromide, or benzyl bromide. Tri- and tetrapeptide libraries made up using 52 different L-, D-, and unnatural amino acid derivatives were permethylated, perallylated, and perbenzylated.[29] The library formats of a permethylated tripeptide SCL **3** and peralkylated tetrapeptide SCLs **4–5** are shown in Table 3. In particular, the amide groups of a tetrapeptide PS-SCL (O_1XXX, XO_2XX, XXO_3X, $XXXO_4$) were chemically modified as described above to enable the identification of individual, biologically active tetrapepti-

Table 3. Peptidomimetic SCLs Generated from Resin-Bound Tri- and Tetrapeptide Libraries through Peralkylation and/or Reduction

	Library Format[a]	*Chemical Transformation*	*Amino Acids in O/X*	*Total # of Compounds*
3	OXX	permethylation	52/52	140,608 (52^3)
4	OXXX	peralkylation **4a** allyl **4b** benzyl	52/52	7,311,616 (52^4)
5	OXXX XOXX XXOX XXXO	peralkylation **5a** allyl **5b** benzyl	52/52	7,311,616 (52^4)
6	OXXX	peralkylation & reduction **6a** methyl **6b** ethyl **6c** benzyl	52/52	7,311,616 (52^4)
7	OXXX XOXX XXOX XXXO	peralkylation & reduction **7c** benzyl	52/52	7,311,616 (52^4)
8	OXXX	reduction	52/52	7,311,616 (52^4)
9	OXXX XOXX XXOX XXXO	reduction	52/52	7,311,616 (52^4)

Note: [a] O, defined position; X, mixture position.

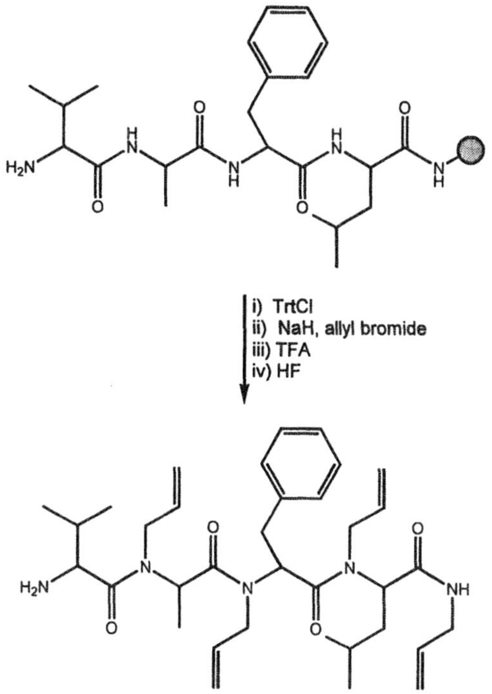

i) TrtCl
ii) NaH, allyl bromide
iii) TFA
iv) HF

Figure 3. Perallylation of the amide nitrogens in a MBHA resin-bound tetrapeptide used as control for the synthesis of **4a** (Table 3).

domimetics in a single screening assay. The synthesis of a perallylated model tetrapeptide representing SCL **4a** is outlined in Figure 3.[30] The α-amino groups were protected with the bulky triphenylmethyl (trityl,Trt) group to avoid modification of the N-terminal amine during manipulation of the amide groups in the resin-bound compounds. The reactions were carried out under an anhydrous nitrogen atmosphere, and the peptide resins were treated with a 20-fold excess of lithium *t*-butoxide (LiOtBu) in tetrahydrofuran to form the amide anions. Following removal of excess base, the alkylating agent (60-fold excess) in dimethyl sulfoxide was reacted with individual resin-bound compounds or libraries. The alkylation reaction mixture was then removed and the base and alkylation treatments repeated to drive the alkylation reaction to completion. Following the removal of the trityl protecting group using 2% trifluoroacetic acid (TFA) in dichloromethane, the peptidomimetic compounds having free N-terminal amino groups were cleaved from their solid support using hydrogen fluoride.

Modifications in the side-chain functionalities vary depending on the groups used to protect the amino acid side chains and the alkylating agent used to modify the amide nitrogens. For example, a number of the amino acid side-chain functionalities underwent alterations during permethylation of model resin-bound dipeptides. Modifications found for amino acid side chains are comparable with the modifications described above for the permethylation (Section 2.1). The hydroxy group of tyrosine was alkylated or remained unmodified depending on the type of side chain protection used and type of alkyl group to be attached. Methylation occurs when tyrosine is protected with the 2-bromobenzyloxycarbonyl group, whereas it remains unmodified using 2,6-dichlorobenzyl. Any benzyl ether formed during the alkylation with benzyl bromide is cleaved during the acidic cleavage of the compounds from the resin. Amino acid derivatives such as aspartic acid, glutamic acid, cysteine, and histidine were not used in the synthesis of the tri- and tetrapeptide libraries due to multiple side reactions occurring during subsequent peralkylation. Over-alkylation was sometimes observed upon treatment of glycine or β-alanine.

The diversity of the peralkylated tetrapeptide SCLs was further expanded by exhaustively reducing the carbonyl groups to produce various peralkylated pentamine libraries **6,7** (Table 3) by applying the method described above (Section 2.1). These libraries were successfully prepared from permethylated, perethylated and perbenzylated tetrapeptide libraries. Reduction of unmodified parent tetrapeptide libraries, either in an iterative or a positional scanning format, resulted in pentamine libraries **8,9** (Table 3).

The tri- and tetrapeptide SCLs synthesized using the libraries-from-libraries concept were screened in a range of *in vitro* biological assays. The screening profiles of the four "first-position-defined" libraries (tetrapeptide, perbenzylated tetrapeptide, pentamine, and perbenzylated pentamine) are shown in Figures 4 and 5.[15] These results were obtained with each of these four libraries upon screening in κ-opioid radio-receptor binding[27,28] and microdilution antimicrobial[24,31] assays, respectively. The screening profiles for each library differs in the amino acid side chains defining the most active mixtures from one library to the other. This was anticipated due to the differing character of the compounds. These results illustrate the utility of the libraries-from-libraries concept for the ready expansion of the diversity of a parent library having very different physicochemical properties when compared to the starting library.

2.3. Acylated Triamine SCL

Peptidomimetic compounds of lower molecular mass were envisioned with the design of an N-alkyl, N-acyl amine SCL **10** (Table 4). A dipeptide library having 52 different amino acid residues in each position (2704 dipeptides) and the N-terminal amino group trityl protected was permethylated as described above (Section 2.2). The resin-bound dipeptidomimetics were reduced following trityl removal and then separately N-acylated using 168 different acylating agents. This library was cleaved from the resin using a prolonged hydrogen fluoride treatment due to greater stability of the amine linkage to the resin relative to the normal amide linkage in resin-bound peptides. The reaction scheme is shown in Figure 6.

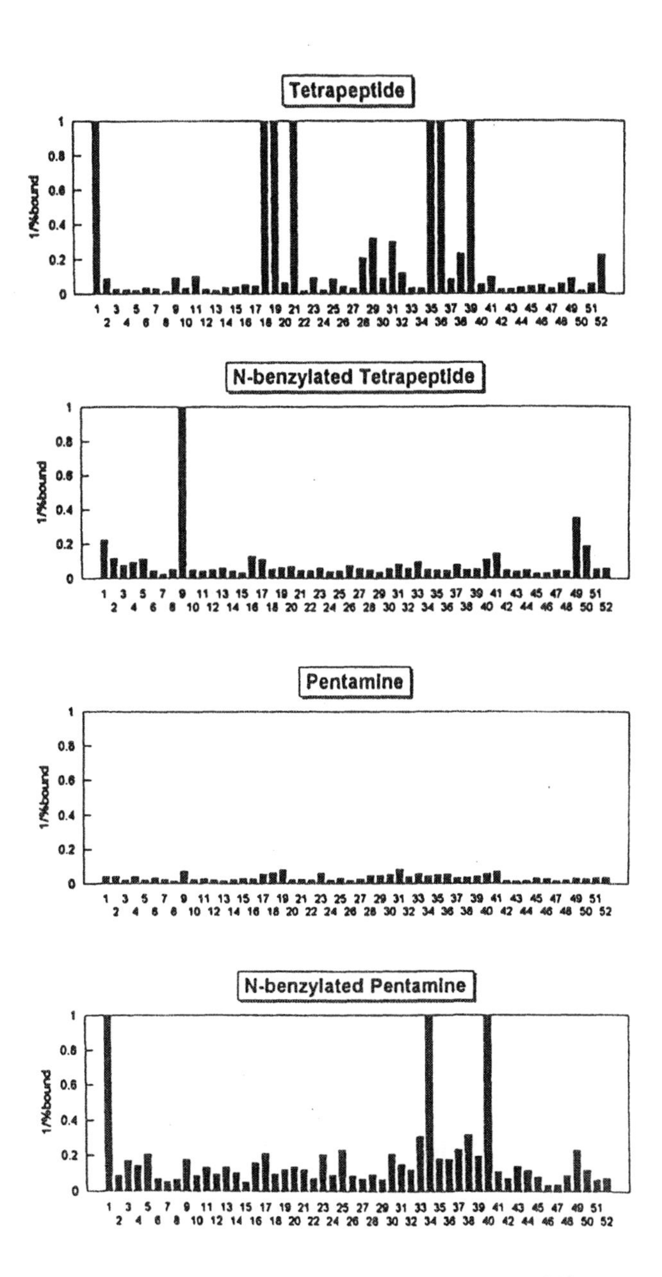

Figure 4. Screening profiles for the four different first positional libraries (O_1XXX) in a κ-opioid radioreceptor binding assay. Each graph represents one of the four libraries; each bar represents the activity of one of the 52 mixtures within the library (x-axis: the 52 different amino acids in the O-position). (Reprinted with permission from Houghten, R. A., Blondelle, S. E., Dooley, C. T., Dörner, B., Eichler, J. and Ostresh, J. M. *Mol. Div.* **1996**, *2*, 41. Copyright 1996 ESCOM Science Publishers B.V.)

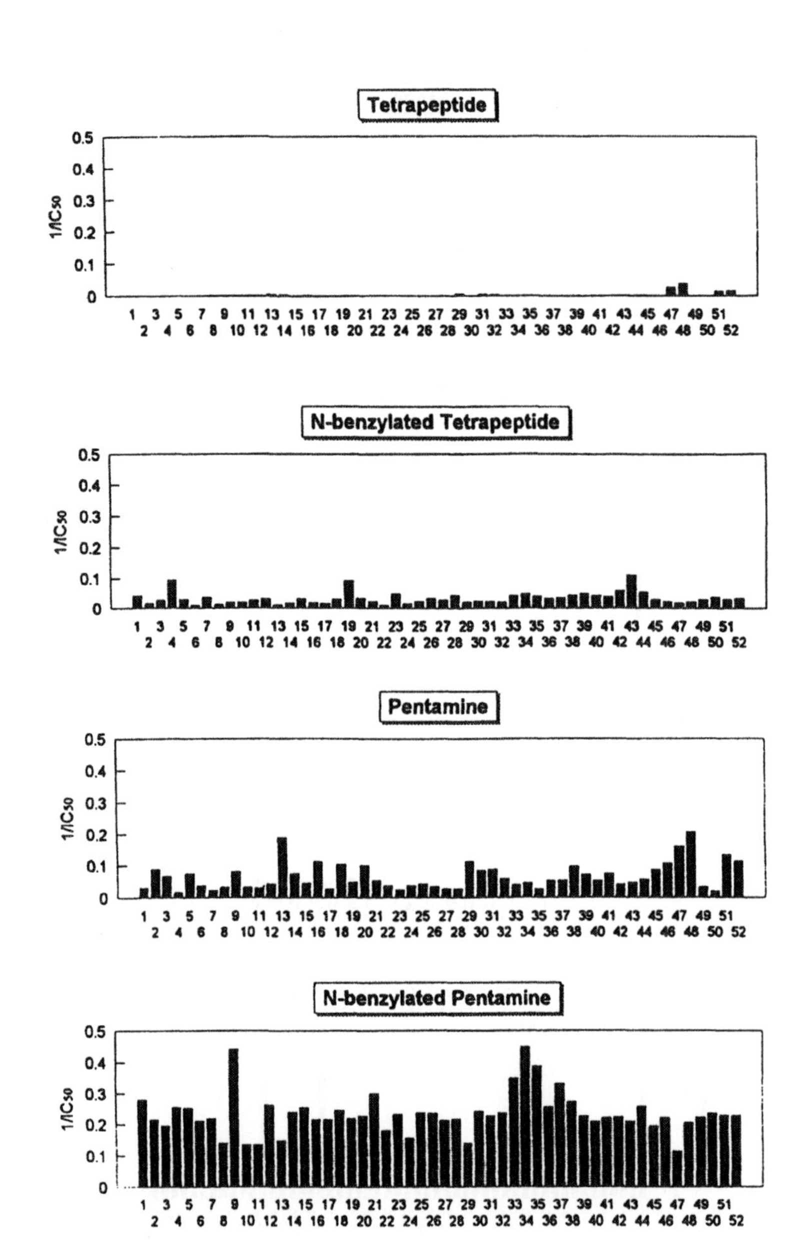

Figure 5. Screening profiles for the four different first positional libraries (O_1XXX) in a microdilution antimicrobial assay. Each graph represents one of the four libraries; each bar represents the activity of one of the 52 mixtures within the library (x-axis: the 52 different amino acids in the O-position). (Reprinted with permission from Houghten, R. A., Blondelle, S. E., Dooley, C. T., Dörner, B., Eichler, J. and Ostresh, J. M. *Mol. Div.*, **1996**, *2*, 41. Copyright 1996 ESCOM Science Publishers B.V.)

Table 4. Acylated Triamine SCL Generated from a Resin-Bound Dipeptide Library

	Library Format[a]	*Chemical Transformation*	*Amino Acids in O/X*	*Total # of Compounds*
10	OXX	peralkylation & reduction & acylation	164/52	443,456 (164 x 52^2)

Note: [a] O, defined position; X, mixture position.

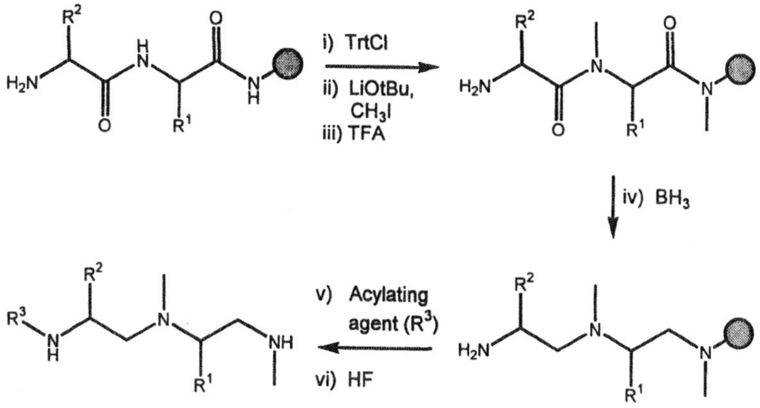

Figure 6. Synthesis of the acylated triamine SCL **10** (Table 4).

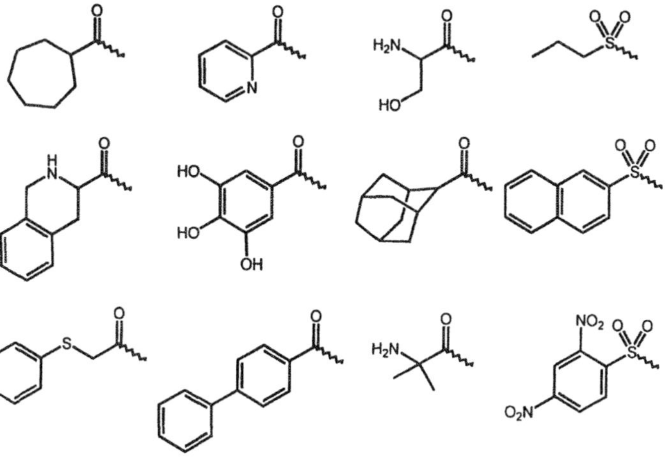

Figure 7. Examples of acyl groups (R^3) used in the synthesis of the acylated triamine SCL **10** (Figure 6, Table 4).

The 168 acyl reagents were composed of 15 sulfonic acids, 62 protected amino acids, and 91 aryl- and alkyl- carboxylic acids. The structures of exemplary acyl building blocks are shown in Figure 7. The compounds included in this peptidomimetic library have an average molecular mass of 425 daltons.

2.4. Dipeptidomimetic SCLs

A dipeptidomimetic SCL **11** (Table 5) of 57,500 compounds was prepared.[32] This library has a dipeptide scaffold with the amide hydrogen replaced with five different alkyl groups (methyl, ethyl, allyl, benzyl, or 2-naphthylmethyl). The dipeptidomimetic SCL was prepared in an iterative format (general formula shown in Figure 8, where R^1 and R^2 represent defined positions and R^3 and R^4 represent mixture positions). Forty-six different amino acids were incorporated into the first X position (R^3), while 50 different amino acids were incorporated into the first O position (R^1). The amide alkyl groups in the second X (R^4) and second O positions (R^2) were methyl, ethyl, allyl, benzyl, and\or 2-naphthylmethyl. This peptidomimetic SCL consists of 250 mixtures (50 amino acids × 5 alkyl groups), each of which is composed of 230 compounds (46 amino acids × 5 alkyl groups), yielding a total number of 57,500 dipeptidomimetics and was prepared using the DCR method.[8] The stepwise synthesis was carried out on the solid phase by alternating amino acid attachment and alkylation of the previously formed amide bond as outlined in Figure 8. Alkylation of the amide bond between the C-terminal amino acid and the MBHA linker was found to significantly decrease the stability of the amide–resin linkage to acidolytic conditions. Therefore, Fmoc chemistry for the incorporation of amino acids was used with MBHA resin as the solid support. The five alkyl halides [methyl iodide, ethyl iodide, allyl bromide, benzyl bromide, and 2-(bromomethyl)naphthalene] were reacted as described above using repeated alkylation treatments. For each alkylation step, individual compound control resins were added to obtain individual compounds for analysis and as controls to study the completeness of the differing alkylation reactions.

Following cleavage from the solid support, the soluble peptidomimetic library mixtures have been screened successfully (i.e., active individual dipeptidomimetics identified) in solution in radioreceptor binding, antimicrobial, and enzyme inhibition assays. Deconvolution of the highly active mixtures, which includes two

Table 5. Alkylated dipeptidomimetic SCLs

	Library Format[a]	*Chemical Transformation*	*Amino Acids in O/X*	*Total # of Compounds*
11	OOXX	stepwise peralkylation	50/46	57,500 (50 x 5 x 46 x 5)
12	OOXX XXOO	Stepwise peralkylation	46/46	52,900 (46 x 5 x 46 x 5)

Note: [a] O, defined position; X, mixture position.

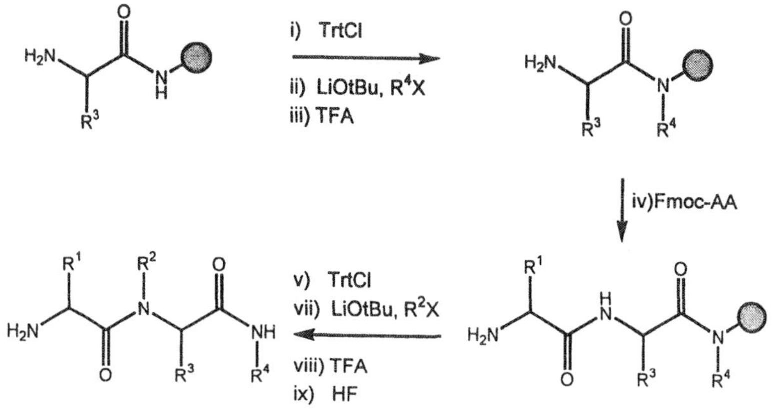

Figure 8. Synthesis of the alkylated dipeptidomimetic SCL **11** (Table 5; R^4X, R^2X: alkyl halides).

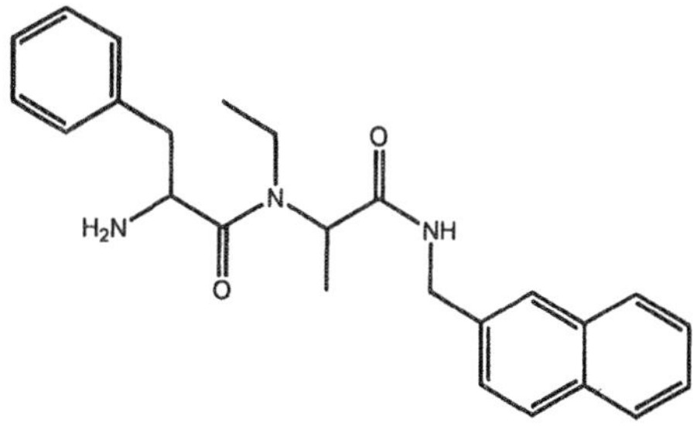

Figure 9. Structure of an μ-opioid receptor selective dipeptidomimetic (IC_{50} = 3 nM) identified from **11** (Figure 8, Table 5).

screening and synthesis steps, was performed to yield compounds with varying activities in the different biological assays. An example for an active and selective dipeptidomimetic ligand for the μ-opioid receptor is shown in Figure 9. This library has also been prepared in the positional scanning format **12** (Table 5).

3. CONCLUSIONS

Synthetic methods have been developed to successfully generate a wide variety of peptidomimetic SCLs of different size and character, obtained through chemical modification of solid-phase-bound precursors using alkylating agents such as benzyl bromide, methyl iodide, etc. Libraries of polyamines and peralkylated polyamines were generated by exhaustive reduction of the peptide SCLs and peptidomimetic SCLs. The concept of the peptidomimetic SCLs presented can be readily extended to include other alkyl groups, longer sequences, etc. A range of other chemical reactions, such as thioamide formation or reductive alkylation of the free N-terminal amino group using a wide variety of aldehydes, can be envisioned to readily yield new diversities having very different physicochemical characteristics and to further expand the range of unique peptidomimetic and small molecule organic SCLs.

The versatility of the SCL concept described here is reflected by the range of libraries with different formats and compositions that can be readily generated. Their broad utility relies on the fact that all these different compound classes are free in solution and in turn can be screened in a wide variety of bioassays.

The concepts described and the ease with which they can be carried out greatly extend the molecular diversities and create novel diversities available for drug discovery and basic research.

ACKNOWLEDGMENTS

We thank Eileen Weiler for her editorial assistance. This work was funded by NSF grant CHE-9520142 (RAH), NIH grant DA09410-01 (RAH) and by Trega Biosciences, Inc., San Diego, CA.

REFERENCES

1. Pinilla, C.; Appel, J.; Blondelle, S.; Dooley, C.; Dörner, B.; Eichler, J.; Ostresh, J.; Houghten, R. A. *Biopolymers (Peptide Science)* **1995**, *37*, 221.
2. Gallop, M. A.; Barrett, R. W.; Dower, W. J.; Fodor, S. P. A.; Gordon, E. M. *J. Med. Chem.* **1994**, *37*, 1233.
3. Thompson, L. A.; Ellman, J. A. *Chem. Rev.* **1996**, *96*, 555.
4. Gordon, E. M.; Barrett, R. W.; Dower, W. J.; Fodor, S. P. A.; Gallop, M. A. *J. Med. Chem.* **1994**, *37*, 1385.
5. Terrett, N. K.; Gardner, M.; Gordon, D. W.; Kobylecki, R. J.; Steele, J. *Tetrahedron* **1995**, *51*, 8135.
6. Kiely, J. S.; Dörner, B.; Ostresh, J. M.; Dooley, C.; Houghten, R. A. In *High Throughput Screening: The Discovery of Bioactive Substances*; Devlin, J. Ed; Marcel-Dekker, New York, 1997, p 155.
7. Merrifield, R. B. *J. Am. Chem. Soc.* **1963**, *85*, 2149.
8. Houghten, R. A.; Pinilla, C.; Blondelle, S. E.; Appel, J. R.; Dooley, C. T.; Cuervo, J. H. *Nature* **1991**, *354*, 84.
9. Pinilla, C.; Appel, J. R.; Blanc, P.; Houghten, R. A. *Biotechniques* **1992**, *13*, 901.
10. Lam, K. S.; Salmon, S. E.; Hersh, E. M.; Hruby, V. J.; Kazmierski, W. M.; Knapp, R. J. *Nature* **1991**, *354*, 82.
11. Furka, Á.; Sebestyén, F.; Asgedom, M.; Dibó, G. *Int. J. Pept. Protein Res.* **1991**, *37*, 487.
12. Ostresh, J. M.; Winkle, J. H.; Hamashin, V. T.; Houghten, R. A. *Biopolymers* **1994**, *34*, 1681.
13. Eichler, J.; Houghten, R. A. *Biochemistry* **1993**, *32*, 11035.

14. Ostresh, J. M.; Husar, G. M.; Blondelle, S. E.; Dörner, B.; Weber, P. A.; Houghten, R. A. *Proc. Natl. Acad. Sci. USA* **1994**, *91*, 11138.
15. Houghten, R. A.; Blondelle, S. E.; Dooley, C. T.; Dörner, B.; Eichler, J.; Ostresh, J. M. *Mol. Div.* **1996**, *2*, 41.
16. Griffith, M. C.; Dooley, C. T.; Houghten, R. A.; Kiely, J. S. In *Molecular Diversity and Combinatorial Chemistry: Libraries and Drug Discovery*; Chaiken, I. M.; Janda, K. D., Eds.; Americal Chemical Society, Washington, DC, 1996, p 50.
17. Houghten, R. A. *Proc. Natl. Acad. Sci. USA* **1985**, *82*, 5131.
18. Pinilla, C.; Appel, J. R.; Houghten, R. A. In *Current Protocols in Immunology*; Coligan, J. E.; Kruisbeek, A. M.; Margulies, D. H.; Shevach, E. M.; Strober, W., Eds.; John Wiley & Sons, New York, 1994; p 9.8.1.
19. Challis, B. C.; Challis, J. A. In *The Chemistry of Amides*; Zabicky, J. Ed.; Interscience, New York, 1970, p 731.
20. Vilkas, E.; Lederer, E. *Tetrahedron Lett.* **1968**, *26*, 3089.
21. Ostresh, J. M.; Blondelle, S. E.; Dörner, B.; Houghten, R. A. *Methods Enzymol.* **1996**, *267*, 220.
22. Houghten, R. A.; Bray, M. K.; DeGraw, S. T.; Kirby, C. J. *Int. J. Pept. Protein Res.* **1986**, *27*, 673.
23. Ostresh, J. M.; Hamashin, V. T.; Husar, G. M.; Dörner, B.; Houghten, R. A. In *Peptides 94: Proceedings of the 23rd European Peptide Symposium*; Maia, H. L. S., Ed.; ESCOM, Leiden, 1995; p 416.
24. Blondelle, S. E.; Pérez-Payá, E.; Houghten, R. A. *Antimicrob. Agents Chemother.* **1996**, *40*, 1067.
25. Cuervo, J. H.; Weitl, F.; Ostresh, J. M.; Hamashin, V. T.; Hannah, A. L.; Houghten, R. A. In *Peptides 94: Proceedings of the 23rd European Peptide Symposium*; Maia, H. L. S., Ed.; ESCOM, Leiden, 1995, p 465.
26. Dooley, C. T.; Houghten, R. A. *Analgesia* **1995**, *1*, 400.
27. Dooley, C. T.; Chung, N. N.; Schiller, P. W.; Houghten, R. A. *Proc. Natl. Acad. Sci. USA* **1993**, *90*, 10811.
28. Pasternak, G. W.; Wilson, H. A.; Snyder, S. H. *Mol. Pharmacol.* **1975**, *11*, 340.
29. Dörner, B.; Ostresh, J. M.; Husar, G. M.; Houghten, R. A. In *Peptides 94: Proceedings of the 23rd European Peptide Symposium*; Maia, H. L. S., Ed.; ESCOM: Leiden, 1995; p 463.
30. Dörner, B.; Ostresh, J. M.; Husar, G. M.; Houghten, R. A. *Methods Mol. Cell. Biol.* **1996**, *6*, 17.
31. Blondelle, S. E.; Takahashi, E.; Weber, P. A.; Houghten, R. A. *Antimicrob. Agents Chemother.* **1994**, *38*, 2280.
32. Dörner, B.; Husar, G. M.; Ostresh, J. M.; Houghten, R. A. *Bioorg. Med. Chem.* **1996**, *4*, 709.

PEPTIDOMIMETIC LIGANDS FOR SRC HOMOLOGY-2 DOMAINS

Charles J. Stankovic, Mark S. Plummer, and Tomi K. Sawyer

Advances in Amino Acid Mimetics and Peptidomimetics
Volume 1, pages 127-163

ISBN: 0-7623-0200-3

ABSTRACT

The design, synthesis, structure–activity relationship (SAR), and drug delivery studies of an emerging new class of peptidomimetics that specifically bind to Src Homology-2 (SH2) domains is described. Such SH2 domain ligands may be therapeutically useful in a number of disease states where the signal transduction of a particular SH2 domain-containing protein can be modulated by blocking its interactions with phosphoproteins. This area of signal transduction research remains, however, quite exploratory and cellular "proof-of-concept" studies with effective compounds are only now beginning to be advanced. In this review we detail the use of X-ray crystallography and molecular modeling in the structure-based design of peptide and peptidomimetic ligands for the Src SH2 domain. This work began with an 11 amino acid phosphopeptide, Glu-Pro-Gln-pTyr-Glu-Glu-Ile-Pro-Ile-Tyr-Leu, and through an iterative process of design and synthesis led to several series of tripeptides, dipeptides, and a series of ureido-type peptidomimetics that contains only a single amino acid. Selected ligands from these series were further modified to provide phosphatase-resistant analogues. Finally, acyloxymethyl ester prodrug derivatives of these stable analogues were prepared and successfully utilized to demonstrate the cellular delivery of these ligands.

1. INTRODUCTION

Since the first report of their structure and role in signal transduction pathways, SH2 domains have become the focus of an intense research effort, which, based on the number of X-ray and NMR structures reported from industrial laboratories, seems to have attracted the interests of most of the pharmaceutical industry. This review is intended to summarize the advancements in the design, synthesis, and study of peptide and peptidomimetic ligands for one of these SH2 domains, specifically, the Src SH2 domain.

The design of high-affinity small molecule ligands for the Src SH2 domain has been a goal for ourselves at Parke-Davis, as well as workers at Glaxo. Both groups have used a well-orchestrated combination of synthetic chemistry, molecular modeling, X-ray crystallography, NMR spectroscopy, as well as *in vitro* and cell-based assays to obtain a wealth of information about how these ligands bind to the Src SH2 domain. While the design of high-affinity ligands has proven to be a reasonable goal, efforts to extend this design to molecules that are cellularly active has proven to be much more challenging. Herein, we review the methodologies and strategies which have proven most successful in this rational drug design effort, and some of the less-fruitful tactics as well.

2. BACKGROUND

The design, chemistry, and biological evaluation of peptidomimetic analogues of phosphotyrosine (pTyr) containing naturally occurring proteins or synthetic peptides, which bind to intracellular targets such as SH2 proteins (e.g., Src, Syp, Grb2; see Table 1), have advanced rapidly over the past few years. Blockade of SH2 protein interactions with cognate phosphorylated proteins provides a new strategy to identify novel therapeutic agents, which selectively modulate the signal transduction pathways in cells related to cancer, osteoporosis, and a plethora of other diseases (Table 1).[1,2]

As illustrated in Figure 1, the phosphorylation state of a tyrosine residue is tightly regulated by specific tyrosine kinases and phosphatases. Furthermore, this

Table 1. Examples of SH2 Protein Targets of Therapeutic Interest[1,2]

Target Protein (Domain Structure)	*Disease State(s)*
Src (SH3-SH2-Kinase)	Cancer, Osteoporosis
Lyn (SH3-SH2-Kinase)	AIDS, Allergy, Asthma
Syk (SH3-SH2-Kinase)	Allergy, Asthma
Zap-70 (SH2-SH2-Kinase)	Autoimmune Disease
Syp (SH2-SH2-Phosphatase)	Anemia
STATs (DNA-binding-SH3-SH2)	Inflammatory Diseases
Grb2 (SH3-SH2-SH3)	Cancer, CM-Leukemia
p85 (SH3-SH2-SH2)	Cancer
Shc (SH2-PTB-Phosphatase)	Cancer, Erythroleukemia
Hck (SH3-SH2-Kinase)	AIDS
Gap (SH2-SH3-SH2-PH-GAP)	Cancer
Bcr/Abl (SH3-SH2-Kinase)	Chronic Mylogenous Leukemia

Figure 1. Regulation of the phosphorylation state of tyrosine.

phosphorylation state can control the catalytic activity of the parent protein and/or its localization inside the cell. At the cellular level, such phosphoprotein "intermediates" provide for temporal and compartmentalized regulation of signal transduction pathways that are initiated by growth factors, cytokines, antigens, and other external stimuli.

As shown in Figure 2, a growing number of signal transduction pathways have been identified which orchestrate downstream biological activities. These pathways often begin with an initial signal from cell membrane-localized effectors (e.g., growth factor receptors). Many of these signaling pathways are directly phosphotyrosine-dependent through further tyrosine phosphorylation[3,4] (e.g. Src) or tyrosine dephosphorylation[5-10] (e.g. Syp). Alternatively, these signals can be mediated indirectly via the binding of SH2 domain-containing proteins to phosphotyrosine-containing proteins; for example, Shc and Grb2, resulting in Ras activation, or PLCγ and PI3K, resulting in lipid metabolism. The control of these pathways by interruption of the phosphotyrosine cascade has been widely investigated due to the expected utility for specific therapeutic intervention.[11-13]

As detailed below, drug discovery research has been focused on the design of potent, selective, metabolically stable and cellularly effective SH2-domain targeted peptidomimetics. The key objective to most of these studies has been to establish the "proof-of-concept" showing that a particular SH2 domain-containing protein (or family of homologues) is a bonafide therapeutic target.

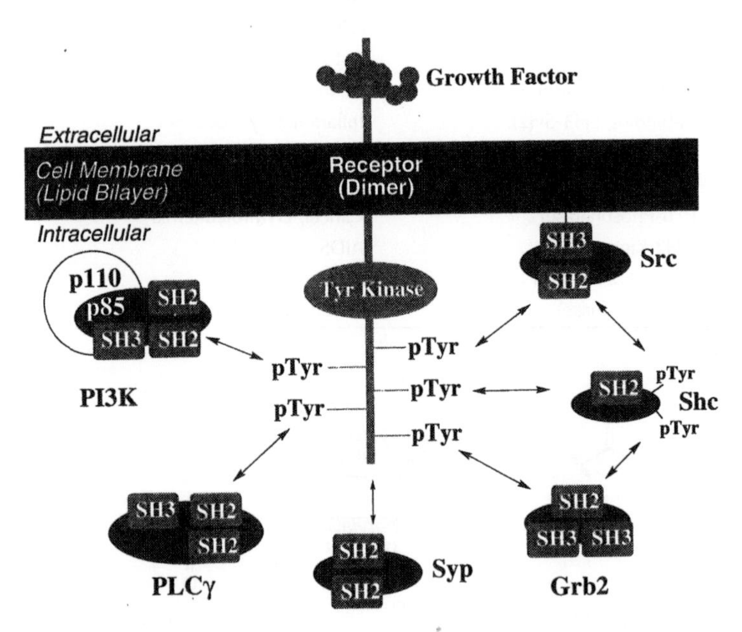

Figure 2. Model for cell signaling via a growth factor receptor and various SH2 domains.

2.1. SH2 Protein Targets: Primary Structure and Cognate Phosphoproteins

SH2 domains are noncatalytic motifs of approximately 100 amino acids, which are being found in an increasing number of signal transduction proteins. As indicated above, these domains bind to pTyr residues of other proteins in a sequence-dependent manner. Such binding often provides for the localization and molecular recognition needed to propagate pTyr-based signal transduction. For our purposes, we have chosen to focus on the Src and Abl SH2 domains, the primary structures of which are shown below (Figure 3). Although, all SH2 domains share a common three-dimensional fold (*vide infra*), the sequence homology between various SH2 domains can be quite variable, which ultimately leads to their binding to pTyr-containing proteins with sequence specificity.[14,15] However, all SH2 domains share several conserved regions, the most important of which is the FLVRES sequence in the phosphate-binding pocket.

It is important to point out that SH2 domains are just one of many domains involved in signal transduction. Others include Src Homology 3 (SH3) domains, which bind to polyproline sequences,[17,18] Pleckstrin Homology (PH) domains, which bind to the β/γ-subunits of G proteins,[19] and PhosphoTyrosine Binding (PTB) domains, which although they share no sequence or structural homology to SH2 domains also bind to phosphotyrosines in proteins.[20-22] SH2 domains are of particular interest since the three-dimensional structures were first determined for two members of this signal transduction domain family—Abl by NMR spectroscopy[23] and Src by X-ray crystallography.[24,25] Since that time, the structure of nu-

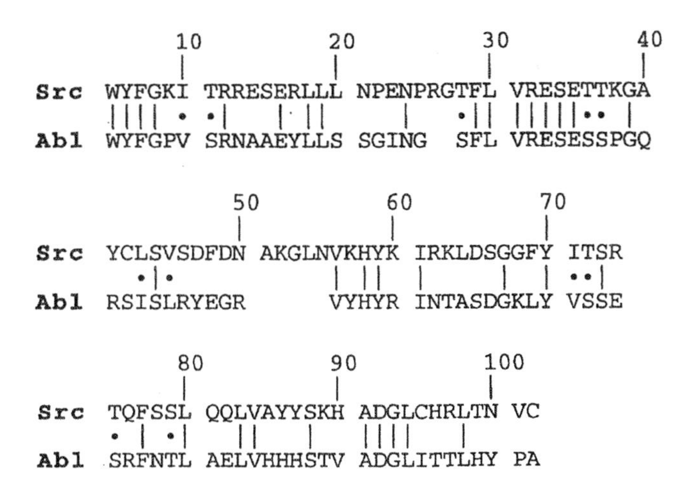

Figure 3. Sequence alignment of the Src and Abl SH2 domains. Numbering is from Holland et al.[16] and is used throughout this review. Identical residues are indicated by "|", and conservative replacements are indicated by "•".

merous SH2 domains have been solved by X-ray crystallographic or NMR techniques in both liganded and apo-protein form (*vide infra*).

The first evidence that SH2 domains bind pTyr residues in a sequence-specific manner was the recognition that not all SH2 domains bind to the same tyrosine autophosphorylation site of a receptor tyrosine kinase.[26] Thus, the amino acid sequence surrounding a particular autophosphorylation site provided an initial "cognate" sequence for the preparation of pTyr-containing peptides, which hopefully would exhibit binding selectivity to that particular SH2 domain.[27] In fact, this approach has successfully provided potent peptide ligands for a number of SH2 domains. For example, both SH2 domains of the p85 subunit of PI3 kinase[28-31] bind effectively to an 18-mer bis-phosphopeptide surrounding pTyr315 and pTyr322 of the hamster middle T antigen. In addition, mono-pTyr-containing phosphopeptides are also effective ligands for the individual SH2 domains of p85/PI3 kinase. Thus, the N-terminal SH2 domain of PI3 kinase binds to 11-mers surrounding pTyr740 and pTyr751 from human PDGF receptor, and to an 11-mer derived from insulin receptor substrate-1 (pTyr608).[31] Interaction of the Grb2 SH2 domain with the activated EGF receptor occurs specifically at an autophosphorylation site (Tyr1068).[32] Based on this discovery, a 13 amino acid peptide derived from the sequence surrounding Tyr1068 in the EGF receptor has been shown to bind to the Grb2 SH2 domain.[33] Phosphorylation of phospholipase Cγ1 (PLCγ1) is dependent upon autophosphorylation of five tyrosine residues at the C-terminus of the EGF receptor.[34] Specifically, PLCγ1 first associates with the EGF receptor via its two SH2 domains then becomes a substrate for phosphorylation by the tyrosine kinase domain. The SH2 domains of PLCγ1 have been shown to preferentially associate with an 11-mer peptide that encompasses pTyr992 of the EGF receptor.[35]

The tandem SH2 domains from the phosphatase Syp have been shown to bind to a 13-mer peptide derived from PDGF receptor[36] (pTyr 1009) and a bis-pTyr containing peptide from the insulin receptor substrate-1 (pTyr1172-pTyr1222).[37] Shc, which contains one SH2 domain, has been found to bind to two phosphopeptides, a 12-mer and an 11-mer, both derived from EGF receptor, at the autophosphorylation sites pTyr992 and pTyr1173, respectively.[38] The tandem SH2 domains of Zap70 have been found to bind to a bis-pTyr containing 17-mer derived from the tyrosine-based activation motifs (TAMs) present in the cytoplasmic tails of the T-cell antigen receptor and CD3 chains.[39] Finally, Src and Src tyrosine kinase family members bind to peptides derived from PDGF receptor (pTyr579 and pTyr581),[40] EGF receptor (pTyr992),[41] a C-terminal auto-regulatory phosphotyrosine (pTyr527 in Src),[42] and a hamster middle T-antigen-derived phosphopeptide (pTyr324).[43,44]

2.2. Synthetic Phosphopeptide Libraries: SH2 Domain Specificity

In addition to the identification of peptides which bind to various SH2 domains by extrapolation from known cognate phosphoproteins (*vide supra*), a more powerful and systematic method for determining sequence-specificity requirements for

SH2 domains has been reported which utilizes phosphopeptide libraries.[45-48] One example of this method relies on a phosphotyrosine peptide library that was degenerate in the P+1 (pTyr+1), P+2, and P+3 positions. These positions were chosen based on the previously reported consensus sequence pTyr-Met/Val-X-Met observed for PI3K,[29,49] and based on the observed binding interactions seen in the first X-ray structure reported by Waksman et al. (*vide infra*).[24,25] Using this method, Songyang and Cantley et al. have determined the preferred amino acids C-terminal to the pTyr necessary for binding to numerous SH2 domains,[45-47] selected examples of which are shown in Table 2.

The above peptide library studies show that the N-terminal SH2 domain of the p85 subunit of PI3 kinase has an optimal sequence of pTyr-Met-Xxx-Met, confirming the consensus sequence previously identified.[29,49] This sequence matches that identified as the p85-binding site on the PDGF receptor (pTyr740) and on hamster middle T antigen (pTyr315). Likewise, this library identified pTyr-Glu-Glu-Ile as the preferred sequence for the Src family members (Src, Fyn, Lck, Yes, Lyn, Yrk, Hck, Blk, and Fgr). This is the exact sequence previously identified from the binding site on hamster polyoma middle T antigen (pTyr324). A key advantage of this methodology is that it provides the "preferred" binding sequence for any given SH2 domain. Such a sequence might represent a possible biological association not yet discovered, but which, through protein sequence database searches, may easily be identified.

2.3. Three-Dimensional Structure of the Src SH2 Domain

The impetus for us and others to synthesize small ligands with reduced peptide character and overall charge is based on the biological relevance of SH2 domains,

Table 2. Sequences Specificities for Selected SH2 Domains[45-47]

Gly-Asp-Gly-pTyr - Xxx - Yyy - Zzz -Ser-Pro-Leu-Leu-Leu
P P+1 P+2 P+3

SH2 Domain	*P+1*	*P+2*	*P+3*
Src	Glu, Asp, Thr	Glu, Asn, Tyr	Ile, Met, Leu
Abl	Glu, Ile, Val	Asn, Glu, Asp	Pro, Val, Leu
PLC-γ1 (N)	Leu, Ile, Val	Glu, Asp	Leu, Ile, Val
PLC-γ2 (C)	Val, Ile, Leu	Ile, Leu	Pro, Val, Ile
Grb2	Gln, Tyr, Val	Asn	Tyr, Gln, Phe
Shc	Ile, Glu, Tyr	(Any)	Ile, Leu, Met
p85 PI3K (N)	Met, Ile, Val, Glu	(Any)	Met
Syk (N)	Gln, Thr, Glu	Glu, Gln, Thr	Leu
Syk (C)	Thr	Thr	Ile, Leu, Met
Lck	Glu, Thr, Gln	Glu, Asp	Ile, Val, Met

as well as the significant three-dimensional structural information recently achieved using X-ray crystallography and NMR spectroscopy. The 3D structures of several SH2 domains, including both apo-proteins and complexes with phosphopeptide ligands, are summarized in Table 3. Most recently the X-ray crystal structures of the full-length proteins of both Src[50] and Hck,[51] including the SH2 domain, the SH3 domain, the kinase domain, and the C-terminal regulatory tail, have been determined. These structures show for the first time how these regulatory subunits (SH2 and SH3) interact with the kinase domain, and offer new insights into the function and regulation of the catalytic activity of these proteins.

As mentioned above, the first SH2 domain X-ray structures were of Src.[24,25] This work provided a detailed molecular map and the likely common topographical features of all SH2 domains. Structurally, all SH2 domains studied thus far have a central anti-parallel β-sheet "core" flanked on each side by an α-helix. In the reported Src structures the phosphopeptide ligands bind perpendicular to the β-sheet core and typically interact with two well-defined binding pockets: the pTyr (P) pocket and hydrophobic binding pocket (P+3) located approximately 12.5 Å apart. Based on these observations, a minimally sized phosphopeptide ligand of at least four amino acid residues is required to span these two pockets (assuming an extended

Table 3. Src Homology-2 Protein Structural Studies

SH2 Protein	*Structural Method*	*References*
Abl SH2 (Apo-protein)	NMR Spectroscopy	23,52
Src SH2 (Apo/Complex)	X-ray Crystallography NMR Spectroscopy	24,25,50,53
Lck SH2 (Complex)	X-ray Crystallography	54-57
PLCγ SH2 (Complex) C-Terminal Domain	NMR Spectroscopy	58
Syp SH2 (Apo/Complex) N-Terminal Domain	X-ray Crystallography	59,60
Zap-70 SH2 (Complex) N- and C-Terminal Domain	X-ray Crystallography	61
p85 SH2 (Apo/Complex) N-Terminal Domain	X-ray Crystallography NMR Spectroscopy	62,63
Syk SH2 (Complex) C-Terminal Domain	NMR Spectroscopy	64
Hck SH2 (Apo)	X-ray Crystallography	51
Shc SH2 (Complex)	X-ray Crystallography NMR Spectroscopy	65
Grb2 SH2 (Complex)	X-ray Crystallography	66-68

conformation). Analysis of the structure of the Src SH2 domain complexed to the high-affinity phosphopeptide containing the pTyr-Glu-Glu-Ile (pYEEI) sequence reveals that most of the peptide is quite solvent exposed, with the only buried portions being the pTyr and the P+3 Ile side chains.[24,25] The only other important direct protein-ligand contact is that of the P+1 NH to His58 carbonyl hydrogen bond.

The pTyr-binding pocket in the Src SH2 domain contains four positively charged residues (Arg12, Arg32, His58, and Lys60). The most critical of these residues is Arg32 found in the FLVRES sequence (Figure 3), which forms two hydrogen bonds with the phosphate oxygen atoms of the pTyr residue (see Figure 4). In Src, the mutation of Arg32 essentially abolishes all binding of pTyr containing

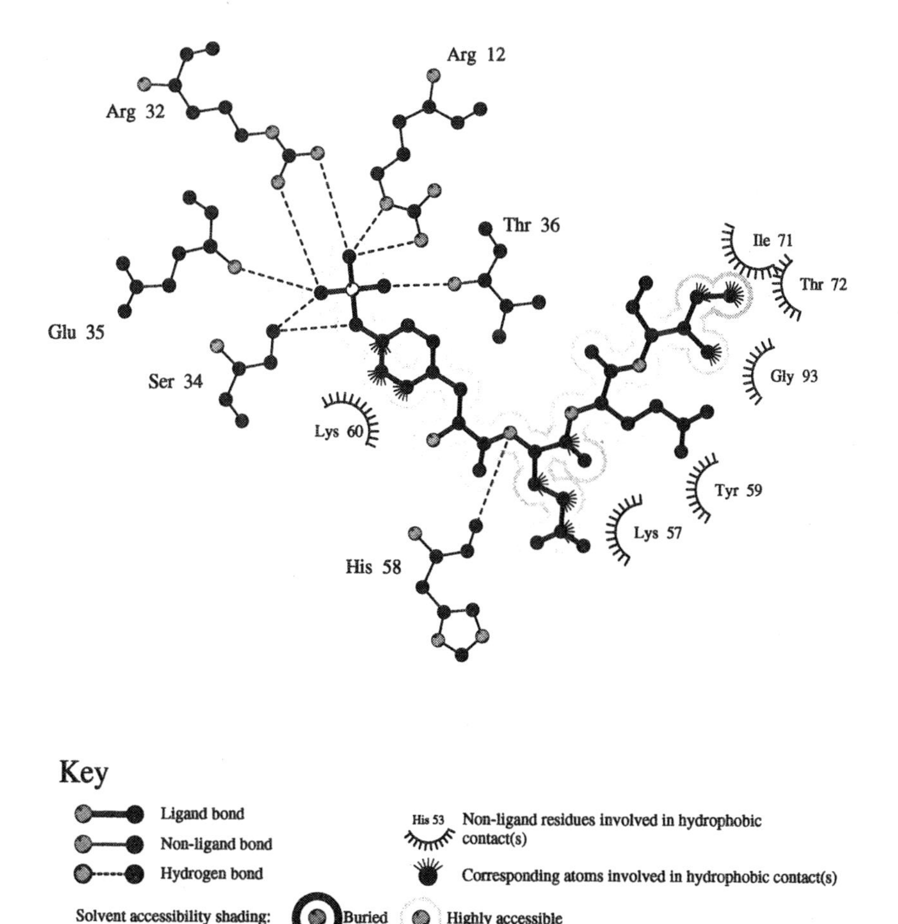

Figure 4. Selected binding site interactions with pYEEI sequence from the Waksman X-ray structure.[24,25] Figure generated using LIGPLOT.[70]

ligands to the SH2 domain;[14] however, the phosphate is also involved in numerous other hydrogen-bonding interactions with Thr36, Glu35, and Ser34, as well as hydrophobic contacts with the sidechain of Lys60. In the X-ray structure of the Src SH2 domain bound with the high-affinity peptide, the P+3 Ile side chain binds in a hydrophobic pocket, which in part is formed by two loop regions, and is comprised of residues Tyr59, Ile71, Gly93, and Thr72.[69]

3. SRC SH2 DOMAIN TARGETED DRUG DESIGN

As mentioned above, we have chosen to focus this review mainly on the design of Src SH2 ligands. The choice of Src was both a consequence of it being the first SH2 domain to be structurally determined by X-ray crystallography,[24,25] thus providing a solid basis for a structure-based design effort, and because of the possible role of Src in the development of cancer or osteoporosis.[71-74] In addition, although several other SH2 domains appear to be interesting therapeutic targets, especially Grb2, Lck, and Zap70, significant structure activity relationships have only been reported for Src SH2 domain ligands.

3.1. Phosphopeptide-SH2 Domain Binding Affinities

In our studies directed at the discovery of potent peptidomimetic ligands for the Src SH2 domain, we have used three methodologies to assess binding affinity: (1) competitive displacement of a radiolabeled 11-mer from a Src SH2 domain–GST fusion protein immobilized on sepharose beads by a test ligand,[75] (2) competitive displacement of ^{35}S labeled Src–GST SH2 domain from a phosphorylated cytoplasmic domain of the PDGF receptor by a test ligand,[75] and (3) isothermal titration calorimetric analysis of direct binding of a test ligand to the Src SH2 domain.[76] Typical affinities for ligands binding to the Src SH2 domain in our assays (ranging from 11-mer peptides to dipeptides) have been in the 0.1 to 50 μM range. It should be noted that caution may be necessary when comparing absolute binding-affinity values between different research groups in this area, even when the same SH2 domain is being studied, since different assays are frequently employed. Particular attention must be paid to values determined by surface resonance, since avidity effects resulting from dimerization of the GST–SH2 domains may be responsible for overestimates of affinities.[77]

3.2. Pentapeptide Lead Series (Glaxo)

The first report of employing structure-based design principles with the aim of modifying the pTyr-Glu-Glu-Ile sequence preferred by the Src SH2 domain appeared in 1994.[78] This elaborate study began with standard truncation studies of pTyr containing peptides from FAK (pTyr397), Src (pTyr527), and the pTyr-Glu-

Glu-Ile sequence from hamster middle T antigen. From these studies the pentapeptide Ac-pTyr-Glu-Glu-Ile-Glu (**1**, see Figure 5) was found to be optimal, with further truncation resulting in significantly decreased activity. This same study also reported the co-crystal structure of this pentapeptide with the Src SH2 domain, which was found to bind in a manner reminiscent of the Waksman[25] structure for the pTyr-Glu-Glu-Ile containing 11-mer described previously (Figure 4).

A systematic study replacing each residue in the pentapeptide was conducted, starting with pTyr replacements. Of the 27 pTyr replacements reported, including some designed to make additional specific interactions with the phosphate-binding pocket, only three replacements retained affinity within 10-fold of their parent pTyr analogue. These included the difluorophosphonate **2**, a thiophosphate **3**, and phosphorylated *p*-hydroxytetrahydoisoquinoline **4** (Figure 5). (A more thorough discussion of pTyr replacements including non-phosphate modifications is detailed later in this chapter.) While these replacements offer some advantages in structural novelty or cellular stability, these analogues are still expected to be dianions at physiological pH and thus show little cellular penetration.

Systematic replacement of the P+1 Glu residue in the pentapeptide indicated that a number of amino acids are well tolerated at this position including: Gln, Ser, Asp, His, Trp, Phe, norvaline (Nva), and Tyr. This reflects the fact that the Glu side chain only weakly interacts with the protein, as discerned in the X-ray structure. Replacement of this Glu with Ala provides a 14-fold reduction in potency, while D-Ala reduced potency 400-fold. The only replacement for the P+1 Glu that retained full affinity was pTyr.

Similarly, replacement of the P+2 Glu in the pentapeptide Ac-pTyr-Glu-Glu-Ile-Glu again indicated a number of amino acids were well tolerated in this position including: Gln, Ser, Asp, His, Trp, Phe, norvaline (Nva), and Tyr. Once again, the Ala for Glu substitution reduces binding 14-fold and, as expected, the D-Ala substitution dramatically drops affinity 57-fold. Interestingly, incorporation of a 1-naphthylalanine (Nal) moiety provides a compound that is equipotent to the Glu-containing pentapeptide. But most surprisingly, when the D-Nal group was incorporated in the P+2 position it caused only a twofold loss in potency as compared to

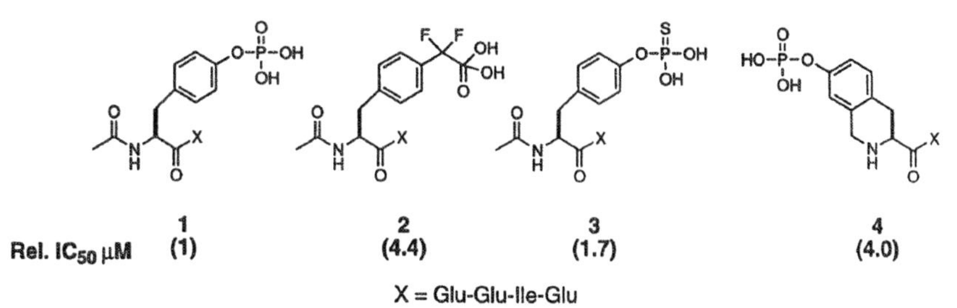

	1	2	3	4
Rel. IC_{50} μM	(1)	(4.4)	(1.7)	(4.0)

X = Glu-Glu-Ile-Glu

Figure 5. Key pTyr replacements from Glaxo in the pentapeptide series.

the 57-fold decrease seen with D-Ala. This unexpected tolerance for large hydrophobic D-amino acids at the P+2 position was subsequently utilized by Glaxo researchers to produce novel tripeptide ligands for the Src SH2 domain (*vide infra*).

Systematic replacement of the P+3 Ile residue, whose side chain interacts directly with the hydrophobic binding pocket, indicates many hydrophobic amino acids (e.g. Leu, homophenylalanine (Hph), norleucine (Nle), norvaline (Nva), or Phe) are well-tolerated and effect a ≤ fivefold decrease in potency. Substitution of Ile by Ala results in a 20-fold reduction in potency indicating that small hydrophobic groups, which cannot fill the P+3 pocket, are insufficient to maintain affinity. Substitution of polar residues such as Gln and Asn for the P+3 Ile reduces affinity 17- and 90-fold respectively.

Interestingly, a highly potent pentapeptide, Ac-pTyr-pTyr-pTyr-Ile-Glu (17-fold more potent than Ac-pTyr-Glu-Glu-Ile-Glu) was discovered in the above study. Unfortunately, this potent pentapeptide is too large and highly charged (-8) to be useful in cellular studies. Interestingly, truncation of the pentapeptide Ac-pTyr-pTyr-pTyr-Ile-Glu to the tripeptide Ac-pTyr-pTyr-pTyr, in which the important hydrophobic P+3 Ile has been removed, results in only a fourfold decrease in potency. Evidently, the highly charged P+1 and P+2 groups of this tripeptide compensate for the loss of the hydrophobic P+3 Ile residue, since this tripeptide is still fourfold more potent than the original parent pentapeptide Ac-pTyr-Glu-Glu-Ile-Glu.

Finally, an attempt was made to span between the critical pTyr and P+3 Ile residues using simple, primarily acyclic, linking groups. This approach was expected to be successful based on the X-ray structure, which shows that the P+1 and P+2 residue side chains do not make specific contacts with the SH2 domain. However, binding affinities for this prototype series of analogues were found to decrease 190- to 860-fold versus pentapeptide **1** (cf. compounds **5–9**, Table 4). These results suggested that ligand binding to the Src SH2 domain was more complex than a simple "two-pronged plug" concept. Nonetheless, a better understanding of the nature of

Table 4. P to P+3 Pocket Spanning Linkers from Glaxo Group[78]

		Relative Potency (μM)
1	Ac-pTyr-Glu-Glu-Ile-Glu	1
5	Ac-pTyr—NH—(linker)—C(=O)—Ile-Glu	190
6	Ac-pTyr—NH—(linker)—C(=O)—Ile-Glu	790
7	Ac-pTyr—NH—(linker)—C(=O)—Ile-Glu	390
8	Ac-pTyr—NH—(linker)—C(=O)—Ile-Glu	860
9	Ac-pTyr—N(piperidine)—C(=O)—Ile-Glu	190

the ligand–Src SH2 binding event could, in principle, result in the design of specific, non-peptide templates or scaffolds which span between the pTyr and P+3 binding pockets and still provide ligands with good binding affinities.[79]

3.3. Tripeptide Lead Series (Glaxo and Parke-Davis)

Pentapeptide to Tripeptide Transformation

In an effort to further reduce the size of their lead structures, Glaxo researchers modified the pentapeptide Ac-pTyr-Glu-Glu-Ile-Glu (**1**) to create a novel tripeptide lead series containing a homologated C-terminal carboxamide group.[80] The logic used for the transformation of a pentapeptide to a tripeptide reasoned that the P+2 through P+3 Glu-Ile sequence of **1** could be viewed as a γ-amino acid with a D-configuration, bearing a hydrophobic side chain that could bind to the Src SH2 P+3 pocket (Figure 6). The effect of incorporation of such a D-amino acid moiety could be anticipated based on the previous P+2 substitution studies employing D-amino acids *(vide supra)*.[78] Although many such tripeptides were designed and tested, the modified tripeptide **10** was found to have the highest affinity (IC_{50} = 1.7 μM) and possesses an overall charge of -3 (Figure 6).

At Parke-Davis, we employed an independent design logic to advance a similar series of tripeptides incorporating D-amino acid carboxamides (but without homologation as compared to the Glaxo series) at the P+2 position, as illustrated in Figure 6. In this design strategy, the carboxamide of the C-terminal D-amino acid was suggested from modeling to intramolecularly hydrogen bond to the carbonyl oxygen atom of the pTyr residue, thus promoting the preorganization of the peptide in a conformation preferred for binding. Briefly, structure–activity relationships for this series of compounds[81,82] indicate that aromatic-, alkyl-, and cycloalkyl-containing D-amino acids are well tolerated in the P+2 position (Table 5). Incorporation of a D-homocyclohexylalanine (Hcy) P+2 group provided the tripeptide **11**, resulting in a fourfold increase in affinity when compared to the aromatic analogue containing a D-homophenylalanine (Hph), **12**. The cyclohexyl group provides a superior hydrophobic interaction with the P+3 pocket. However, modeling suggests that the cyclohexyl ring of **11** and phenyl ring of **12** only cap the pocket, unlike the side chain of the isoleucine in the pentapeptide **1** which binds more deeply in the pocket. As exemplified by analogues **13** and **14**, N-methylation of the D-amino acid also proved favorable, presumably by inducing a conformation preferred for binding. Finally, replacement of the hydrolyzable pTyr group of **11** by phosphonodifluoromethyl phenylalanine (F_2Pmp),[83,84] provided a metabolically stable tripeptide lead **15** that was only fivefold less potent than the original 11-mer.

X-ray Structure of Tripeptide-Src SH2 Complex (Parke-Davis)

To gain a better understanding of the structural requirements for the binding of these tripeptides, and to confirm and further refine our modeling studies, the tripeptide Ac-pTyr-Glu-D-Hcy-NH_2 (**11**) was selected for X-ray crystallographic analy-

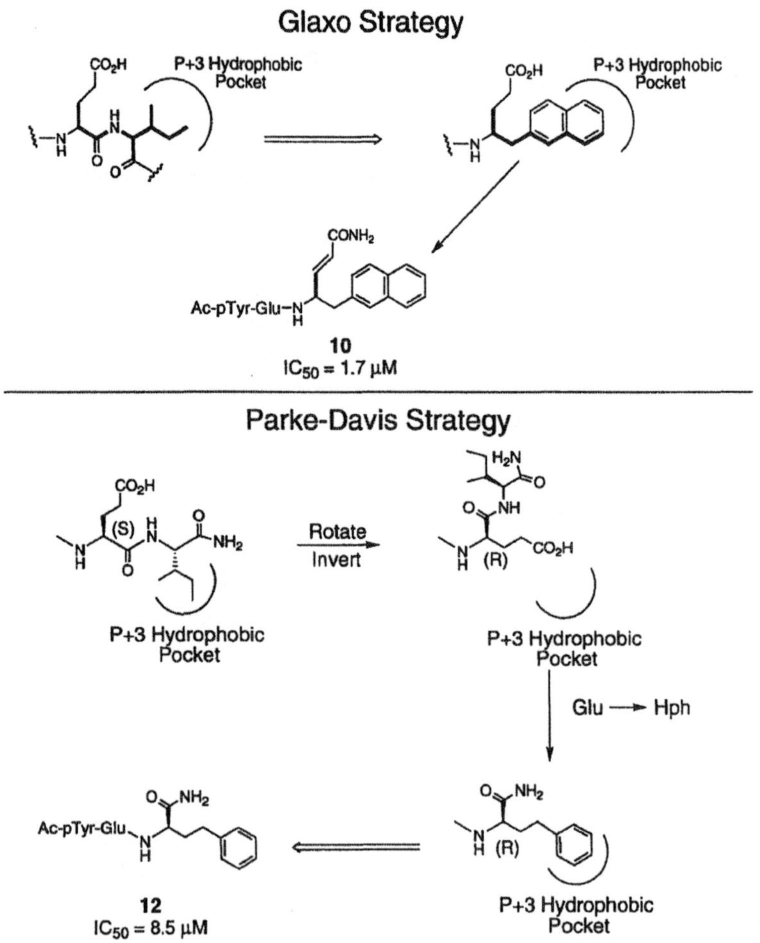

Figure 6. Design of P+2 D-amino acid tripeptides.

sis. Crystals of the complex of Src SH2 with tripeptide **11** diffracted to 2.1-Å resolution,[16] and the X-ray structure shows the ligand bound to the protein in a manner similar to that seen for the Waksman high-affinity peptide ligand.[25] As predicted by modeling, the phosphate moiety was observed to bind in the P pocket and the tripeptide backbone adopts a somewhat extended conformation, perpendicular to the central β-sheet core of the SH2 domain (Figure 7).[75] The P+2 cyclohexyl group interacts, as predicted, to the P+3 pocket via hydrophobic interactions, but its contacts are with the surface of the pocket only. In fact, this group appears not to be held in a single rotational conformation, and the X-ray data suggest at least one other secondary conformation for the ring. Elsewhere, tripeptide **11** makes a number of

Table 5. P+2 D-Amino Acid Tripeptide SAR

	Peptide[a]	IC_{50} (μM)
11	Ac-pTyr-Glu-D-Hcy-NH_2	1.8
12	Ac-pTyr-Glu-D-Hph-NH_2	8.5
13	Ac-pTyr-Glu-D-N(Me)Phe-NH_2	2.2
14	Ac-pTyr-Glu-D-N(Me)Cha-NH_2	1.5
15	Ac-$_{F2}$Pmp-Glu-D-Hcy-NH_2	4.9

Note: [a] Cha = cyclohexylalanine, Hcy = homocylohexylalanine

specific contacts with the protein including: the carbonyl group of the N-terminal acetyl group of **11** hydrogen bonds to the Arg12 side chain of the SH2 domain; the phosphate interacts with the P pocket as described for the high-affinity peptides;[25] and the backbone NH of the P+1 residue forms the expected hydrogen bond with the backbone His58 of the SH2 domain.

Interestingly, the P+2 carboxamide moiety does not interact with the protein, but is projected toward solvent, thus suggesting that this group may be expendable. However, one of two bound water molecules (neither of which is shown) is positioned between the nitrogen of the carboxamide and the carbonyl of the pTyr residue, perhaps helping to intramolecularly stabilize the bound inhibitor conformation. Another bound water molecule exists between the carbonyl oxygen of the tripeptide's P+1 residue and the Lys60 backbone NH of the Src SH2. This

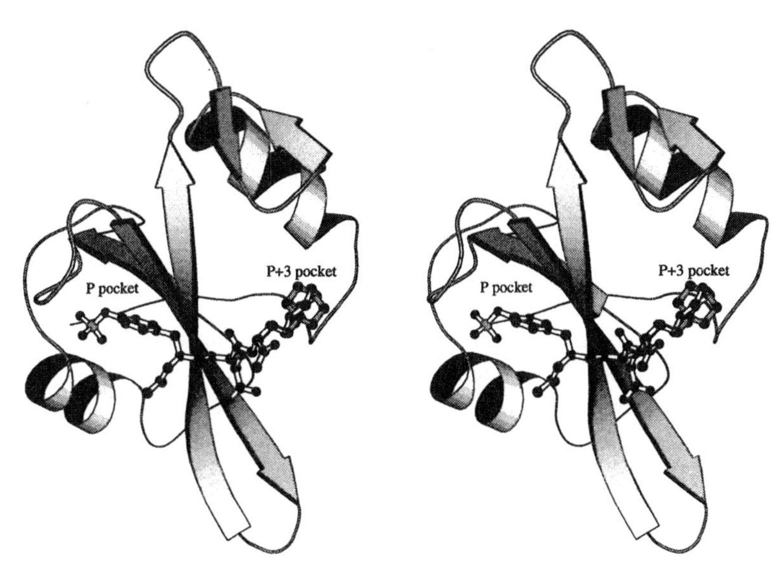

Figure 7. A stereoview of the structure of D-tripeptide **11** bound to Src SH2 domain. Both observed rotamers of the sidechain cyclohexyl ring are shown. This figure was generated with MOLSCRIPT.[85]

structural water was also seen in the Waksman high-affinity peptide ligand complex with the Src SH2 domain.[25]

3.4. Dipeptide Lead Series (Parke-Davis)

Inspection of the X-ray structure of the tripeptide **11** indicated that the C-terminal carboxamide group did not directly hydrogen bond to the pTyr carbonyl oxygen atom. Thus, the carboxamide group may not have a role in providing the expected preorganization of the peptide binding conformation. To test this possibility, a series of dipeptide analogues, **16–23**, were synthesized (Table 6). Indeed, removal of the carboxamide group from tripeptide **12** gave the dipeptide **16** (IC_{50} = 10.0 μM), which was essentially equipotent to the parent tripeptide **12** (IC_{50} = 8.5 μM). Truncation of the C-terminal phenylpropylamide moiety of **16** to the phenethylamide analog **17** (IC_{50} = 9.8 μM) was also an effective modification. Efforts next focused on modifying and substituting the P+2 aromatic ring with the goal of increasing the potency of the dipeptide lead series. The aromatic substitution, which provided the greatest increase in potency, was a *m*-hydroxyacetic acid group as in **18** (IC_{50} = 2.5 μM). Modeling suggested that this acidic group interacts with Arg74 in a manner similar to the P+4 Glu in the Glaxo pentapeptide.[78]

Similar to the tripeptide lead series, potency could be enhanced in the dipeptide analogue series by *N*-methylation, as exemplified by compound **19** (IC_{50} = 3.5 μM), which was threefold more potent than **16**. Reduction of the P+2 aromatic ring in **19** gave the *N*-methyl cyclohexylpropyl amide analogue **20** (IC_{50} = 0.8 μM). This combined the more hydrophobic P+3 moiety with the favorable conformational constraints of *N*-methylation, resulting in a fourfold increased potency. Reintroduction of the chiral center adjacent to the nitrogen as in **21** (IC_{50} = 3.5 μM) provides almost a threefold increase in potency when compared with **17** (IC_{50} = 9.8 μM), presumably by inducing a rotamer favorable for binding. Unfortunately, this C-methyl group was incompatible with other potency enhancing modifications such as N-methylation due to a methyl/methyl *gauche* interaction in the rotamer preferred for binding (cf. **22**, IC_{50} = 2.4 μM). However, by linking both methyl groups into a pyrrolidine ring, these two modifications could be combined constructively to provide our most potent dipeptide analogue **23** (IC_{50} = 0.6 μM).

Relative to the dipeptide **16**, we next focused on replacement of the acidic P+1 Glu residue with uncharged amino acids (Table 7).[86] The X-ray structure of the tripeptide **11**[16] shows that the Cβ and Cα of the P+1 Glu side chain make van der Waals contacts with a small hydrophobic patch on the surface of the protein, while the carboxylate is more solvent exposed and extends toward Lys57. To further explore binding interactions with this small hydrophobic patch, we substituted the P+1 Glu of **16** with norvaline (Nva) and aminobutyric acid (Abu) to give **24** and **25**, respectively, which were each essentially equipotent with **16**. Furthermore, replacement of the Glu with other amino acids such as Ala (**27**) or Val (**28**) had a modest effect on binding, but substitution with Gly (**29**) resulted in fourfold decreased

Table 6. SAR of Dipeptides with Varying C-Terminal Hydrophobic Groups

	Peptide	*IC_{50} (μM)*
12	Ac-pTyr-Glu-NH–CH($CONH_2$)–CH_2CH_2–Ph	8.5
16	Ac-pTyr-Glu-NH–$(CH_2)_3$–Ph	10.0
17	Ac-pTyr-Glu-NH–$(CH_2)_2$–Ph	9.8
18	Ac-pTyr-Glu-NH–$(CH_2)_3$–C_6H_4–O–CH_2–CO_2H	2.5
19	Ac-pTyr-Glu-N(Me)–$(CH_2)_3$–Ph	3.5
20	Ac-pTyr-Glu-N(Me)–$(CH_2)_3$–cyclohexyl	0.8
21	Ac-pTyr-Glu-NH–CH(Me)–CH_2–Ph	3.5
22	Ac-pTyr-Glu-N(Me)–CH(Me)–CH_2–Ph	2.4
23	Ac-pTyr-Glu-N(pyrrolidine)–CH_2CH_2–cyclohexyl	0.6

binding affinity, presumably due to increased rotational freedom and decreased hydrophobicity. The only P+1 Glu replacement which provided increased binding affinity relative to **16** was analogue **29** (IC_{50} = 3.9 μM); the P+1 pTyr residue may interact with Lys57 thus causing the increase in affinity. Dipeptide **29** is also reminiscent of the potent pentapeptide analogue Ac-pTyr-pTyr-pTyr-Ile-Glu and the tripeptide Ac-pTyr-pTyr-pTyr as previously reported by Glaxo researchers and discussed above.[78] Overall, the effects of P+1 substitutions in this series of dipeptides were less pronounced when compared to the identical substitutions in the Glaxo pentapeptide series. For example, Ala and Gly substitutions effected a 1.5- and 4.5-fold decreased affinity in the dipeptide series (Table 7), respectively, while the same substitutions in the Glaxo pentapeptide series resulted in 14- and 90-fold decreased affinity relative to P+1 Glu.[78]

Combining the *N*-methylcyclohexylpropyl amide, a potent C-terminal amide capping group, with the uncharged P+1 Abu substitution provided **30** (IC_{50} = 5.4 μM), which was about twofold more potent than the parent dipeptide **16**, and has an overall charge of only -2. This reduction in the overall charge on our molecules was viewed to be a positive step toward our goal of ultimately delivering these compounds to cellular targets *(vide infra)*.

Table 7. SAR of Dipeptides with P+1 Modifications

	Peptide	*IC_{50} (μM)*
16	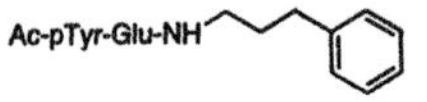 Ac-pTyr-Glu-NH	10
24	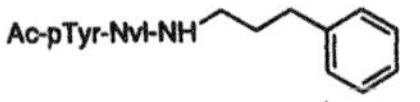 Ac-pTyr-Nvl-NH	12
25	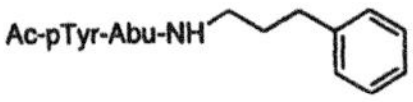 Ac-pTyr-Abu-NH	10
26	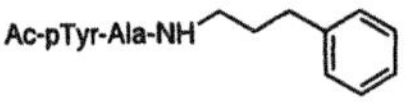 Ac-pTyr-Ala-NH	15
27	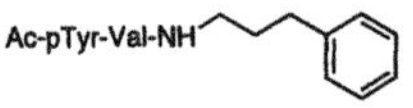 Ac-pTyr-Val-NH	30
28	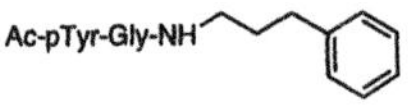 Ac-pTyr-Gly-NH	46
29	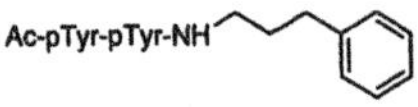 Ac-pTyr-pTyr-NH	3.5
30	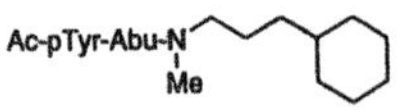 Ac-pTyr-Abu-N Me	5.4

3.5. Novel Phosphotyrosine Replacements (Parke-Davis)

To further reduce the peptide character of the above dipeptide Src ligands shown in Table 6 and 7, we examined replacements of the tyrosine residue. Tyrosine replacements[87] were incorporated into the pTyr-Glu-Glu-Ile-Glu sequence to allow convenient comparison to previous studies on phosphate replacements by the Glaxo researchers.[78] The parent compound **1**, Ac-pTyr-Glu-Glu-Ile-Glu, was quite potent ($IC_{50} = 0.50$ μM) in our assay. Removal of the acetylated α-amino group in **1** provided **31** which showed a fivefold diminished binding affinity. Compounds **32–34** were designed to constrain the freely rotating tyrosine side chain of **31** by incorporation of either an α-methyl group, *ortho*-substituted dimethyl groups on the phenyl ring, or a *trans* olefin. However, none of these strategies provided increased potency relative to compound **31**. Molecular modeling suggested that introduction of an aromatic ring *alpha* to the carbonyl group of **31** might augment the interaction of the aromatic π-system with the guanidinium group of Arg12.[25] Illustrative of this concept are compounds **35** and **36** (each existing as diastereomeric pairs). Unfortunately, no additional potency gain was realized with **35a** ($IC_{50} = 1.9$ μM) when compared with **31** ($IC_{50} = 2.8$ μM). A hydroxyl group was introduced (**36a** and **36b**) to explore the possibility of gaining an additional hydrogen bond with Arg12, but again no significant potency gain was obtained as compared with **31**. Finally, replacement of the acetylated α-amino group in **1** by an α-carboxymethyl group provided compound **37** (IC_{50} = 0.35 μM), which was equipotent to the

Table 8. SAR of Phosphopeptides with Novel pTyr Replacements

	Peptide		IC_{50} (μM)
1			0.6
31			2.8
32			20
33			4.5
34			5.5
35a		Isomer 1	1.9
35b		Isomer 2	9.3
36a		Isomer 1	2.1
36b		Isomer 2	3.3
37			0.4

Ac-pTyr analogue. Molecular modeling of **37** suggests that the additional carboxy group forms a bifurcated hydrogen bond with Arg12. Since no increase in potency was realized by replacing the tyrosine residue in Ac-pTyr-Glu-Glu-Ile-Glu, as seen in Table 8, these replacements were never incorporated into the dipeptide Src SH2 ligands.

3.6. Novel Peptidomimetic Lead Series (Parke-Davis)

Continuing the above study to replace the acetylated α-amino group with the α-carboxymethyl moiety **37**, we further advanced this design concept by creating a novel ureido-type peptidomimetic lead series (Table 9). This strategy was based on replacement of the α-carbon atom of the pTyr mimetic **37** with a nitrogen, which simultaneously eliminates chirality and induces conformational rigidity (upon coupling with the P+1 amino acid). As a prototype test of this concept, the potent dipeptide **20** (Table 9, $IC_{50} = 0.8$ μM) was modified to provide the peptidomimetic **38**, which contains only one amino acid (i.e., Glu at P+1). Indeed, peptidomimetic

Table 9. SAR of Peptidomimetics with Ureido-Type pTyr Modifications

Peptide	IC_{50} (μM)
20	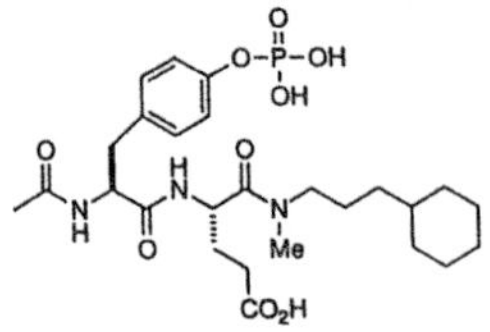 0.8
38	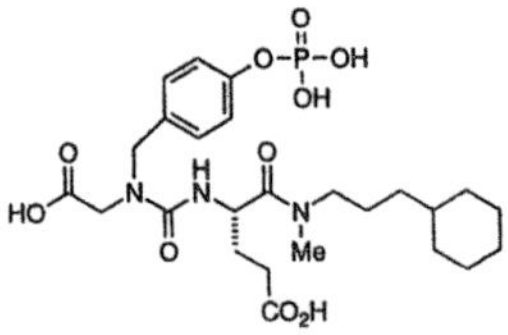 7.0
39	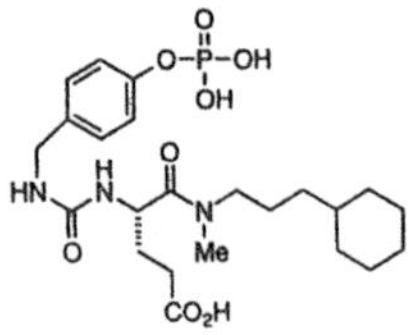 12

38 was determined to bind to the Src SH2 domain with an $IC_{50} = 7.0$ μM, which was about 10-fold lower in potency than the parent dipeptide **20**. Interestingly, replacement of the carboxymethyl moiety of **38** by a hydrogen atom gave **39**, and did not markedly effect binding affinity.

X-ray Structure of Ureido-Type Peptidomimetic-Src SH2 Complex

Cocrystallization of peptidomimetic **38** with the Src SH2 domain gave a 2.4-Å X-ray structure,[88] which reveals a binding mode somewhat different than that previously reported for Src SH2 peptide ligands, including the tripeptide **11** described above.[16] Although peptidomimetic **38** binds with the phosphate group in the P pocket and the cyclohexyl group in the hydrophobic P+3 pocket, as expected, the conformation and placement of the intervening backbone of the ligand is quite unique relative to previously reported Src SH2 domain peptide ligands (Figure 8). In particular, the orientation of the phenylphosphate ring in the P pocket is nearly perpendicular to that observed in the structures of all the previously reported pTyr-containing peptide ligand complexes.[16,25,78] Nevertheless, the phosphate group of **38** superimposes well with that of the pTyr-containing ligands. Also, the N-terminal carboxymethyl group is completely exposed to solvent instead of hydrogen-bonded to Arg12, as predicted from modeling of **38**. The most surprising difference between the modeled and the X-ray structures for the bound ureido ligand is the existence of a *cis*-amide linkage between the P+1 Glu and the C-

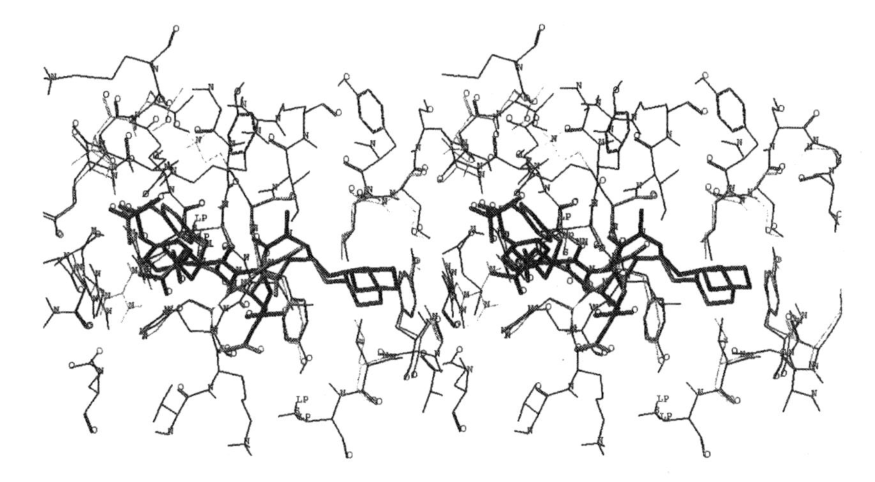

Figure 8. Stereoview overlay of the X-ray crystal structure of the tripeptide **11** (gray) and the ureido analog **38** (black) bound to the Src SH2 domain. *Note*: The dominant rotamer for the cyclohexyl group is shown for the tripeptide.

terminal *N*-methyl-cyclohexylpropyl group. As a result of this unexpected configuration, the structural water molecule (not shown), which usually mediates a hydrogen bond between Lys60 backbone and the P+1 ligand carbonyl of pTyr-containing ligands,[24,25,78] is displaced by the P+1 Glu carbonyl oxygen of **38** to provide a novel direct protein–ligand contact. The hydrogen bond between His58 and the p+1 NH is preserved in the X-ray structure of **38**, similar to all the other reported peptide complexes, as is the solvent-exposed environment of the P+1 Glu carboxylate. Finally, the C-terminal cyclohexyl ring of **38** does not bind as deeply into the P+3 pocket as does the preferred Ile residue as related to the high-affinity phosphopeptide (~pTyr-Glu-Glu-Ile~) as described by Waksman et al.[24,25]

4. PURSUIT OF CELLULAR ACTIVITY

4.1. The Problem with Phosphates and Phosphonates

Through the use of structure-based drug design and traditional medicinal chemistry, we have simplified the original parent 11-mer phosphopeptide lead down to a series of prototype peptidomimetics containing one or two amino acids, the best of which maintain binding affinities in the low μM range. However, all of these analogues require a phosphate group to maintain good binding affinity. In the absence of a phosphate group the binding affinity of these ligands is completely lost. This requirement for a phosphate group poses two serious problems for the further development of these compounds beyond the stage of *in vitro* binding assays. The first is that the targeted SH2 domains are intracellular, and thus these compounds must be able to cross cell membranes and penetrate into cells. However, most phosphates are expected to be dianions at physiological pH, and in general highly charged molecules do not readily cross cell membranes by simple diffusion. Second, even if these phosphorylated compounds could be delivered into cells, it is expected that specific and nonspecific phosphatases inside most cells would cause rapid dephosphorylation, rendering these ligands inactive.

4.2. Phosphate and Phosphotyrosine Replacements

Literature precedent suggests that, in general, a number of other functional groups might be able to replace the phosphate group and still maintain effective binding affinity to the target of interest.[89-96] However, in the case of the phosphotyrosine–SH2 domain interaction, the phosphate group would appear to be especially critical to binding. As mentioned previously, Glaxo researchers have reported an extensive number of phosphate replacements, including a number of *para*-functionalized phenylalanines to replace the phosphotyrosine in the pentapeptide Ac-pTyr-Glu-Glu-Ile-Glu (Table 10).[78] However, most of these analogues were completely inactive or significantly less effective in their assay. Even the rather conservative replacement of the dianionic phosphonate by the monoanionic sulfonate effected a 300-fold loss in activity. The only functional groups which effectively replace the phosphate of pTyr to date are phosphonates (e.g., F_2Pmp as in **2**), and the dianionic malonates as in **46**. Both of these groups were first introduced by

Table 10. SAR of Phosphopeptides with pTyr Replacements[78,100]

	X	Relative IC_{50}
1	OPO_3H_2	1
2	$CF_2PO_3H_2$	4.4
3	$OPSO_2H_2$	1.7
40	$CH_2PO_3H_2$	40
41	CH_2SO_3H	300
42	$CH_2CH_2COCOCH_3$	520
43	CH_2CH_2COCHO	370
44	NO_2	>1000
45	CH_2CO_2H	940
46	$OCH(CO_2H)_2$	8*

Note: * Relative to Ac-QpYEEIP. From Ye and Burke et al.[100]

Burke and coworkers,[83,84,97-99] and at least in the case of F_2Pmp have been extensively used by us and others to provide phosphatase-resistant replacements for pTyr *(vide infra)*. In the case of the malonate group, this analogue is obviously resistant to phosphatases, but is still expected to be dianionic at physiological pH. It was anticipated that this analogue would provide easier access to prodrug analogues, which might provide better cellular activity. However, the success of this approach has not yet been reported.[100-103]

The use of phosphonates as replacements for phosphates in biological systems is well established,[89,91] and in general, these phosphonates retain significant potency while providing complete resistance to degradation via phosphatases. In the case of SH2 domain ligands there seems to be an extra requirement that these phosphonates have a low second pK_a similar to that of the phosphate of pTyr. Thus, *alpha* hetero-substituted phosphonates are more effective ligands than the parent methylene phosphonates, with the difluoromethylenephosphonate (F_2Pmp) derivative being most preferred.[97,98] The use of F_2Pmp was pioneered by Burke and coworkers[83,84,99] and has subsequently become a somewhat standard replacement for pTyr in SH2-targeted phosphopeptide ligand drug design and lead discovery.[39,78,82,104,105]

In our case, we have incorporated F_2Pmp into a number of our lead series to evaluate its role in the binding and selectivity for Src over other SH2 domains. For this purpose we have used the two assays described previously *(vide supra)*, the results of which are shown in Table 11 for selected compounds. In this table SSRC and SABL refer to data from the radioligand displacement assay of ^{125}I-labeled 11-mer from Src or Abl. Likewise Src and Abl refer to data from the protein association assay which measures the displacement of ^{35}S-labeled SH2–GST fusion proteins of

Table 11. Binding and Specificity Data for pTyr, F_2Pmp, and Phe(PO_3H_2) Containing Peptides

		IC_{50} (μM)[b]			
		^{125}I Peptide		*35_S Protein*	
	Peptide[a]	*SSRC*	*SABL*	*Src*	*Abl*
11	Ac-pTyr-Glu-D-Hcy-NH_2	1.8	4.9	7.0	6.4
15	Ac-F_2Pmp-Glu-D-Hcy-NH_2	4.9	1.2	27	1.4
20	Ac-pTyr-Glu-NMe-(Pr-Chx)	0.8	4.0	7.4	18
47	Ac-F_2Pmp-Glu-NMe-(Pr-Chx)	3.4	1.7	30	2.1
48	Ac-Phe(PO_3H_2)-Glu-NMe-(Pr-Chx)	9.3	14	ND[c]	ND
30	Ac-pTyr-Abu-NMe-(Pr-Chx)	5.5	15	26	46
49	Ac-F_2Pmp-Abu-NMe-(Pr-Chx)	7.3	1.6	~100	4.1
50	Ac-Phe(PO_3H_2)-Abu-NMe-(Pr-Chx)	6.1	22	>100	17

Notes: [a] Hcy = homocyclohexylalanine, Pr-Chx = cyclohexylpropyl, Abu = *alpha*-amino-butyric acid, Phe(PO_3H_2) = 4-phosphonophenylalanine.
[b] The IC_{50}'s were determined as described in Shahripour et al. [87]
[c] Not determined.

Src and Abl from the phosphorylated intracellular domain of the PDGF receptor by our ligands.

Several observations can be made from this data and from that observed for other analogues (data not shown). First, incorporation of F_2Pmp generally decreases the affinity toward Src (e.g. **15**, **47**, and **49**), as was expected from previous reports.[98] However, incorporation of F_2Pmp generally favors binding to Abl over Src. This was especially true in the protein association assays (Src and Abl) where this reversal in selectivity ultimately provided compounds which were >20-fold selective for Abl over Src. Also included in Table 11 are data for a series of arylphosphonate analogues of the same dipeptides (i.e. **48** and **50**). Although these compounds generally exhibit lower binding affinities for Src, they are included based on their projected utility in our cell delivery strategy *(vide infra)*.

4.3. Strategies for the Design of Cellular-Active Compounds

Although such F_2Pmp analogues provide excellent resistance to phosphatases, and thus possess good cellular stability, they still exist as dianions at physiological pH, and are unlikely to show significant cellular penetration. At this point we contemplated three options for making our dipeptidyl Src SH2 ligands suitable for cellular penetration. The first and ideal choice would be to find singularly charged, or preferably nonionic, replacements for pTyr. However both our unpublished results and those reported by the Glaxo group[78] (Table 10) suggested that discovery of such a pTyr replacement would be extremely challenging. The second alternative would be to replace the phosphate group with a reactive functional group, which might selectively react with potential nucleophiles in the phosphotyrosine recognition pocket.[106] An example of this approach has been reported by the Glaxo group,[107] who used the keto-aldehyde **43** and the diketone **42** which were designed to form a dihydroxyimidazoline adduct with the conserved Arg32 of the FLVRES sequence (Figure 9). However, no significant binding was observed (cf. **42** and **43** in Table 10).

Finally, a third alternative would be to use a prodrug approach to mask the charges on the phosphonate with bioreversible ester groups (Figure 10). We reasoned that this approach would give us a rapid means to provide cellularly active analogues, which could, in turn, be used for key biological "proof-of-concept" studies designed to identify potential therapeutic targets for these compounds.

4.4. The Prodrug Approach to Cellular Delivery

The use of prodrugs for the delivery of charged compounds is a well established strategy in medicinal chemistry, and many methods for accomplishing this goal have been reported and reviewed.[108] Of particular interest are several recent reviews on nucleotide and phosphate/phosphonate prodrugs.[109-111] Based on the information gained from these reviews and the above stated problems with finding pTyr re-

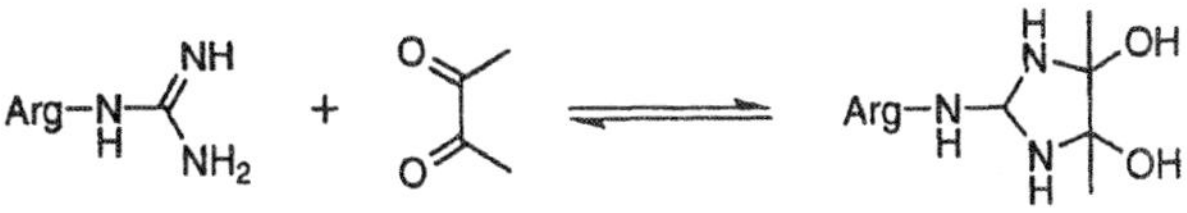

Figure 9. Model reaction for arginine reacting with dicarbonyl trap.

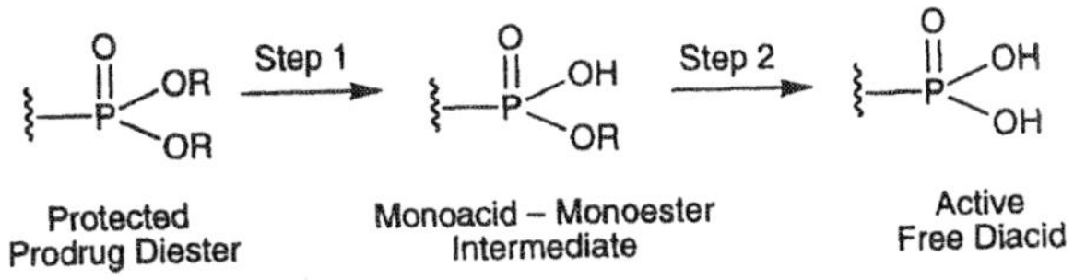

Figure 10. Phosphonate ester prodrug model.

placements, we focused most of our initial efforts on producing cellularly active analogues via the use of prodrugs of some of our best compounds.

Given that the *in vitro* binding affinities of our lead analogues were in the 0.5–10 μM range, and the lack of precedented *in vivo* models to evaluate the effect of SH2 ligands on cellular systems, we anticipated difficulties in assessing the effect of our compounds in cellular assays based on a simple biological readout (e.g. DNA synthesis or cell proliferation). To circumvent these problems we established an analytical assay for measuring the uptake of our prodrug analogues into a model cell system, thus providing an independent and unambiguous assessment of both their ability to penetrate cells and their ability to reconvert giving the parent Src SH2 ligands.

For our initial studies two compounds, **49** and **50**, were selected (Figure 11). When protected as their phosphonate diethyl esters (**51** and **53**), these compounds were completely uncharged and relatively hydrophobic and, as such, were expected to show good cellular penetration by simple diffusion across the cell membrane. Compound **49** was one of our most active dipeptides (IC_{50} = 7.3 μM) and, although compound **50** was significantly less potent, it was included in our initial studies based on the anticipated ease of synthesis of prodrug esters in this series *(vide infra)*.

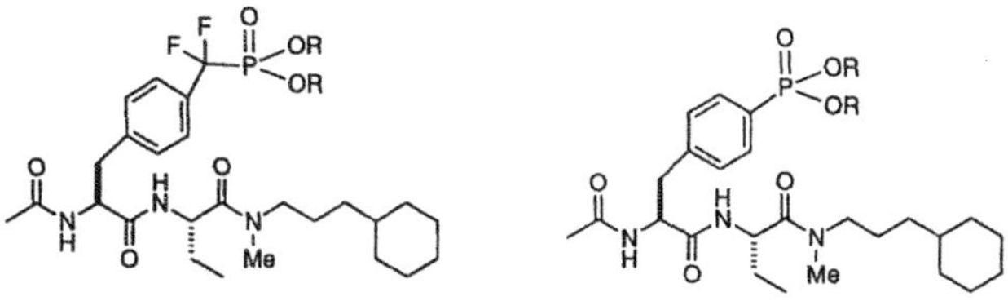

Figure 11. Model diethyl ester prodrugs and their parent diacids.

Our initial cellular studies used the CACO-2 cell transport model[112-114] to evaluate the cell permeability of compounds **49** and **50**. From these studies we observed that not only did these compounds readily cross this cell monolayer, but they were also readily accumulated in these cells.[115,116] Based on these results we decided to investigate the direct cellular uptake of our compounds into Balbc3T3 cells. This fibroblast cell line was chosen since it would also be used in subsequent "proof-of-concept" studies based on cellular assays designed to evaluate the ability of these compounds to interrupt various Src-dependent signal transduction pathways. To measure the uptake of these compounds we developed a simple HPLC method which allowed for the rapid separation and quantification of the prodrug bis-esters, the intermediate monoesters, and the parent diacids.[115] This HPLC method was particularly robust, allowing us to monitor the uptake of the prodrug analogues in several series of our best ligands.

The first step in this prodrug approach investigated the uptake of simple alkyl esters (Figure 11). Although these esters were not expected to reconvert to the parent diacids, they did provide stable diesters and were readily available. The uptake studies for **51** and **53** clearly demonstrated the potential utility of this prodrug approach (Table 12). Both of these simple diesters readily penetrate into Balbc3T3

Table 12. Cellular Uptake Data for Dipeptide Prodrugs

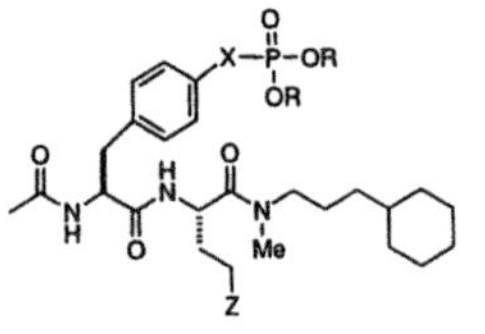

	Z	*X*	*OR*	*OR′*	*Uptake Rate (nmol/min-sq cm)*[a]
51	H	CF_2	OEt	NA[b]	0.3
49	H	CF_2	OH	NA	<0.0026
52	H	CF_2	OH, OEt	NA	<0.002
53	H	—	OEt	NA	0.092
50	H	—	OH	NA	<0.002
54	CO_2R'	CF_2	OEt	OtBu	3.2
55	CO_2R'	CF_2	OEt	OH	0.001
47	CO_2R'	CF_2	OH	OH	<0.002
56	CO_2R'	—	OEt	OtBu	0.52
48	CO_2R'	—	OH	OH	<0.002
57	H	—	OPh	NA	>LOQ[c]

Notes: [a] Normalized to a donor concentration of 100 μM. The uptake rates were determined as reported in Surendran et al. see ref. [115,116]

[b] NA = Not Applicable

[c] Limit of quantitation.

cells and accumulate to significant levels. Noteworthy is that the HPLC method used is limited by detector sensitivity, which in this case leads to a lower limit of detection of ~ 0.25 μM for the injected solution. This in turn leads to an approximated detection limit of 100 μM to 1 mM for the cell-associated or "intracellular" concentration. However, we have found that this is not a problem for many of our compounds, since many have been found to accumulate to these high levels in our cell model systems.

Based on the success of these initial studies, our focus turned to assessing the generality of this methodology using several other analogues (Table 12). These studies demonstrate that the fully protected analogues (**51**, **53**, **54**, and **56**) show good to excellent uptake. However, as expected, the parent di- and triacids (**49**, **50**, **47**, and **48**) did not show any uptake within our detection limits. This latter result was also confirmed by the lack of cellular/biological activity observed for any of the parent di- or triacids (data not shown). Interestingly, however, interpretation of the uptake for several monoacids was less obvious. The monocarboxylate **55** showed a weak, but measurable, uptake into cells and this may reflect the ability of the protonated form of the carboxylic acid to cross the cell membrane. The phosphonate monoacid **52**, however, did not show any uptake, indicating that the more acidic phosphonate group is unlikely to cross the cell membrane in its anionically charged state.

Although the ethyl esters (**51**, **53**, **54**, and **56**) provided good uptake they did not, as expected, exhibit any reconversion to the parent diacids. The use of phosphonate diesters as prodrugs is further complicated by the need to remove two prodrug ester groups thus revealing the desired active parent diacid. Although, in general, the two esters groups of phosphonate diester prodrugs are usually the same, the prodrug diester and the intermediate monoester are chemically different and removal of the first and second esters must be considered as two separate and different deprotection steps. Prodrugs can be removed either passively, through their inherent chemical instability in water at physiological pH, or actively by enzymes. For phosphonate diesters, the second hydrolysis step is thought to be the easiest, since conversion of the monoester/monoacid to the diacid can generally be accomplished enzymatically by common phosphodiesterases in the cell. In this regard, a phosphonate monoester is equivalent to a phosphate diester (Figure 12), which are ubiquitous in cells and serve as the normal substrates for a variety of phosphodiesterases.

Thus, the rate-determining step in the reconversion of phosphonate diester prodrugs is the removal of the first ester. The problem with most simple alkyl esters is that the derived phosphonate diesters are chemically stable towards hydrolysis at physiological pH, and are enzymatically inert due to the lack of any phosphotriesterases in most cells. Two general strategies have been employed to resolve this problem. The first strategy is to use esters which are more chemically labile, so that they spontaneously hydrolyze at a useful rate at physiological pH.[117-122] The second strategy is to replace the difficult first step of hydrolyzing a phosphonate diester with a more readily achievable enzymatic transformation.[120-134]

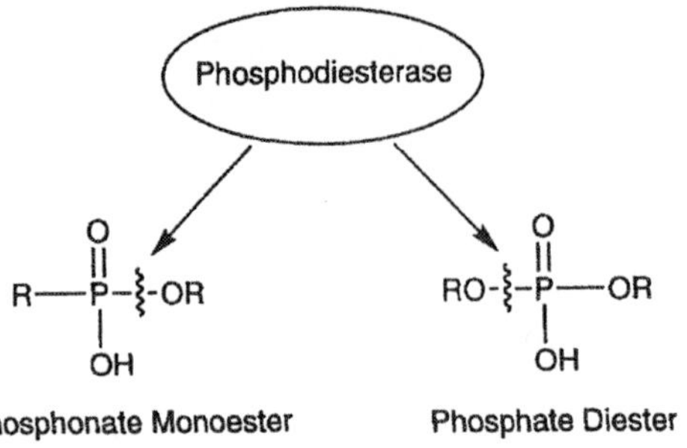

Phosphodiesterase

Phosphonate Monoester

Phosphate Diester

Figure 12. Equivalence of phosphate and phosphonate monoacids as substrates for phosphodiesterases.

Our first attempt to prepare prodrugs for our compounds focused on the use of chemically more labile phosphonate esters. Initially we planned to use the arylphosphonate template as exemplified by **53** to develop a series of prodrug esters based on the expected ease of preparation of these phosphonate esters via the route outlined in Scheme 1.[135] For simple phosphite esters such as ethyl, this route cleanly provided the desired product in excellent yield, however attempts to extend this methodology to esters with better prodrug potential (phenyl or trichloroethyl)[117-120] were generally unsuccessful. In the case of the phenyl ester derivative, only very low yields of the desired product were obtainable, but incorporation of this analogue into our arylphosphonate series could be achieved to provide **57**. Unfortunately, this compound was extremely water insoluble, and although some cellular uptake could be observed, quantitation was not possible, and no reconversion was observed (Table 12).

Due to the difficulties with preparing prodrug esters in the arylphosphonate series, as well as the lower binding affinities of the parent diacids in our *in vitro* assays, we returned to the synthesis of prodrugs in the F_2Pmp series. We further decided to explore the use of the second prodrug strategy described above, which was to replace the slow phosphonate diester cleavage step with a more readily achieved enzymatic transformation. A review of the literature indicated that the best prodrug for phosphates and phosphonates were esters of acyloxymethyl type.[121-126]

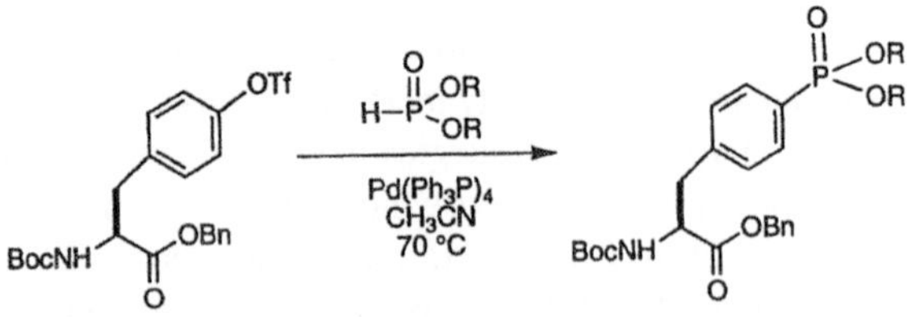

$Pd(Ph_3P)_4$, CH_3CN, 70 °C

Scheme 1. Synthesis of arylphosphonate diester prodrug precursors.

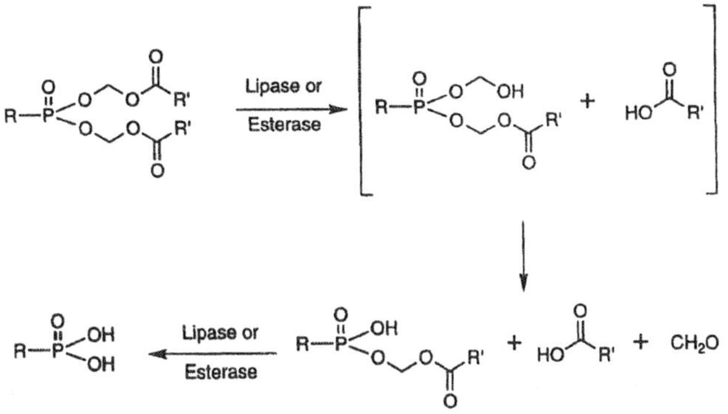

Figure 13. Proposed mechanism for the hydrolysis of a bis-acyloxymethyl ester of a generic phosphonate.

The advantage of these esters is that they replace the slow first hydrolysis step in the reconversion of simple phosphonate esters with a more facile enzymatic hydrolysis of a carboxylate ester by either lipases or esterases in the cell (Figure 13).

Such esters have found extensive use in the preparation of nucleoside prodrugs and have been shown to provide effective oral absorption of several different compound classes.[121-126] Although, originally described for use in Ampicillin prodrugs,[136] they were subsequently exploited by Bodor in his "soft drug" approach,[137-139] and then further championed by Farquhar for their use in the development of orally active nucleoside analogues.[125,140] Most importantly, these esters were generally easy to prepare and provided bioreversible prodrugs for many different structural subtypes. Therefore, it was anticipated that this generality would also extend to our F_2Pmp derivatives.

In our case, subjecting any of the F_2Pmp analogues (X = CF_2) to the standard synthetic methodology yielded only the monoacyloxymethyl ester **A**, even under forcing conditions (Scheme 2). However, when this method was applied to the arylphosphonate series (X = bond), both the mono- **A** and bis-esters **B** could be obtained.

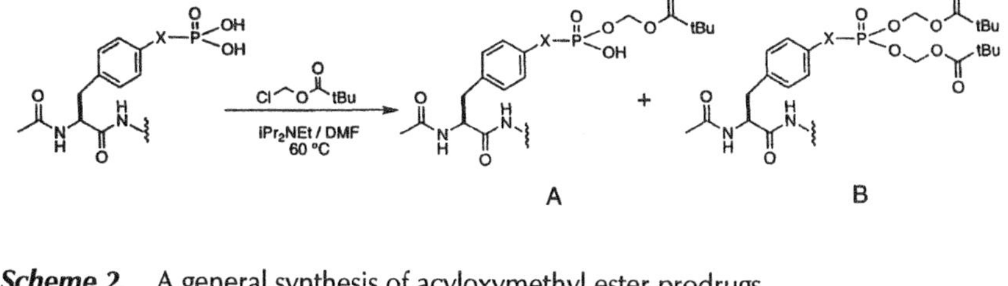

Scheme 2. A general synthesis of acyloxymethyl ester prodrugs.

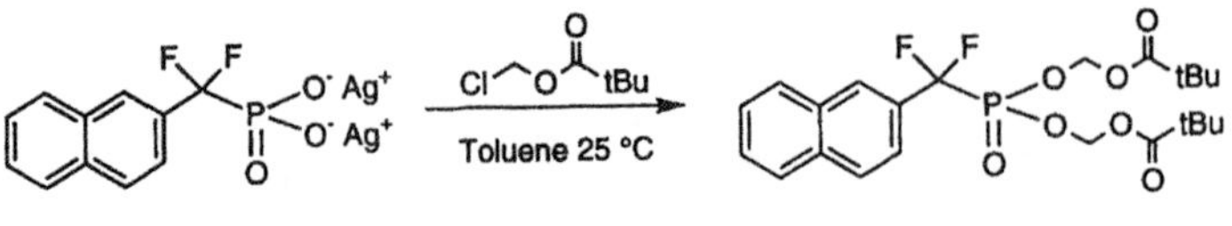

58

Scheme 3. Synthesis of a model difluorophosphonate prodrug diester.

We reasoned that the expected lower pK_a[97,98] of our F_2Pmp derivatives makes them less nucleophilic and, thus, more difficult to alkylate a second time to form the bis-esters. The only method which successfully generated the bis-esters in the F_2Pmp series was to use the bis-silver salt of the phosphonate in conjunction with an acyloxymethyl iodide.[124,125] This worked well in model systems to provide the bis-ester **58** (Scheme 3), but application of this method to our peptide analogues has proven problematic due to low yields and difficulties performing these reaction on small scales.

Fortunately, when the acyloxymethyl esters were tested in our cellular uptake assay a number of interesting results were obtained (Table 13). First, as expected, the fully protected di- and triesters in the arylphosphonate series (**60** and **62**) each showed excellent uptake rates and reasonable reconversion to the parent di- and tri-acids. Unexpectedly, however, the monoacyloxymethyl esters in the F_2Pmp series, **63** and **64**, each also showed moderate uptake and good reconversion to the parent phosphonate diacids as well.

Based on the lack of uptake of the monoethyl ester **52** (Table 12) and the normal lack of uptake of charged molecules, the observation that F_2Pmp monoacyloxy-methyl esters **63** and **64** show good uptake was particularly intriguing. Namely, the ability to use a monoacyloxymethyl ester may provide water soluble prodrugs with effective cellular uptake, and therefore, offer promising candidates for use in further signal transduction-based cellular models.

In contrast to the F_2Pmp series, the monoacyloxymethyl esters in the arylphosphonate series (**59** and **61**) did not show any detectable uptake. On first inspection, this result was rather puzzling, but further analysis suggests that the differences in uptake between the F_2Pmp and arylphosphonate monoacyloxymethyl esters may be correlated to their differences in hydrophobicity, as judged by their cLogPs[141,142] and independently by their log k'_{IAM} on an immobilized artificial membrane-phosphotidyl choline (IAM.PC) HPLC column.[116,143] However, further analysis employing more compounds will be needed to more fully evaluate this hypothesis.

Thus, the use of F_2Pmp as a pTyr replacement has proven to be uniquely advantageous to our goal of producing cellularly active SH2 targeted peptidomimetic ligands, by providing prodrugs with good cellular penetration, but which are still water soluble. Furthermore, the use of the above prodrug analogues, especially the acyloxymethyl ester type, have provided us with compounds which readily penetrate into cells and reconvert to give the active parent compounds. Current studies are focused on extending these preliminary uptake experiments to the use of cellu-

Table 13. Cellular Uptake and Reconversion Data for Dipeptide Prodrugs

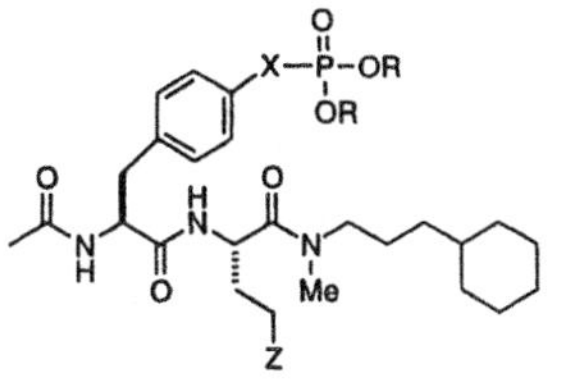

	Z	*X*	*OR*[a]	*OR′*	*Uptake Rate*[b]	*Reconversion*
59	H	—	OCH_2OPiv, OH	NA[c]	<0.002	ND[d]
60	H	—	OCH_2OPiv	NA	6.6 ± 0.76 (n=3)	Yes
61	CO_2R'	—	OCH_2OPiv, OH	OCH_2OPiv	<0.002	ND
62	CO_2R'	—	OCH_2OPiv	OCH2OPiv	$>>>0.002$[e]	Yes
63	H	CF_2	OCH_2OPiv, OH	NA	0.01 ± 0.001 (n=3)	Yes
64	CO_2R'	CF_2	OCH_2OPiv, OH	OCH_2OPiv	$>>0.002$[e]	Yes

Notes: [a] Piv = pivaloate.
[b] In (nmol/min-sq cm) normalized to a donor concentration of 100 μM. The uptake rates were determined as reported in Surendran et al.[115,116]
[c] Not applicable.
[d] Not detected.
[e] Uptake rate was much greater than our detection limit, but overlapping peaks preclude quantitation.

lar assays which are designed to reflect effects of these compounds on signal transduction pathways, as indicated by interruption of signaling via the PDGF receptor, or interruption of DNA synthesis, or their effects on osteoclast apoptosis or bone resorption. Results from these studies will be reported in due course.

5. CONCLUSIONS

In this review we have described the design, chemistry and biological evaluation of phosphopeptides and peptidomimetic analogues as prototype drugs for targeting the Src SH2 domain. The design of potent and selective peptidomimetic ligands which may block phosphoprotein–SH2 protein interactions in cells may provide a novel approach to future therapeutic agents targeting a variety of diseases. To date, the design of such drugs has been very challenging. These challenges include transforming a highly charged peptide into a cell penetrable small molecule. This review detailed a significant step forward which advanced several prototype series of peptidomimetics that show promising affinities and selectivities for Src SH2. In this regard, our effort focused on Src SH2 targeted drug design, since X-ray crystallography first succeeded for the Src SH2 domain and since Src (and Src family tyrosine kinases) are of intense interest in many therapeutic areas.

In summary, the iterative transformation of phosphopeptide ligands for the Src SH2 domain into dipeptides and peptidomimetics has been achieved. Such compounds have been designed to provide molecular recognition with the Src SH2 domain by virtue of binding at the P (pTyr) site and the P+3 hydrophobic pocket. Such structure-based design strategies have been aided by X-ray crystallography, which has provided the three-dimensional structures of key lead compounds (e.g. the tripeptide and the ureido-type peptidomimetic).[16,88] Furthermore, an additional effort has been focused on the pTyr residue itself to advance metabolically stable and effective mimetics which may provide cellular penetration. Such work has led to the discovery of lead compounds having prodrug modifications of the phosphotyrosine mimetic F_2Pmp. As described in this review, a convergent strategy to exploit structure-based design of dipeptide and peptidomimetic ligands for the Src SH2 domain and replacement of the phosphate moiety of pTyr has provided prodrug derivatives useful in cellular "proof-of-concept" studies.

ACKNOWLEDGMENTS

We would like to thank all of our collaborators who have contributed to this project, but especially Debra Holland and Ron Rubin for their contributions to the X-ray studies reported herein, and to Beth Lunney for providing all the modeling support for this project. We further would like to thank Narayanan Surendran and Barbra Stewart for providing all the cell uptake data for our compounds, and Jim Fergus for running all of the *in vitro* assays. We would also like to thank John Blankley for providing the cLoyPs used in these studies. Finally, we would like to acknowledge the assistance of all of the above in the preparation and critical review of this manuscript.

REFERENCES

1. Botfield, M. C.; Green, J. *Annu. Rep. Med. Chem.* **1995**, *30*, 227.
2. Smithgall, T. E. *J. Pharmacol. Toxicol. Methods* **1995**, *34(3)*, 125.
3. Liu, D.; Wang, L. H. *J. Biomed. Sci.* **1994**, *1(2)*, 65.
4. Levitzki, A.; Gazit, A. *Science* **1995**, *267(5205)*, 1782.
5. Hunter, T. *Cell* **1995**, *80, 225.*
6. Kassim, S. K.; Wiener, J. R.; Berchuck, A.; Bast, R. C. *Cancer Mol. Biol* **1994**, *1(2)*, 133.
7. Barford, D.; Jia, Z.; Tonks, N. K. *Nat. Struct. Biol.* **1995**, *2(12)*, 1043.
8. Barford, D. *Curr. Opin. Struct. Biol.* **1995**, *5(6)*, 728.
9. Feng, G. S.; Pawson, T. *Trends Genet.* **1994**, *10(2)*, 54.
10. Tonks, N. K. *Adv. Pharmacol.* **1996**, *36*.
11. Bridges, A. J. *Chemtracts: Org. Chem.* **1995**, *8(2)*, 73.
12. Brugge, J. S. *Science* **1993**, *260(5110)*, 918.
13. Levitzki, A. *Curr. Opin. Cell Biol.* **1996**, *8(2)*, 239.
14. Bibbins, K. B.; Boeuf, H.; Varmus, H. E. *Mol. Cell. Biol.* **1993**, *13(12)*, 7278.
15. Marengere, L. E. M.; Songyang, Z.; Gish, G. D.; Schaller, M. D.; Parsons, J. T.; Stern, M. J.; Cantley, L. C.; Pawson, T. *Nature* **1994**, *369(6480)*, 502.
16. Holland, D. R. et al. Manuscript in preparation .

17. Ren, R.; Mayer, B. J.; Cicchetti, P.; Baltimore, D. *Science* **1993**, *259(5098)*, 1157.
18. Yu, H.; Chen, J. K.; Feng, S.; Dalgarno, D. C.; Brauer, A. W.; Schreiber, S. L. *Cell* **1994**, *76(5)*, 933.
19. Touhara, K.; Inglese, J.; Pitcher, J. A.; Shaw, G.; Lefkowitz, R. J. *J. Biol. Chem.* **1994**, *269(14)*, 10217.
20. Eck, M. J. *Structure* **1995**, *3(5)*, 421.
21. Kavanaugh, W. M.; Turck, C. W.; Williams, L. T. *Science* **1995**, *268(5214)*, 1177.
22. Eck, M. J.; Dhe-Paganon, S.; Trub, T.; Nolte, R. T.; Shoelson, S. E. *Cell* **1996**, *85(5)*, 695.
23. Overduin, M.; Rios, C. B.; Mayer, B. J.; Baltimore, D.; Cowburn, D. *Cell* **1992**, *70(4)*, 697.
24. Waksman, G.; Kominos, D.; Robertson, S. C.; Pant, N.; Baltimore, D.; Birge, R. B.; Cowburn, D.; Hanafusa, H.; Mayer, B. J.; Overduin, M.; Resh, M. D.; Rios, C. B.; Silverman, L.; Kuriyan, J. *Nature* **1992**, *358(6388)*, 646.
25. Waksman, G.; Shoelson, S. E.; Pant, N.; Cowburn, D.; Kuriyan, J. *Cell* **1993**, *72(5)*, 779.
26. Cantley, L. C.; Auger, K. R.; Carpenter, C.; Duckworth, B.; Graziani, A.; Kapeller, R.; Soltoff, S. *Cell* **1991**, *64(2)*, 281.
27. Birge, R. B.; Hanafusa, H. *Science* **1993**, *262(5139)*, 1522.
28. Carpenter, C. L.; Auger, K. R.; Chanudhuri, M.; Yoakim, M.; Schaffhausen, B.; Shoelson, S.; Cantley, L. C. *J. Biol. Chem.* **1993**, *268(13)*, 9478.
29. Piccione, E.; Case, R. D.; Domchek, S. M.; Hu, P.; Chaudhuri, M.; Backer, J. M.; Schlessinger, J.; Shoelson, S. E. *Biochemistry* **1993**, *32(13)*, 3197.
30. Herbst, J. J.; Andrews, G.; Contillo, L.; Lamphere, L.; Gardner, J.; Lienhard, G. E.; Gibbs, E. M. *Biochemistry* **1994**, *33(32)*, 9376.
31. Sun, X. J.; Crimmins, D. L.; Myers, M. G., Jr.; Miralpeix, M.; White, M. F. *Mol. Cell. Biol.* **1993**, *13(12)*, 7418.
32. Batzer, A. G.; Rotin, D.; Urena, J. M.; Skolnik, E. Y.; Schlessinger, J. *Mol. Cell. Biol.* **1994**, *14(8)*, 5192.
33. Lemmon, M. A.; Ladbury, J. E.; Mandiyan, V.; Zhou, M.; Schlessinger, J. *J. Biol. Chem.* **1994**, *269(50)*, 31653.
34. Sorkin, A.; Helin, K.; Waters, C. M.; Carpenter, G.; Beguinot, L. *J. Biol. Chem.* **1992**, *267(12)*, 8672.
35. McNamara, D. J.; Dobrusin, E. M.; Zhu, G.; Decker, S. J.; Saltiel, A. R. *Int. J. Pept. Protein Res.* **1993**, *42(3)*, 240.
36. Huyer, G.; Li, Z. M.; Adam, M.; Huckle, W. R.; Ramachandran, C. *Biochemistry* **1995**, *34(3)*, 1040.
37. Pluskey, S.; Wandless, T. J.; Walsh, C. T.; Shoelson, S. E. *J. Biol. Chem.* **1995**, *270(7)*, 2897.
38. Zhou, M.-M.; Harlan, J. E.; Wade, W. S.; Crosby, S.; Ravichandran, K. S.; Burakoff, S. J.; Fesik, S. W. *J. Biol. Chem.* **1995**, *270(52)*, 31119.
39. Wange, R. L.; Isakov, N.; Burke, T. R., Jr.; Otaka, A.; Roller, P. P.; Watts, J. D.; Aebersold, R.; Samelson, L. E. *J. Biol. Chem.* **1995**, *270(2)*, 944.
40. Alonso, G.; Koegl, M.; Mazurenko, N.; Courtneidge, S. A. *J. Biol. Chem.* **1995**, *270(17)*, 9840.
41. Luttrell, D. K.; Lee, A.; Lansing, T. J.; Crosby, R. M.; Jung, K. D.; Willard, D.; Luther, M.; Rodriguez, M.; Berman, J.; Gilmer, T. M. *Proc. Natl. Acad. Sci. USA* **1994**, *91(1)*, 83.
42. Roussel, R. R.; Brodeur, S. R.; Shalloway, D.; Laudano, A. P. *Proc. Natl. Acad. Sci. USA* **1991**, *88(23)*, 10696.
43. Payne, G.; Stolz, L. A.; Pei, D.; Band, H.; Shoelson, S. E.; Walsh, C. T. *Chem. Biol.* **1994**, *1(2)*, 99.
44. Payne, G.; Shoelson, S. E.; Gish, G. D.; Pawson, T.; Walsh, C. T. *Proc. Natl. Acad. Sci. USA* **1993**, *90(11)*, 4902.
45. Songyang, Z.; Shoelson, S. E.; Chaudhuri, M.; Gish, G.; Pawson, T.; Haser, W. G.; King, F.; Roberts, T.; Ratnofsky, S.; Lechleider, R. J.; Neel, B. G.; Birge, R. B.; Fajardo, J. E.; Chou, M. M.; Hanafusa, H.; Schaffhausen, B.; Cantley, L. C. *Cell* **1993**, *72(5)*, 767.
46. Songyang, Z.; Cantley, L. C. *Trends Biochem. Sci.* **1995**, *20(11)*, 470.

47. Songyang, Z.; Shoelson, S. E.; McGlade, J.; Olivier, P.; Pawson, T.; Bustelo, X. R.; Barbacid, M.; Sabe, H.; Hanafusa, H.; Yi, T.; Ren, R.; Baltimore, D.; Ratnofsky, S.; Feldman, R. A.; Cantley, L. C. *Mol. Cell. Biol.* **1994**, *14(4)*, 2777.
48. Mueller, K.; Gombert, F. O.; Manning, U.; Grossmueller, F.; Graff, P.; Zaegel, H.; Zuber, J. F.; Freuler, F.; Tschopp, C.; Baumann, G. *J. Biol. Chem.* **1996**, *271(28)*, 16500.
49. Fantl, W. J.; Escobedo, J. A.; Martin, G. A.; Turck, C. W.; Del Rosario, M.; McCormick, F.; Williams, L. T. *Cell* **1992**, *69(3)*, 413.
50. Xu, W.; Harrison, S. C.; Eck, M. J. *Nature* **1997**, *385(6617)*, 595.
51. Sicheri, F.; Moarefi, I.; Kuriyan, J. *Nature* **1997**, *385(6617)*, 602.
52. Overduin, M.; Mayer, B.; Rios, C. B.; Baltimore, D.; Cowburn, D. *Proc. Natl. Acad. Sci. USA* **1992**, *89(24)*, 11673.
53. Xu, R. X.; Word, J. M.; Davis, D. G.; Rink, M. J.; Willard, D. H., Jr.; Gampe, R. T., Jr. *Biochemistry* **1995**, *34(7)*, 2107.
54. Eck, M. J.; Shoelson, S. E.; Harrison, S. C. *Nature* **1993**, *362(6415)*, 87.
55. Eck, M. J.; Atwell, S. K.; Shoelson, S. E.; Harrison, S. C. *Nature* **1994**, *368(6473)*, 764.
56. Mikol, V.; Baumann, G.; Keller, T. H.; Manning, U.; Zurini, M. G. M. *J. Mol. Biol.* **1995**, *246(2)*, 344.
57. Tong, L.; Warren, T. C.; King, J.; Betageri, R.; Rose, J.; Jakes, S. *J. Mol. Biol.* **1996**, *256(3)*, 601.
58. Pascal, S. M.; Singer, A. U.; Gish, G.; Yamazaki, T.; Shoelson, S. E.; Pawson, T.; Kay, L. E.; Forman-Kay, J. D. *Cell* **1994**, *77(3)*, 461.
59. Lee, C.-H.; Kominos, D.; Jacques, S.; Margolis, B.; Schlessinger, J.; Shoelson, S. E.; Kuriyan, J. *Structure* **1994**, *2(5)*, 423.
60. Eck, M. J.; Pluskey, S.; Trub, T.; Harrison, S. C.; Shoelson, S. E. *Nature* **1996**, *379(6562)*, 277.
61. Hatada, M. H.; Lu, X.; Laird, E. R.; Green, J.; Morgenstern, J. P.; Lou, M.; Marr, C. S.; Phillips, T. B.; Ram, M. K.; Theriault, K.; Zoller, M. J.; Karas, J. L. *Nature* **1995**, *377(6544)*, 32.
62. Nolte, R. T.; Eck, M. J.; Schlessinger, J.; Shoelson, S. E.; Harrison, S. C. *Nat. Struct. Biol.* **1996**, *3(4)*, 364.
63. Booker, G. W.; Breeze, A. L.; Downing, A. K.; Panayotou, G.; Gout, I.; Waterfield, M. D.; Campbell, I. D. *Nature* **1992**, *358(6388)*, 684.
64. Narula, S. S.; Yuan, R. W.; Adams, S. E.; Green, O. M.; Green, J.; Philips, T. B.; Zydowsky, L. D. *Structure* **1995**, *3(10)*, 1061.
65. Mikol, V.; Baumann, G.; Zurini, M. G. M.; Hommel, U. *J. Mol. Biol.* **1995**, *254(1)*, 86.
66. Guilloteau, J. P.; Fromage, N.; Ries-Kautt, M.; Reboul, S.; Bocquet, D.; Dubois, H.; Faucher, D.; Colonna, C.; Ducruix, A.; Becquart, J. *Proteins: Struct. Funct. Genet.* **1996**, *25(1)*, 112.
67. Maignan, S.; Guilloteau, J.-P.; Fromage, N.; Arnoux, B.; Becquart, J.; Ducruix, A. *Science* **1995**, *268(5208)*, 291.
68. Rahuel, J.; Gay, B.; Erdmann, D.; Strauss, A.; Garcia-Echeverria, C.; Furet, P.; Caravatti, G.; Fretz, H.; Schoepfer, J.; Gruetter, M. G. *Nat. Struct. Biol.* **1996**, *3(7)*, 586.
69. Waksman, G. *Bull. Inst. Pasteur* **1994**, *92(1)*, 19.
70. Wallace, A. C.; Laskowski, R. A.; Thorton, J. M. *Prot. Eng.* **1995**, *8(2)*, 127.
71. Levitzki, A. *Anti-Cancer Drug Des.* **1996**, *11(3)*, 175.
72. Hall, T. J.; Schaeublin, M.; Missbach, M. *Biochem. Biophys. Res. Comm.* **1994**, *199(3)*, 1237.
73. Lowe, C.; Yoneda, T.; Boyce, B. F.; Chen, H.; Mundy, G. R.; Soriano, P. *Proc. Natl. Acad. Sci. USA* **1993**, *90(10)*, 4485.
74. Boyce, B. F.; Yoneda, T.; Lowe, C.; Soriano, P.; Mundy, G. R. *J. Clin. Invest.* **1992**, *90(4)*, 1622.
75. Plummer, M. S.; Lunney, E.; Para, K. S.; Shahripour, A.; Stankovic, C. J.; Humblet, C.; Fergus, J. H.; Marks, J. S.; Herrera, R.; Hubbell, S.; Saltiel, A.; Sawyer, T. K. *Bioorg. Med. Chem.* **1997**, *5(1)*, 41.
76. Brouillette, C. Unpublished results .
77. Ladbury, J. E.; Lemmon, M. A.; Zhou, M.; Green, J.; Botfield, M. C.; Schlessinger, J. *Proc. Natl. Acad. Sci. USA* **1995**, *92(8)*, 3199.

78. Gilmer, T.; Rodriguez, M.; Jordan, S.; Crosby, R.; Alligood, K.; Green, M.; Kimery, M.; Wagner, C.; Kinder, D.; Charifson, P.; Hassel, A. M.; Willard, D.; Luther, M.; Rusnak, D.; Sternbach, D. D.; Mehrotra, M.; Peel, M.; Shampine, L.; Davis, R.; Robins, J.; Patel, I. R.; Kassel, D.; Burkhart, W.; Moyer, M.; Bradshaw, T.; Berman, J. *J. Biol. Chem.* **1994**, *269(50)*, 31711.
79. Lunney, E. A.; Para, K. S.; Rubin, J. R.; Humblet, C.; Fergus, J. H.; Marks, J. S.; Sawyer, T. K. *J. Am. Chem. Soc.* In press.
80. Rodriguez, M.; Crosby, R.; Alligood, K.; Gilmer, T.; Berman, J. *Lett. Pept. Sci.* **1995**, *2(1)*, 1.
81. Plummer, M. S.; Lunney, E. A.; Para, K. S.; Prasad, J. V. N. V.; Shahripour, A.; Singh, J.; Stankovic, C. J.; Humblet, C.; Fergus, J. H.; Marks, J. S.; Sawyer, T. K. *Drug Design and Discovery* **1996**, *13*, 75.
82. Shahripour, A.; Plummer, M. S.; Lunney, E. A.; Vara Prasad, J.; Singh, J.; Para, K. S.; Stankovic, C. J.; Eaton, S. R.; Marks, J. S.; Decker, S. J.; Herrera, R.; Hubbell, S.; Saltiel, A. R.; Sawyer, T. K. *Novel Phosphotyrosine and Hydrophobic D-Amino Acid Replacements in the Design of Peptide Ligands for pp60src SH2 Domain*; Kaumaya, P. J. P.; Hodges, R. S. Eds.; Mayflower Scientific, 1995, pp 394.
83. Smyth, M. S.; Burke, T. R., Jr. *Tetrahedron Lett.* **1994**, *35(4)*, 551.
84. Smyth, M. S.; Burke, T. R., Jr. *Org. Prep. Proc. Int.* **1996**, *28(1)*, 77.
85. Kraulis, P. J. *J. App. Cryst.* **1991**, *24*, 946.
86. Shahripour, A.; Para, K. S.; Plummer, M. S.; Lunney, E.; Stankovic, C. J.; Holland, D. R.; Rubin, J. R.; Humblet, C.; Fergus, J. H.; Marks, J. S.; Saltiel, A. R.; Sawyer, T. K. *Bioorg. Med. Chem. Lett.* **1997**, *7(9)*, 1107.
87. Shahripour, A.; Plummer, M. S.; Lunney, E.; Para, K. S.; Stankovic, C. J.; Rubin, J. R.; Humblet, C.; Fergus, J. H.; Marks, J. S.; Herrera, R.; Hubbell, S. E.; Saltiel, A. R.; Sawyer, T. K. *Bioorg. Med. Chem. Lett.* **1996**, *6(11)*, 1209.
88. Plummer, M. S.; Holland, D. R.; Shahripour, A.; Lunney, E. A.; Fergus, J. H.; Marks, J. S.; McConnell, P.; Mueller, W. T.; Sawyer, T. K. *J. Med. Chem.* In Press.
89. Engel, R. *Chem. Rev.* **1977**, *77(3)*, 349.
90. Blackburn, G. M.; Perree, T. D.; Rashid, A.; Bisbal, C.; Lebleu, B. *Chem. Scr.* **1986**, *26(1)*, 21.
91. Blackburn, G. M. *Chem. Ind.* **1981**, *(5)*, 134.
92. Guida, W. C.; Elliott, R. D.; Thomas, H. J.; Secrist, J. A., III; Babu, Y. S.; Bugg, C. E.; Erion, M. D.; Ealick, S. E.; Montgomery, J. A. *J. Med. Chem.* **1994**, *37(8)*, 1109.
93. Sikorski, J. A.; Miller, M. J.; Braccolino, D. S.; Cleary, D. G.; Corey, S. D.; Font, J. L.; Gruys, K. J.; Han, C. Y.; Lin, K. C.; Pansegrau, P. D.; Ream, J. E.; Schnur, D.; Shah, A.; Walker, M. C. *Phosphorus, Sulfur Silicon Relat. Elem.* **1993**, *76*, 375.
94. Miller, M. J.; Anderson, K. S.; Braccolino, D. S.; Cleary, D. G.; Gruys, K. J.; Han, C. Y.; Lin, K. C.; Pansegrau, P. D.; Ream, J. E.; Sammons, D.; Sikorski, J. A. *Bioorg. Med. Chem. Lett.* **1993**, *3(7)*, 1435.
95. Miller, M. J.; Braccolino, D. S.; Cleary, D. G.; Ream, J. E.; Walker, M. C.; Sikorski, J. A. *Bioorg. Med. Chem. Lett.* **1994**, *4(21)*, 2605.
96. Wissner, A.; Kohler, C. A.; Goldstein, B. M. *J. Med. Chem.* **1985**, *28(9)*, 1365.
97. Smyth, M. S.; Ford, H., Jr.; Burke, T. R., Jr. *Tetrahedron Lett.* **1992**, *33(29)*, 4137.
98. Burke, T. R., Jr.; Smyth, M. S.; Otaka, A.; Nomizu, M.; Roller, P. P.; Wolf, G.; Case, R.; Shoelson, S. E. *Biochemistry* **1994**, *33(21)*, 6490.
99. Otaka, A.; Burke, T. R., Jr.; Smyth, M. S.; Nomizu, M.; Roller, P. P. *Tetrahedron Lett.* **1993**, *34(44)*, 7039.
100. Ye, B.; Akamatsu, M.; Shoelson, S. E.; Wolf, G.; Giorgetti-Peraldi, S.; Yan, X.; Roller, P. P.; Burke, T. R., Jr. *J. Med. Chem.* **1995**, *38(21)*, 4270.
101. Ye, B.; Burke, T. R., Jr. *Tetrahedron Lett.* **1995**, *36(27)*, 4733.
102. Burke, T. R., Jr.; Ye, B.; Akamatsu, M.; Ford, H.; Yan, X.; Kole, H. K.; Wolf, G.; Shoelson, S. E.; Roller, P. P. *J. Med. Chem.* **1996**, *39(5)*, 1021.
103. Akamatsu, M.; Ye, B.; Yan, X.; Kole, H. K.; Burke, T. R., Jr.; Roller, P. P. *Pept. Chem.* **1996**, 369.

104. Stankovic, C. J.; Surendran, N.; Lunney, E.; Plummer, M. S.; Para, K. S.; Shahripour, A.; Humblet, C.; Fergus, J. H.; Marks, J. S.; Herrera, R.; Hubbell, S. E.; Saltiel, A. R.; Stewart, B. H.; Sawyer, T. K. , *Bioorg. Med. Chem. Lett.* **1997**, *7(14)*, 1909.
105. Kelly, M. A.; Liang, H.; Sytwu, I.-I.; Vlattas, I.; Lyons, N. L.; Bowen, B. R.; Wennogle, L. P. *Biochemistry* **1996**, *35(36)*, 11747.
106. Jagoe, C. T.; Kreifels, S. E.; Li, J. *Bioorg. Med. Chem. Lett.* **1997**, *7(2)*, 113.
107. Mehrotra, M. M.; Sternbach, D. D.; Ridriguez, M.; Charifson, P.; Berman, J. *Bioorg. Med. Chem. Lett.* **1996**, *6(16)*, 1941.
108. Testa, B.; Caldwell, J. *Med. Res. Rev.* **1996**, *16(3)*, 233.
109. Jones, R. J.; Bischofberger, N. *Antiviral Res.* **1995**, *27(1-2)*, 1.
110. Krise, J. P.; Stella, V. J. *Adv. Drug Delivery Rev.* **1996**, *19(2)*, 287.
111. Friis, G. J.; Bundgaard, H. *Eur. J. Pharm. Sci.* **1996**, *4(1)*, 49.
112. Hidalgo, I. J.; Raub, T. J.; Borchardt, R. T. *Gastroenterology* **1989**, *96(736-749)*.
113. Schasteen, C. S.; Donovan, M. G.; Cogburn, J. N. *J. Cont. Rel.* **1992**, *21*, 49.
114. Stewart, B. H.; Chan, O. H.; Lu, R. H.; Reyner, E. L.; Schmid, H. L.; Hamilton, H. W.; Steinbaugh, B. A.; Taylor, M. D. *Pharm. Res.* **1995**, *12(5)*, 693.
115. Surendran, N.; Stankovic, C. J.; Stewart, B. H. *J. Chrom. B.* **1997**, *691(2)*, 305.
116. Surendran, N.; et al. Manuscript in preparation.
117. McGuigan, C.; Davies, M.; Pathirana, R.; Mahmood, N.; Hay, A. J. *Antiviral Res.* **1994**, *24(1)*, 69.
118. McGuigan, C.; Devine, K. G.; O'Connor, T. J.; Kinchington, D. *Antiviral Res.* **1991**, *15(3)*, 255.
119. De Lombaert, S.; Blanchard, L.; Berry, C.; Ghai, R. D.; Trapani, A. J. *Bioorg. Med. Chem. Lett.* **1995**, *5(2)*, 151.
120. Chawla, R. R.; Freed, J. J.; Hampton, A. *J. Med. Chem.* **1984**, *27(12)*, 1733.
121. Starrett, J. E., Jr.; Tortolani, D. R.; Russell, J.; Hitchcock, M. J. M.; Whiterock, V.; Martin, J. C.; Mansuri, M. M. *J. Med. Chem.* **1994**, *37(12)*, 1857.
122. Serafinowska, H. T.; Ashton, R. J.; Bailey, S.; Harnden, M. R.; Jackson, S. M.; Sutton, D. *J. Med. Chem.* **1995**, *38(8)*, 1372.
123. Starrett, J. E., Jr.; Tortolani, D. R.; Hitchcock, M. J. M.; Martin, J. C.; Mansuri, M. M. *Antiviral Res.* **1992**, *19(3)*, 267.
124. Srivastva, D. N.; Farquhar, D. *Bioorg. Chem.* **1984**, *12(2)*, 118.
125. Farquhar, D.; Khan, S.; Srivastva, D. N.; Saunders, P. P. *J. Med. Chem.* **1994**, *37(23)*, 3902.
126. Srinivas, R. V.; Robbins, B. L.; Connelly, M. C.; Gong, Y. F.; Bischofberger, N.; Fridland, A. *Antimicrob. Agents Chemother.* **1993**, *37(10)*, 2247.
127. Farquhar, D.; Smith, R. *J. Med. Chem.* **1985**, *28(9)*, 1358.
128. Farquhar, D.; Khan, S.; Wilkerson, M. C.; Andersson, B. S. *Tetrahedron Lett.* **1995**, *36(5)*, 655.
129. Valette, G.; Pompon, A.; Girardet, J.-L.; Cappellacci, L.; Franchetti, P.; Grifantini, M.; Colla, P. L.; Loi, A. G.; Perigaud, C.; et al. *J. Med. Chem.* **1996**, *39(10)*, 1981.
130. Hunston, R. N.; Jones, A. S.; McGuigan, C.; Walker, R. T.; Balzarini, J.; De Clercq, E. *J. Med. Chem.* **1984**, *27(4)*, 440.
131. Puech, F.; Gosselin, G.; Lefebvre, I.; Pompon, A.; Aubertin, A. M.; Kirn, A.; Imbach, J. L. *Antiviral Res.* **1993**, *22(2-3)*, 155.
132. Lefebvre, I.; Perigaud, C.; Pompon, A.; Aubertin, A.-M.; Girardet, J.-L.; Kirn, A.; Gosselin, G.; Imbach, J.-L. *J. Med. Chem.* **1995**, *38(20)*, 3941.
133. Freeman, S.; Irwin, W. J.; Mitchell, A. G.; Nicholls, D.; Thomson, W. *J. Chem. Soc., Chem. Commun.* **1991**, *(13)*, 875.
134. Benzaria, S.; Pelicano, H.; Johnson, R.; Maury, G.; Imbach, J.-L.; Aubertin, A.-M.; Obert, G.; Gosselin, G. *J. Med. Chem.* **1996**, *39(25)*, 4958.
135. Thurieau, C.; Simonet, S.; Paladino, J.; Prost, J.-F.; Verbeuren, T.; Fauchere, J.-L. *J. Med. Chem.* **1994**, *37(5)*, 625.
136. Von Daehne, W.; Frederiksen, E.; Gundersen, E.; Lund, F.; Moerch, P.; Petersen, H. J.; Roholt, K.; Tybring, L.; Godtfredsen, W. O. *J. Med. Chem.* **1970**, *13(4)*, 607.

137. Bodor, N.; Kaminski, J. J.; Selk, S. *J. Med. Chem.* **1980**, *23(5)*, 469.
138. Bodor, N.; Kaminski, J. J. *J. Med. Chem.* **1980**, *23(5)*, 566.
139. Bodor, N.; Woods, R.; Raper, C.; Kearney, P.; Kaminski, J. J. *J. Med. Chem.* **1980**, *23(5)*, 474.
140. Farquhar, D. *Preparation and Testing of Masked Deoxynucleotide Drugs*; Farquhar, D., University of Texas System, USA: U.S, pp 10.
141. cLogP was calculated using CLOGP3, v2, 12, BioByte Corp., Pomona, CA.
142. Leo, A. J. *Methods Enzymol* **1991**, *202A*, 544.
143. Pidgeon, C.; Ong, S.; Liu, H.; Qiu, X.; Pidgeon, M.; Dantzig, A. H.; Munroe, J.; Hornback, W. J.; Kasher, J. S.; Glunz, L.; Szczerba, T. *J. Med. Chem.* **1995**, *38(4)*, 590.

PEPTIDOMIMETIC INHIBITORS OF FARNESYLTRANSFERASE:

AN APPROACH TO NEW ANTITUMOR AGENTS

Yimin Qian, Saïd M. Sebti, and Andrew D. Hamilton

Advances in Amino Acid Mimetics and Peptidomimetics
Volume 1, pages 165-192

ISBN: 0-7623-0200-3

ABSTRACT

Ras is a small, 21-kDa GTPase that plays a critical role in signal transduction by mediating the transmission of signals from external growth factors to the cell nucleus. Mutated forms of Ras are found in 30% of human cancers with high occurrence in colon and pancreatic carcinomas. The structural result of these mutations is to destroy the ability of Ras to hydrolize GTP. As a result the protein is locked in its active, GTP bound form, leading to constitutive activation of the signaling pathways and uncontrolled cell growth. In order to function in signaling, Ras must associate with the plasma membrane. A key step in this translocation is the posttranslational farnesylation of a cysteine residue near the carboxyl-terminal of Ras. The enzyme Ras farnesyltransferase (FTase) recognizes and farnesylates the cysteine residue of the C-terminal CA_1A_2X sequence, present in all Ras proteins. Because of its critical role in processing Ras for its signaling function, this enzyme has become an important target for the design of inhibitors that act as antitumor agents. We, and others, have designed a family of peptidomimics that reproduce features of the carboxyl-terminal tetrapeptide CA_1A_2X sequence. These compounds have been shown to be highly potent inhibitors of FTase and some are capable of selectively blocking Ras processing in a range of Ras-transformed tumor cell lines. Furthermore, certain CA_1A_2X peptidomimetics will block tumor growth in various mouse models with little sign of toxic side effects.

1. INTRODUCTION

Cancer in its most basic form is a breakdown in the control mechanisms of cell growth. In normal cells extracellular growth factors interact with membrane-bound receptors and switch on a signal transduction cascade that leads to cell division. In cancerous tissues these signaling mechanisms are disrupted and uncontrolled cell proliferation takes place.

The molecular details of one type of signaling breakdown are becoming clearer as the cancer-causing mechanism of the oncogene, *ras*, unfolds.[1] Three *ras* genes have been identified in mammals as encoding for four small GTP-binding proteins (H-, N-, K_A- and K_B- Ras). These Ras proteins have 21-kDa molecular masses and weak GTPase activity that is at the core of their role as a switch in signal transduction. The currently understood function for Ras in signal transduction (shown schematically in Figure 1) is in mediating the transmission of signals from external growth factors to the cell nucleus. In particular, extracellular growth factors (e.g. platelet-derived growth factor–PDGF) bind to and dimerize a receptor tyrosine kinase in the cell membrane. This dimerization results in auto cross-phosphorylation

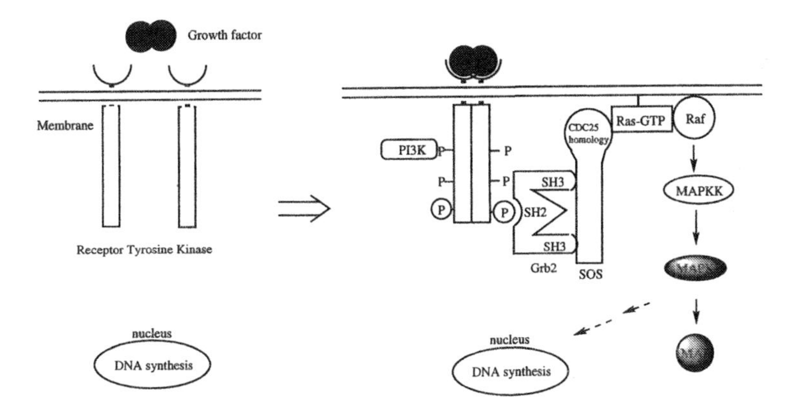

Figure 1. Ras signaling pathway.

of tyrosine residues on the receptor. The resulting phosphotyrosines act as specific binding sites for a number of src homology 2 (SH-2) domain-containing proteins. In the Ras signaling pathway one of the key SH-2-containing proteins is growth factor receptor-binding protein (GRB-2). In addition to an SH-2 domain, GRB-2 contains two SH-3 domains that can bridge to a small protein known as mSOS-1 (mammalian son-of-sevenless-1). The role of mSOS-1 is to bind to Ras and to activate the exchange of GDP for GTP in the active site of the Ras protein. GTP-bound Ras binds to a serine/threonine kinase called Raf which in turn activates a cascade of mitogen-activated protein (MAP) kinases believed to be involved in the stimulation of DNA synthesis and cell division.[2]

Mutated forms of Ras are found in 30% of human cancers with particularly high prevalence in colon and pancreatic carcinomas. Mutated forms of all four Ras proteins (H-, N-, K_A- and K_B-) have been found in tumors, but of these K_B-Ras is the most frequent in human cancers. The most common amino acid mutations are found at positions -12, 13, and 61, all near the nucleotide-binding pocket of the enzyme. These structural changes destroy the GTPase activity of Ras and cause the protein to be locked in its GTP bound form.[3] This activated form of Ras is fully capable of binding to Raf and other effectors, and activating the signaling pathways that leads to uncontrolled tumor growth.

2. FARNESYLTRANSFERASE

The ability of Ras to transduce biological information from receptor tyrosine kinases to the nucleus requires its association with the plasma membrane. This association is accomplished through a series of posttranslational modifications that increase the hydrophobicity of Ras. The protein is rendered more lipophilic by posttranslational farnesylation of a cysteine residue near the carboxyl-terminus. The key residue

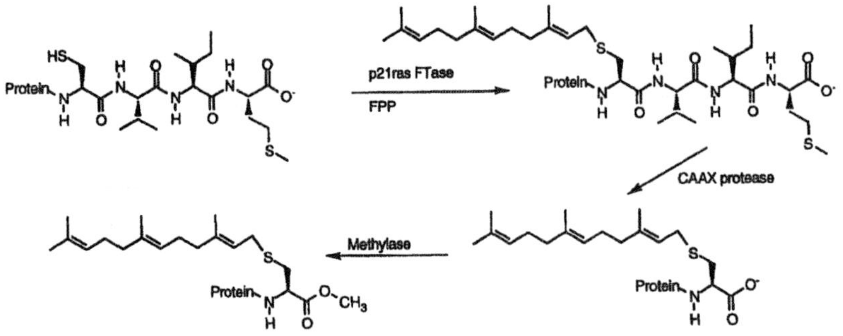

Figure 2. Posttranslation modifications of Ras.

(cysteine 186) is part of the CA_1A_2X carboxyl-terminal tetrapeptide of all Ras proteins, where A_1 and A_2 are aliphatic amino acids and X is methionine or serine. The terminal amino acid of the sequence plays a key role in directing farnesylation, in the case of methionine or serine, or geranylgeranylation in the case of leucine or isoleucine. Cysteine alkylation is then followed by proteolytic cleavage of the AAX tripeptide and lastly methylation of the terminal carboxylic acid before insertion of the protein into the membrane (Figure 2). Of these three posttranslational modifications, farnesylation is the most important. Farnesylation alone is both necessary and sufficient for the membrane-binding and transforming ability of Ras.[4]

The enzyme that recognizes and farnesylates the CA_1A_2M sequence, farnesyltransferase (FTase), is a heterodimer composed of an α-subunit (common to FTase and GGTase-I) and β-subunit that binds both Ras and farnesylpyrophosphate (FPP). The enzyme is a metalloenzyme requiring both Zn^{2+} and Mg^{2+} for activity. Recent data suggest that two forms of this enzyme with different metal requirements are present in several tissues.[5] Little is known about the active site of the enzyme, although some recent transfer NOE studies suggest a turn conformation for the bound peptide substrates.[6]

3. DESIGN OF FTASE INHIBITORS

FTase represents an Achilles' heel in the signal transduction pathway. Disruption of the membrane translocation of mutated (GTP-locked) Ras should inhibit its signaling function and permit the elimination of a molecular switch that had been stuck in the "on" position. Encouraging support for this approach was found with the cholesterol-lowering drug, lovastatin. This inhibitor of HMG-CoA reductase (a key enzyme in both cholesterol and FPP biosynthesis) was shown to block Ras farnesylation and membrane association and to reduce the growth of *ras*-oncogene transformed tumors in nude mice.[7] While high concentrations of lovastatin exert a complex and highly toxic effect on the cell (reducing all isoprenoid levels), these results suggested that more selective inhibitors of Ras farnesylation might be better as antitumor agents.

In 1990 a key observation directed attention firmly towards FTase as a potential target. Goldstein, Brown and coworkers showed[8] that small peptides containing the CA_1A_2X sequence could act as alternative substrates and strong competitive inhibitors of FTase. Even the minimum tetrapeptide, such as CVIM **1** (the carboxyl-terminal sequence of K_BRas), inhibited FTase with an IC_{50} of 200 nM. In certain sequences (e.g. CVFM) potent inhibition ($IC_{50} = 20$ nM) was observed without farnesylation of the cysteine residue. However, while effective at inhibiting purified FTase *in vitro*, no simple tetrapeptide was able to enter cells and be stable enough to disrupt Ras processing. The double edged sword of poor membrane permeability and proteolytic instability disqualified CA_1A_2X peptides as antitumor agents.

Since 1991 the search for *in vivo* active inhibitors of farnesyltransferase has been engaged in earnest.[9] Several approaches have been taken. These include the identification of natural products such as the chaetomellic and zaragozic acids[10] or synthetic long-chain carboxylates[11] that presumably associate with the farnesylpyrophosphate-binding pocket in the enzyme. Others have combined elements of the farnesyl and tetrapeptide substrates in a bi-substrate transition state analogue approach.[12] However, the most active and presently most successful strategy has focused on the design of mimics of the minimal tetrapeptide substrate (CVIM for K_BRas) that is recognized by the β-subunit of FTase. Different peptidomimetic approaches have included the use of pseudopeptides,[13] ethers as amide bond surrogates,[14] benzodiazepine scaffolds,[15] and constrained amino acid derivatives.[16] This review summarizes our efforts in the field with particular emphasis on peptidomimetic design, synthesis, and biological evaluation. Other more comprehensive reviews have recently been published.[17]

4. INHIBITORS FOR FTASE BASED ON SIMPLE DIPEPTIDE MIMETICS

The C-terminal amino acid sequences in Ras proteins contain a CA_1A_2X motif. While tetrapeptides with these sequences are competitive inhibitors of FTase they are ineffective at blocking Ras processing in whole cells due presumably to poor membrane permeability. A systematic study of tetrapeptide sequence requirement for FTase inhibition indicated that methionine is the preferred residue at the C-terminal X position, and cysteine is required for farnesylation. The central A_1A_2 positions were shown to tolerate different aliphatic amino acid residues[8]. When the A_2 position is occupied by an aromatic amino acid, the tetrapeptide binds to the enzyme but cannot be farnesylated.[18]

In seeking to design molecules that can mimic these tetrapeptides and inhibit Ras farnesylation, we conjectured that the central A_1A_2 dipeptide may simply contribute to a hydrophobic interaction and the amide bond may be unimportant in binding to the enzyme. To test this hypothesis, we replaced the central dipeptide Val-Ile in CVIM with a series of non-amino acid spacers. Our hope was that these peptidomimetics would lead not only to effective inhibitors of FTase but also provide us with further information on the conformation of the bound substrate.

The rationale for choosing different spacers is shown in Figure 3. The simplest spacer was based on 5-aminopentanoic acid which separates the terminal cysteine and methionine residues by the same number of atoms as in the original tetrapeptide. However, the flexibility of this molecule suggested that alternative spacers with more rigid backbones would also be needed. One possibility was 3-aminomethylbenzoic acid which retained the same separation of the Cys and Met residues but increased the rigidity of the spacer. This dipeptide mimetic also emulated the well-known use of a *trans* olefin as a peptide amide bond isostere.[19] By using this approach not only are two bond rotations restricted, but also some hydrophobic character of the A_1A_2 region is retained. A further extension of this strategy involved using 4-aminobenzoic acid as a spacer. This group corresponds to a *cis* olefin-amide bond replacement and leads to a further removal of one freely rotating C–C bond.

The advantage of using an aromatic spacer is that the distance between cysteine and methionine can be readily controlled by varying the sites of attachment of the amino acid groups. To study structure–activity relationships, positional isomers of 3-aminomethylbenzoic acid and 4-aminobenzoic acid were also chosen as spacers.

The syntheses of these key compounds are described in Schemes 1, 2, and 3. The amino group in 5-aminopentanoic acid and 3-aminomethylbenzoic acid was introduced by hydrogenation of the corresponding azide which was prepared from nucleophilic substitution of the alkyl halide by sodium azide. Amide bond formation was carried out by using coupling reagents such as dicyclohexylcarbodiimide (DCC) and ethyl dimethylaminopropylcarbodiimide (EDCI) or through mixed anhydrides derived from N-protected amino acid and isobutylchloroformate (IBCF).[20] In the final deprotection step, esters were hydrolyzed by aqueous lithium

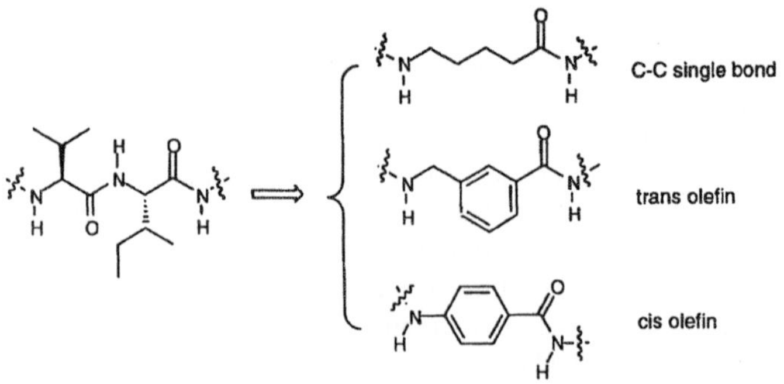

C-C single bond

trans olefin

cis olefin

Figure 3. Replacement of Val-Ile dipeptide with aliphatic and aromatic spacers.

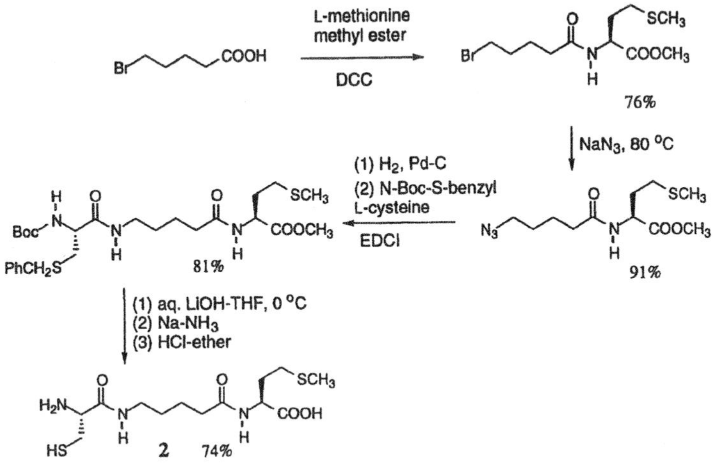

Scheme 1. Preparation of compound **2**.

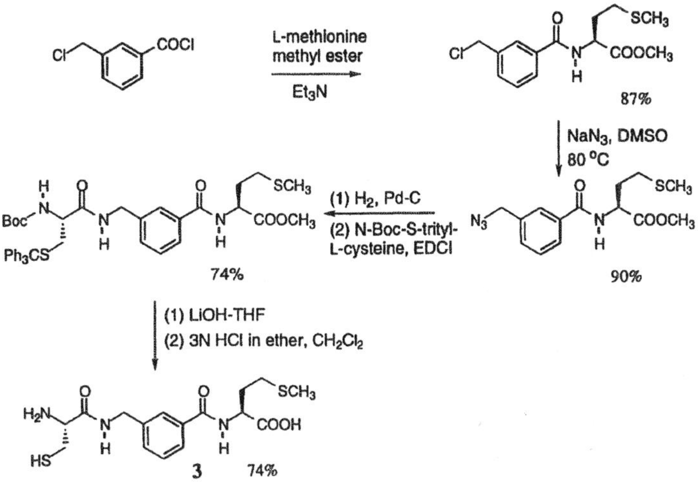

Scheme 2. Preparation of compound **3**.

hydroxide in tetrahydrofuran. The mercapto group in cysteine was either protected as a benzyl or trityl ether. The benzyl thioether was deprotected by sodium in liquid ammonia, while the trityl group was deprotected using hydrogen chloride in ether.

The structures of the first generation of peptidomimetics are shown in Table 1. Inhibition activities of these compounds are indicated by IC_{50} values which correspond to concentrations at which FTase activity is inhibited by 50%. The levels of enzyme activity were determined by the measurement of [^{3}H]-FPP incorporated

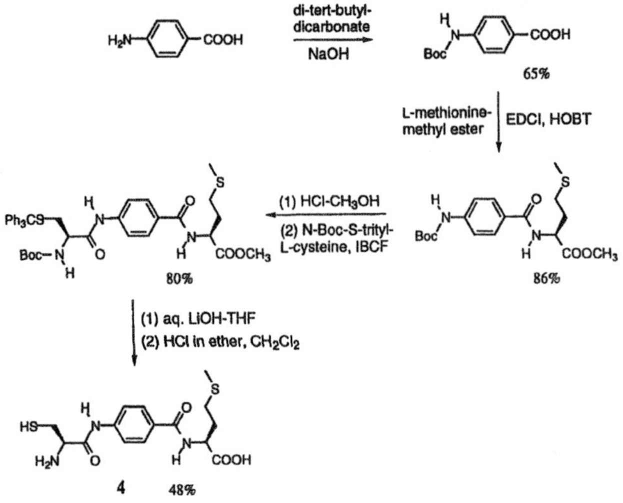

H_2N–C6H4–COOH → (di-tert-butyl-dicarbonate, NaOH) → Boc–NH–C6H4–COOH 65%

L-methionine-methyl ester | EDCI, HOBT

Boc–NH–C6H4–CO–NH–CH($COOCH_3$)–$CH_2CH_2SCH_3$ 86%

(1) HCl-CH_3OH
(2) N-Boc-S-trityl-L-cysteine, IBCF

Ph_3CS / Boc–NH / H / O / $COOCH_3$ 80%

(1) aq. LiOH-THF
(2) HCl in ether, CH_2Cl_2

HS / H_2N / O / H / COOH

4 48%

Scheme 3. Synthesis of peptidomimetic **4**.

Table 1. Structures and Inhibition Activities of Spacer-Based Peptidomimetics

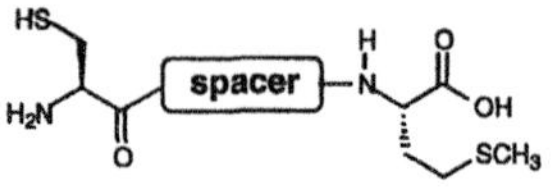

HS / H_2N / O / spacer / H / N / O / OH / SCH_3

Compounds	*Spacer*	*IC_{50} (μM) FTase*	*S-farnesylation*
1	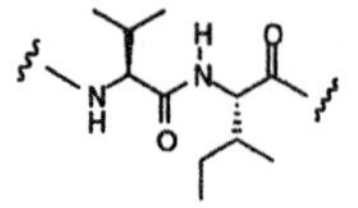 N-H, O, H-N, O	0.34	yes
2	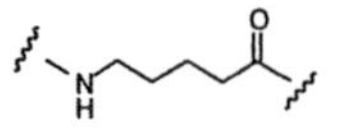 N-H, O	8.0	no
3	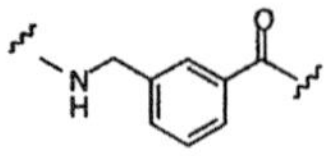 N-H, O	0.38	no
4	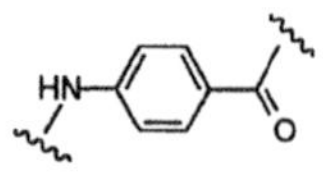 HN, O	0.15	no
5	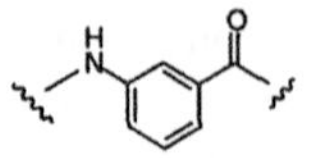 H-N, O	6.5	no
6	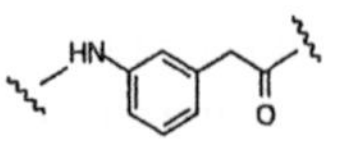 HN, O	7.0	no

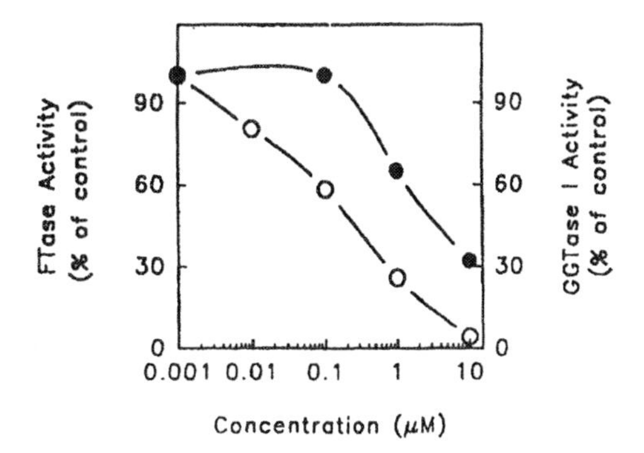

Figure 4. Inhibition of FTase (O) and GGTase-I (•) by peptidomimetic **4** (reproduced with permission).

into H-Ras protein in the presence or absence of inhibitors. The compounds were further characterized for their ability to accept a farnesyl group in the presence of the enzyme.

Table 1 shows that both **3** and **4** have potent inhibition activities against FTase. The inhibition potency of **3** (IC_{50} = 0.38 μM) is similar to that of tetrapeptide **1** (IC_{50} = 0.34 μM), while peptidomimetic **4** is twofold more potent (IC_{50} = 0.15 μM).[21,22] The closely related 3-aminobenzoate isomer **5** is 20-fold less active at inhibiting FTase. A comparison of **6** (IC_{50} = 7.0 μM) with **2** (IC_{50} = 8.0 μM) clearly shows that the simple presence of an aromatic ring (in **6**) does not necessarily improve inhibition. A striking result is seen with **5** which differs from **4** only in the positioning of the amino group. Despite this similarity, **5** is over 40-fold less active than **4**. These results emphasize that correct positioning of cysteine and methionine is critical for potent inhibition activity. Table 1 also shows that unlike tetrapeptide **1** none of the peptidomimetics derived from aromatic or aliphatic spacers is farnesylated by FTase, suggesting that for enzyme turnover more structural features of the substrate are required.

One important question to address in the design of FTase inhibitors is their selectivity for FTase over the closely related enzyme GGTase-I. Peptidomimetic **4** was shown to inhibit FTase and GGTase-I with IC_{50} values of 0.15 μM and 1.5 μM, respectively, corresponding to a 10-fold selectivity (Figure 4).

5. PEPTIDOMIMETICS MIMIC EXTENDED CONFORMATIONS OF TETRAPEPTIDES

A critical question in the design of peptidomimetics concerns the bioactive conformation of the native peptides. In tetrapeptide CVIM, the number of conformations with similar low energies is huge due to many possible bond rotations. Figure 5 shows two extreme conformations in which the flexible tetrapeptide CVIM takes up either an extended or a β-turn structure.

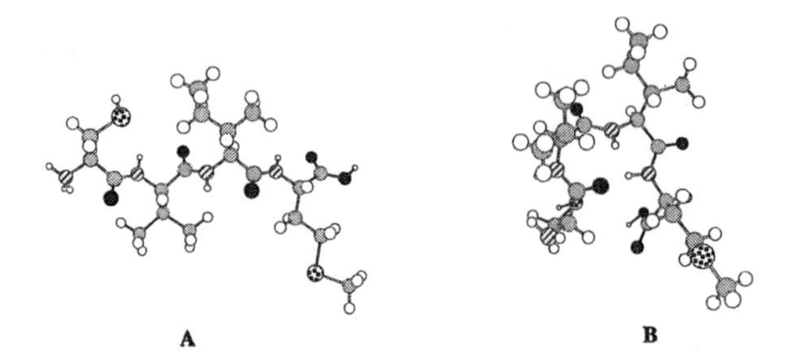

Figure 5. Cys-Val-Ile-Met tetrapeptide in extended (**A**) and β-turn (**B**) conformations.

One hypothesis regarding the Zn^{2+} requirement for Ras binding to FTase is that Zn^{2+} coordinates to carboxylate and thiol groups in the C-terminal tetrapeptide CAAX to form a bidentate complex.[23] This coordination requires that the tetrapeptide adopt a turn structure to bring the carboxylate and thiol groups close together. This turn structure has been observed in heptapeptide KTKCVFM and tetrapeptide CVWM by transferred NOE experiments.[24,25] It is possible that the turn structure is required for FTase-catalyzed farnesyl group transfer. However, the potent inhibition activity of **4** clearly demonstrated that this bidentate coordination is not required for enzyme binding. A β-turn structure to bring carboxylate and thiol groups in a close position is not possible for **4** due to the rigid 4-aminobenzoic acid spacer.

Molecular modeling (using AMBER force field within MacroModel) indicates that the distance between the thiol and carboxylic acid groups in **4** is about 10.5 Å (Figure 6A).[22] Tetrapeptide CVIM in its extended conformation has an analogous distance of 10.8 Å (Figure 5 A). The conformational rigidity of **4** and its inhibition activity suggests that it mimics an extended rather than a turn structure of the tetrapeptide. Furthermore, molecular modeling of compound **5**, a poor inhibitor of FTase, shows that thiol and carboxylate distance is about 6.4 Å, a much shorter dis-

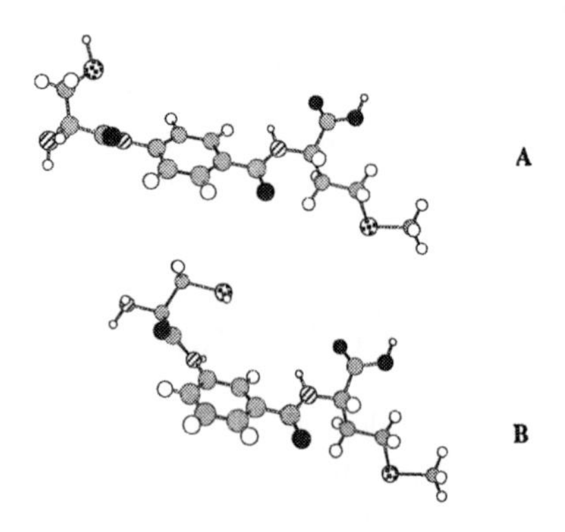

Figure 6. Energy-minimized structures for peptidomimetics **4** (**A**) and 5 (**B**).

tance compared to **4** (Figure 6B). This result suggests a correct distance between cysteine and methionine is important for enzyme binding.

6. PEPTIDE BOND MODIFICATION AND HYDROPHOBIC POCKET PROBING LEAD TO POTENT FTASE INHIBITORS

The success of **4** as a conformationally restricted enzyme inhibitor suggests that the central dipeptide in the CA_1A_2X sequence can be replaced with a structurally defined non-amino acid fragment. Although **4** lacks dipeptide A_1A_2, it still has two amide bonds. On the other hand, the simple aromatic ring in **4** is clearly too small in size when compared with the two side chains from valine and isoleucine. Therefore, two directions were targeted to improve the inhibition potency of **4**. One was a pseudopeptide modification, via the reduction of the cysteine amide bond. Another is modification of the aminobenzoate group to increase the hydrophobicity of the aromatic spacer.

The amide bond linking cysteine and the aromatic spacer was first reduced to a secondary amine which is expected to be stable in the presence of peptidases. Later this secondary amine was methylated to study the importance of possible hydrogen bonding and steric effects. Results from the enzyme inhibition studies indicated that reduction of the cysteine amide bond did not affect enzyme binding significantly, with only a slight reduction in potency being observed (Table 2). This is consistent with the earlier published results which showed that cysteine reduction in tetrapeptide CVFM did not cause any loss in inhibition activity.[26] Furthermore, cysteine amide reduction increased peptidomimetic selectivity for FTase over GGTase-I (Table 2). The selectivity for **7** is 15-fold compared to 10-fold for **4**.

A striking result comes from methylation of the secondary amine. Compound **8** is over 100-fold less active than the parent compound **4**. The loss of activity may be caused by the disrupting of hydrogen bonding to the aniline NH group or may be due to steric effects in the tertiary amine.

Table 2. Peptidomimetics that Lack Cysteine Amide Bond

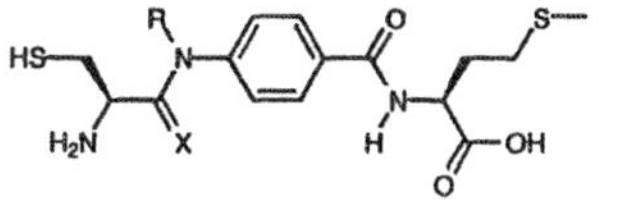

Compounds	*X*	*R*	*FTase* IC_{50} (μM)	*GGTase-I* IC_{50} (μM)
4	O	H	0.15	1.5
7	H, H	H	0.26	4.0
8	H, H	CH_3	20	Not determined

The synthesis of **7** and **8** is shown in Scheme 4. Reductive amination of N,S-protected cysteine aldehyde with 4-aminobenzoylmethionine methyl ester gave the desired peptide isostere.[27] The aminoaldehyde was prepared from the reduction of Weinreb's amide with $LiAlH_4$.[28] To prepare the *N*-methylated pseudopeptide, N-methyl-4-aminobenzoic acid was used as a starting material in the same sequence. Deprotection of the *S*-trityl group was carried out by using TFA in the presence of triethylsilane.[29] Purity of the final compounds was checked by reverse-phase analytical HPLC and structures were confirmed by spectroscopic data. No racemization was observed in the reductive amination step as indicated by HPLC and NMR analysis.

Neither the original tetrapeptide CVIM or zwitterionic **4** showed any activity against whole cells due to their poor membrane permeability. In order to improve the hydrophobicity of the inhibitors we exploited a prodrug strategy in which the free carboxylate was masked as its methyl ester **7b** (Scheme 4). To test the ability of the inhibitors to disrupt cellular processing of Ras, NIH 3T3 cells were first treated with inhibitors at different concentrations and then cells were lysed and immunoprecipitated with a monoclonal antibody Y-13-259 which is specific for Ras protein. The farnesylated and non-farnesylated Ras in the immunoprecipitates were first separated on SDS-PAGE and then transferred to nitrocellulose and immunoblotted with Y-13-259 antibody. The membrane-bound farnesylated (m-p21) and cytosolic non-farnesylated (c-p21) Ras were detected with a peroxidase-conjugated secondary antibody. Results are shown in Figure 7.[30] Cells treated with no inhibitor (Figure 7, lane 2) contained only membrane-bound Ras. As expected, lovastatin (20 μM) treated cells contained both m-p21ras and c-p21ras (Figure 7, lane 1). Compound **3** (lane 3) and **4** (lane 5) in their free carboxylate form were not effective in inhibiting Ras membrane association at concentrations as high as 400 μM. In contrast, the corresponding methyl esters were effective at high concentrations (lane 4, **3** methyl ester, 400 μM; lane 6, **4** methyl ester, 400 μM). The methyl ester of **4** was also effective at 200 μM (lane 7) but not effective at 100 μM (lane 8). Reduction of the amide bond in **4** did not significantly affect *in vitro* activity, but the whole cell activity was greatly improved. Peptidomimetic **7b** was able to inhibit Ras membrane association at 200, 100, 50, and 25 μM (lane 9, 10, 11, and 12). At the concentration of 1 μM, **7b** was not effective (lane 13). Comparison of **4** methyl ester and **7b** clearly suggests that the secondary amine is more stable toward peptidase degradation. It is also possible that polar character of an amide hampers membrane uptake. The low pK_a of the aniline excludes the possibility of protonation on the amino group.

Although compound **7** is as potent as the parent tetrapeptide at inhibiting FTase, structural comparison of CVIM with **7** suggests that the large hydrophobic side chains in valine and isoleucine are not fully reproduced in the 4-aminobenzoic acid spacer. Therefore, increasing the size and hydrophobicity of the 4-aminobenzoic acid spacer is expected to occupy better the FTase substrate-binding pocket. To increase hydrophobicity of the peptidomimetics, the central aromatic ring was substi-

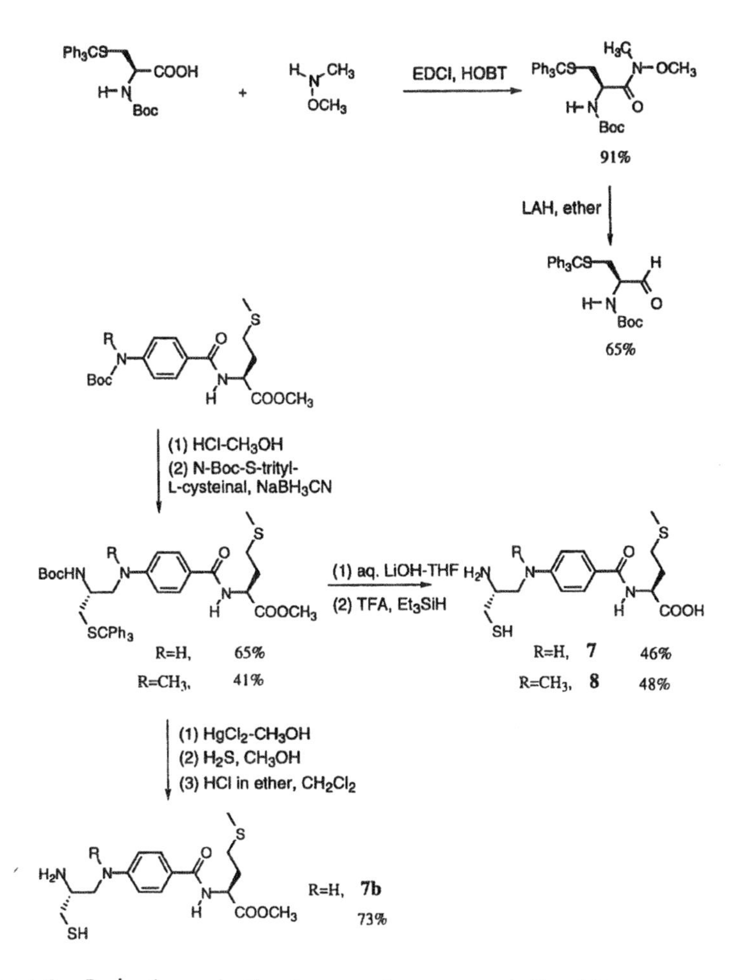

Scheme 4. Reductive amination to generate compounds 7 and **8**.

Figure 7. Reduction of cysteine amide bond in peptidomimetics improves activity of inhibiting Ras processing. (*Lane 1*) lovastatin (20 μM); (*lane 2*) control; (*lane 3*) **3** (400 μM); (*lane 4*) **3** methyl ester (400 μM); (*lane 5*) **4** (400 μM); (*lane 6*) **4** methyl ester (400 μM); (*lane 7*) **4** methyl ester (200 μM); (*lane 8*) **4** methyl ester (100 μM); (*lane 9*) **7b** (200 μM); (*lane 10*) **7b** (100 μM); (*lane 11*) **7b** (50 μM); (*lane 12*) **7b** (25 μM); (*lane 13*) **7b** (1 μM).

tuted by different groups. Two factors were considered in choosing the position and hydrophobicity of the potential substitution groups. First, in the tetrapeptide series the most potent inhibitor is CVFM. This indicates that the side chain benzyl group may have a favorable interaction with the FTase binding pocket. Therefore, different aromatic groups were connected to the spacer. Second, the CVFM sequence suggests a possibly favorable substitution in the aromatic spacer at the 2-position. Moreover, substitution at this position can partially restrict aryl–amide bond rotation. Third, our earlier results showed that when the cysteine amide bond was not reduced, aliphatic substitution at the 3-position of the aromatic spacer was not tolerated, whereas a simple phenyl ring substitution at the 2-position dramatically improved inhibition activity. With this information, phenyl and naphthyl groups were attached to the 2-position of the 4-aminobenzoic acid spacer. Peptidomimetics derived from these improved spacers were assayed and their FTase inhibition activities are listed in Table 3.

Two important conclusions can be drawn from Table 3. First, hydrophobic substitution at the 2-position of the 4-aminobenzoic acid spacer enhances enzyme binding affinity. Compound **9** is 130-fold more potent than parent compound **7**. **FTI-276** has a subnanomolar inhibition potency and is one of the most active FTase inhibitors reported so far. Second, a suitable hydrophobic group in the aromatic spacer can increase the inhibitor's selectivity for FTase over GGTase-I. This selectivity in **7** is only ~ 15-fold, while **FTI-276** has a selectivity of 100-fold. It is also interesting to note that the much larger hydrophobic spacer in **9** does not affect its FTase inhibition potency significantly when compared to a smaller spacer in **FTI-276**. However, the GGTase-I inhibition potency is dramatically increased for this naphthalene containing compound. Although compound **9** is more potent in inhibiting FTase than GGTase-I, its selectivity dropped to fourfold. This may be explained by the presence of a larger hydrophobic-binding pocket in GGTase-I complementing the larger substituent in **9**. The selective inhibition of FTase over GGTase-I is shown in Figure 8. As with the FTase inhibition assay, the GGTase-I inhibition

Table 3. Peptidomimetics Derived from Hydrophobic Substitution and Amide Reduction

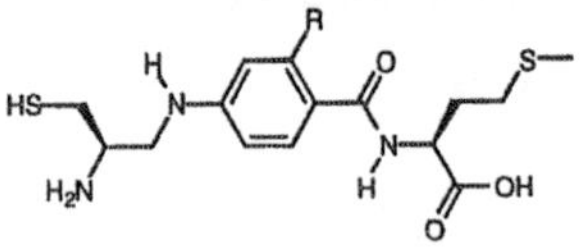

Compound	*R*	*FTase* IC_{50} *(nM)*	*GGTase-I* IC_{50} *(nM)*
7	H	260	4,000
9	1-$C_{10}H_7$	1.9	7.0
FTI-276	C_6H_5	0.5	50

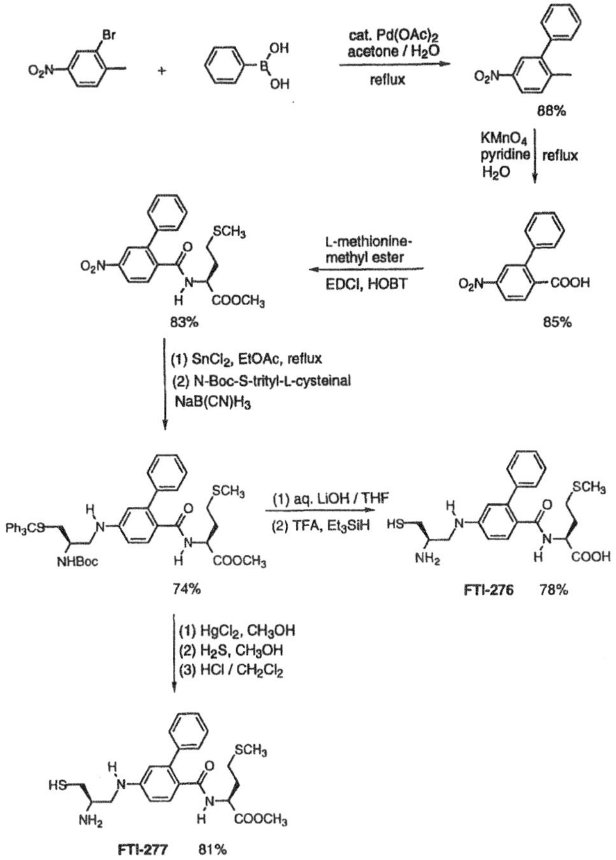

Scheme 5. Synthesis of **FTI-276** and **FTI-277**.

assay was based on the measurement of [^{3}H]-geranylgeranyl group incorporated into H-Ras-CVLL.

Synthesis of **FTI-276** and its methyl ester **FTI-277** is shown in Scheme 5. In the preparation of **FTI-276**, commercially available 2-bromo-4-nitrotoluene was coupled with phenylboronic acid under modified Suzuki coupling conditions.[31] The resulting 2-phenyl-4-nitrotoluene was oxidized to 2-phenyl-4-nitrobenzoic acid by potassium permanganate. Coupling of methionine with this carboxylic acid provided the nitroamide. Reduction of the nitro group by stannous chloride gave the corresponding amine which was coupled with N-Boc-S-trityl-cysteinal under reductive amination conditions to give the fully protected inhibitor.[32] This compound was deprotected, first under basic conditions and then under acid conditions to give **FTI-276**. The final compound was purified by preparative HPLC. **FTI-277**, the methyl ester of **FTI-276**, was also prepared from the same intermediate. The syn-

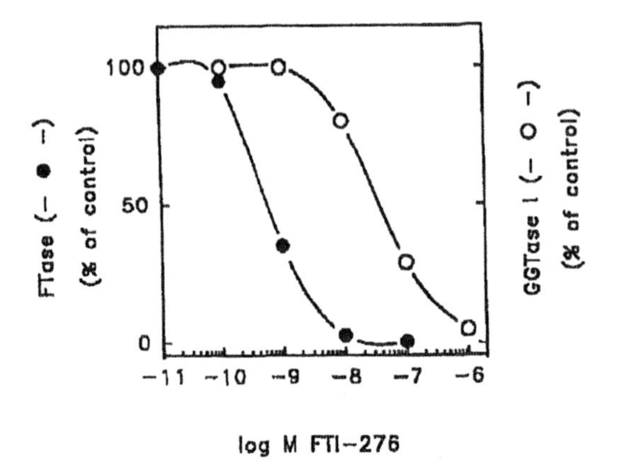

Figure 8. **FTTI-276** selectively inhibits FTase over GGTase-I. Inhibition assays were carried out by determining the ability of **FTI-276** to inhibit the transfer of farnesyl and geranylgeranyl to recombinant H-Ras-CVLS and H-Ras-CVLL, respectively (reproduced with permission)

thesis of **9** involved a modification of this route. Commercially available 2-bromo-4-nitrotoluene was first oxidized to 2-bromo-4-nitrobenzoic acid. This acid was esterified and the ester was coupled with 1-naphthylboronic acid.[33] The coupled product was first hydrolyzed to generate a free carboxylic acid and then coupled with L-methionine methyl ester to form an amide. Reduction of the nitro group followed by reductive amination and final deprotection was carried under the same conditions as for the preparation of **FTI-276**.

7. PEPTIDOMIMETIC FTI-276 SELECTIVELY INHIBITS ONCOGENIC RAS PROCESSING, SIGNALING, AND TUMOR GROWTH

The main reason for generating FTase inhibitors is to inhibit oncogenic Ras farnesylation selectively and efficiently in intact cells. To study Ras-processing inhibition by **FTI-276**, the methyl ester derivative **FTI-277** was prepared. H-RasF cells (NIH 3T3 cells transformed with oncogenic H-Ras-CVLS) were treated with different concentrations of **FTI-277**. Cell lysates were immunoblotted with anti-Ras or anti-Rap1A antibodies. **FTI-277** at concentrations as low as 10 nM inhibited Ras processing but at concentrations as high as 10 μM did not inhibit processing of the geranylgeranylated protein Rap1A. **FTI-277** inhibited Ras processing with an IC_{50} of 100 nM (Figure 9), and therefore represents one of the most potent inhibitors reported to date of Ras farnesylation in whole cells.[34] Interestingly, **FTI-277** was less effective at blocking the processing of NIH 3T3 cells transformed with oncogenic KB-Ras. Recent results suggest that when FTase is inhibited, on-

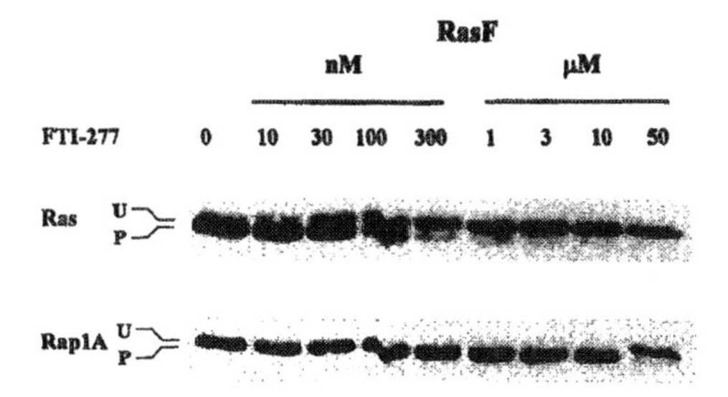

Figure 9. Effects of **FTI-277** on Ras and Rap1A processing.

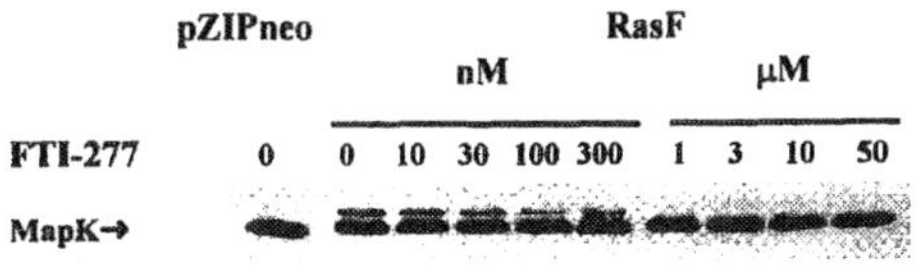

Figure 10. Effects of **FTI-277** on oncogenic Ras activation of MAPK (reproduced with permission).

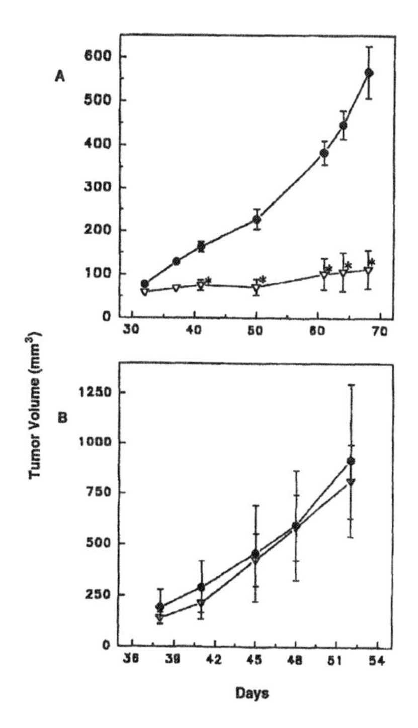

Figure 11. Antitumor effecacy in nude mice of **FTI-276** against human lung carcinomas Calu-1 (A) and NCI-H810 (B). Animals treated with vehicle (*circles*), animals treated with **FTI-276** at 50 mg/kg/day (*triangles*) (reproduced with permission).

cogenic KB-Ras may act as a substrate for GGTase-I and so continue its signaling function.[35]

FTI-277 was further studied for its effects on oncogenic Ras activation of mitogen-activated protein kinase (MAPK), a Ras downstream signaling event (Figure 1). Figure 10 shows that NIH 3T3 cells transfected with pZIPneo (control vector without oncogenic H-Ras) contain only inactive MAPK, but upon transformation with oncogenic H-Ras, MAPK is activated. MAPK activation by oncogenic Ras was inhibited when cells were treated with 300 nM **FTI-277**. Higher concentrations of **FTI-277** (1 μM) completely blocked MAPKactivation (Figure 10).[34]

To evaluate the antitumor efficacy of **FTI-276**, human tumor xenografts in nude mouse models were used. Two human lung carcinoma cell lines were chosen: one (Calu-1) with a K-Ras oncogenic mutation and the other (NCI-H810) with no Ras mutation. Animals were treated with **FTI-276** at the dosage of 50 mg/kg/day.[36] Tumor growth in Calu-1 was dramatically inhibited by **FTI-276** (Figure 11 A), while no inhibition effect was observed in NCI-H810 (Figure 11 B). Interestingly, Calu-1 cells express a mutated Ras, whereas NCI-H810 do not.[36] Although animals were treated once daily for 36 days, no weight loss was observed, and animals appeared to be normal without any evidence of gross toxicity.

8. NON-PEPTIDE CAAX MIMETICS AS POTENT FTASE INHIBITORS

In our first generation of FTase inhibitors, a 3-aminomethylbenzoyl group replaced the central dipeptide Val-Ile. To investigate whether methionine is required for potent inhibition of FTase, a hydrophobic spacer was designed to replace methionine as well as the amide bond (as shown in **10** in Figure 12).

To further study the effect of the cysteine amide bond on inhibition activity, compound **11** was also prepared where the cysteine amide is reduced to a secondary amine (Table 4). Neither **10** nor **11** is active in inhibiting FTase. Since this may be due to the flexibility of the aminomethyl group, more rigid derivatives **12** and **13** (Table 4) were also prepared. The syntheses of these compounds are described in Scheme 6.[37,38]

Although compound **10** and **11** showed very poor activity for FTase inhibition, compound **12** (Table 4) has an IC_{50} of 114 nM, and is a more potent FTase inhibitor

HS H2N N H O O N H SCH3 COOH 3 ⟹ HS H2N N H O COOH 10

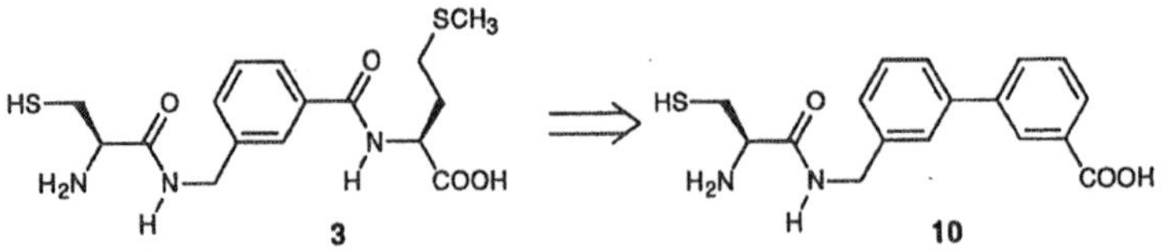

Figure 12. Structural replacement of methionine by an aromatic carboxylic acid.

Table 4. Peptidomimetics Lacking Tripeptide A_1A_2X Inhibit FTase

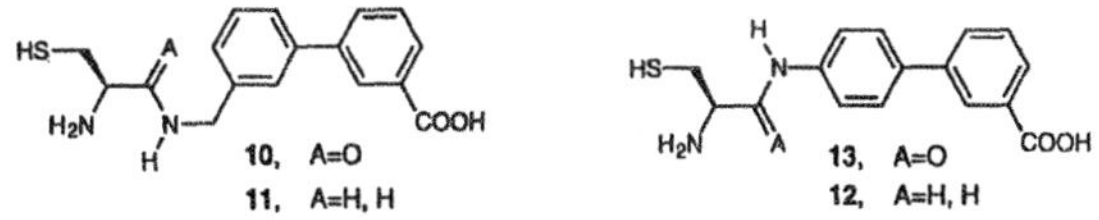

10, A=O
11, A=H, H
13, A=O
12, A=H, H

Compounds	*FTase* *IC_{50} (μM)*	*GGTase-I* *IC_{50} (μM)*	*FTase substrate*
10	100	>100	no
11	11	3	not determined
13	13.5	5100	not determined
12 (FTI-265)	0.11	100	no

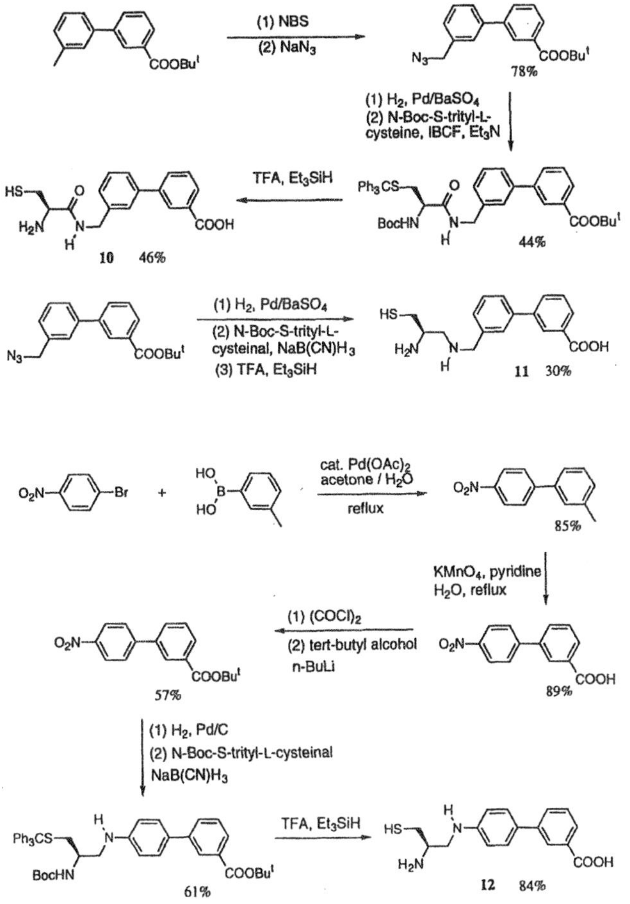

(1) NBS
(2) NaN_3
78%
(1) H_2, Pd/$BaSO_4$
(2) N-Boc-S-trityl-L-cysteine, IBCF, Et_3N
44%
TFA, Et_3SiH
10 46%
(1) H_2, Pd/$BaSO_4$
(2) N-Boc-S-trityl-L-cysteinal, $NaB(CN)H_3$
(3) TFA, Et_3SiH
11 30%
cat. $Pd(OAc)_2$
acetone / H_2O
reflux
85%
$KMnO_4$, pyridine
H_2O, reflux
89%
(1) $(COCl)_2$
(2) tert-butyl alcohol
n-BuLi
57%
(1) H_2, Pd/C
(2) N-Boc-S-trityl-L-cysteinal
$NaB(CN)H_3$
61%
TFA, Et_3SiH
12 84%

Scheme 6. Synthesis of compounds **10**, **11**, and **12**.

than the methionine containing parent compound **7** (Table 2). It is interesting to note that reduction of the cysteine amide bond produces much larger effects in **12** than in the parent compound **7**. The poor activity of compound **13** is probably due to unfavorable conformations of the amide bond form. Reduction of the cysteine amide bond gives a conformationally more flexible molecule which may lead to the accessibility of alternative binding conformations.[38]

The success of **12** as a non-peptide FTase inhibitor raised the question of whether the carboxylic acid substituent was in the optimum position. To answer this, regioisomers of **12** were prepared with methods described in Scheme 6. The inhibition potencies of these isomers are listed in Table 5.

As expected, the meta-substituted carboxylic acid derivative **12** gives the best activity (Table 5). Three important conclusions are revealed from these structure–activity studies. First, the methionine residue is not necessary for potent FTase inhibition. Compound **12** only contains a partial structure from cysteine; all other three amino acids in CVIM are replaced. This result suggests that a key pharmacophore resides in the cysteine part, all other regions can be replaced. In fact, when the thiol group in **12** was changed to an amino group, inhibition activity was totally lost (data not shown). Second, selectivity for FTase over GGTase-I is not necessarily determined by methionine or leucine in the terminal amino acid, as is the case for the parent tetrapeptide CAAX. Whether the tetrapeptide is farnesylated or geranylgeranylated is determined by the last amino acid. Compound **12** lacks this amino acid, but still retains a high selectivity for FTase over GGTase-I. Selectivity is much higher in **12** than in methionine-containing peptidomimetic **7** (Table 2). It is possible that, as substrates, tetrapeptides need a defined amino acid in C-terminus. Table 5 clearly suggests no such requirements are needed for enzyme inhibition. In fact, **12** and other molecules listed in Table 4 and Table 5 are not substrates. Third, a correctly positioned carboxylic acid is required for potent FTase inhibition. Compounds **14** and **15** only differed in their carboxylic acid position from **12**, but showed dramatically different activity in enzyme inhibition. Non-peptide inhibitor **12** is over 40-fold more potent than its isomer **15**, while a smaller difference was observed between **12** and compound **14** (five-fold, Table 5).

Table 5. Carboxylic Acid Position Affects FTase Inhibition

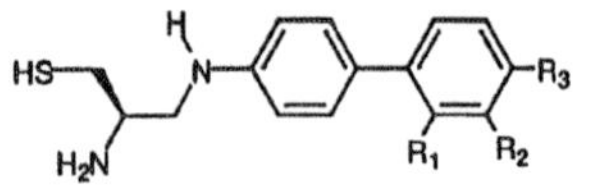

HS, H, N, H_2N, R_1, R_2, R_3

R_2=COOH, R_1=R_3=H, **12 (FTI-265)**
R_1=COOH, R_2=R_3=H, **14**
R_3=COOH, R_1=R_2=H, **15**

Compound	*FTase IC_{50} (μM)*	*GGTase-I IC_{50} (μM)*	*FTase Substrate*
12	0.11	100	no
14	0.54	140	Not determined
15	4.6	>100	no

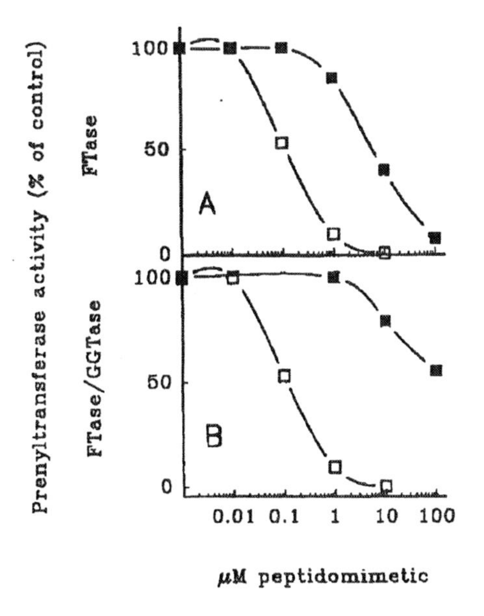

Figure 13. FTase and GGTase-I inhibition studies. (**A**) FTase inhibition by **12** (ÿ) and **15** (þ). (**B**) GGTase-I (þ) and FTase (ÿ) inhibition by nonpepetide inhibitor **12** (reproduced with permission).

Comparison of FTase and GGTase-I inhibition by compounds **12** and **15** is shown in Figure 13. At a concentration of 1 μM, FTase activity was totally inhibited by **12** ,while almost no effect was seen on GGTase-I. This compound was further shown to inhibit Ras processing in whole cells at a concentration of 100 μM without the requirement of masking the carboxylic acid.[37]

An earlier proposal suggested that potent inhibitory activity toward FTase might require inhibitors to take up a β-turn conformation bringing the cysteine thiol and free carboxylate in close proximity to form a bidentate complex with Zn^{2+}.[23] Structurally restricted biphenyl derivative **12** cannot take this turn conformation (Figure 14). An energy-minimized conformation of **12** by using the AMBER force field

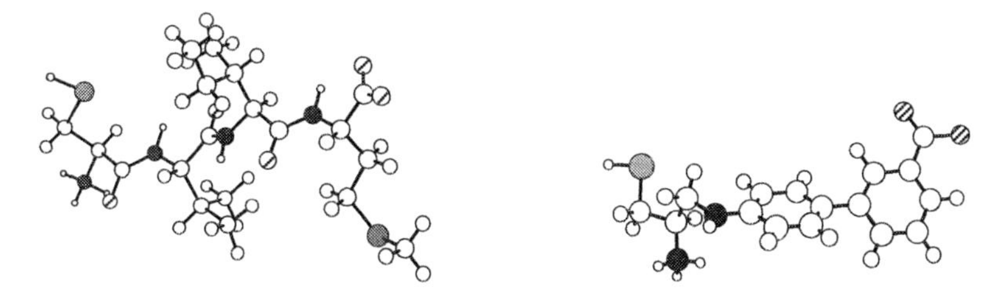

Figure 14. Local energy-minimized structures of CVIM in an extended conformation and non-peptide **12**.

within the MacroModel program in the absence of solvent showed it has a similar thiol–carboxylate distance when compared to the extended conformation of CVIM (SH to COOH distance is about 10.5 Å). This result confirms that potent inhibition of FTase does not require a β-turn conformation. The conformation of CVIM shown in Figure 14 is not the lowest energy conformation. When global energy minimization is applied, CVIM showed a β-turn conformation.

9. MODIFICATIONS IN THE HYDROPHOBIC CHARACTER AND CARBOXYLIC ACID REPLACEMENT

In earlier studies, we had shown that methyl esterification of the methionine residue decreased FTase inhibition potency *in vitro*. Since methionine is removed in **12**, it is important to know whether the carboxylic acid group in **12** can be replaced by a non-carboxylic acid fragment. For this purpose, different groups were connected to the 3'-position to replace the carboxylic acid. Compounds prepared in this series are listed in Table 6.

The syntheses of compounds **16**, **17**, and **18** (Table 6) were analogous to those shown in Scheme 6. Compound **19** was prepared in several steps. Coupling of 1-bromo-2-methoxy-4-nitrobenzene with 3-methylphenylboronic acid followed by oxidation and then *tert*-butyl ester formation gave 2-methoxy-4-nitro-3'-(*tert*-butoxycarbonyl)biphenyl. This nitro compound was reduced and then coupled with N-Boc-S-tritylcysteinal to give the fully protected intermediate which was deprotected to give **19**. Preparation of compound **20** is shown in Scheme 7. Deprotection of the methoxy group under acidic conditions generated a substituted phenol which was alkylated to give the propanoxy nitro derivative. Reduction of the nitro group followed by reductive amination and deprotection gave **20** (Scheme 7).

Table 6. Inhibiton of FTase and GGTase-I by Non-peptide Molecules

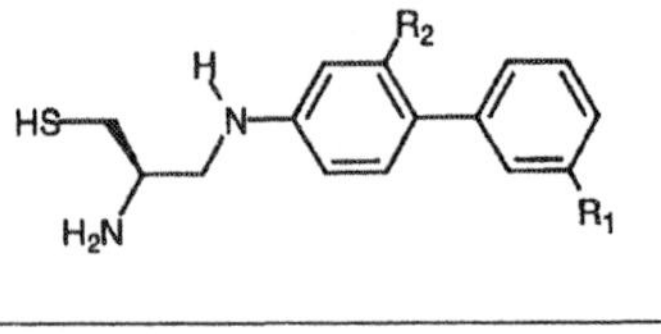

Compound	R_1	R_2	*FTase* IC_{50} *(nM)*	*GGTase-I* IC_{50} *(nM)*
12	COOH	H	114	100,000
16	$COOCH_3$	H	917	>100,000
17	CH_3	H	710	>100.000
18	H	H	1,070	>100,000
19	COOH	OCH_3	37	35,000
20	COOH	$O(CH_2)_2CH_3$	30	2,400

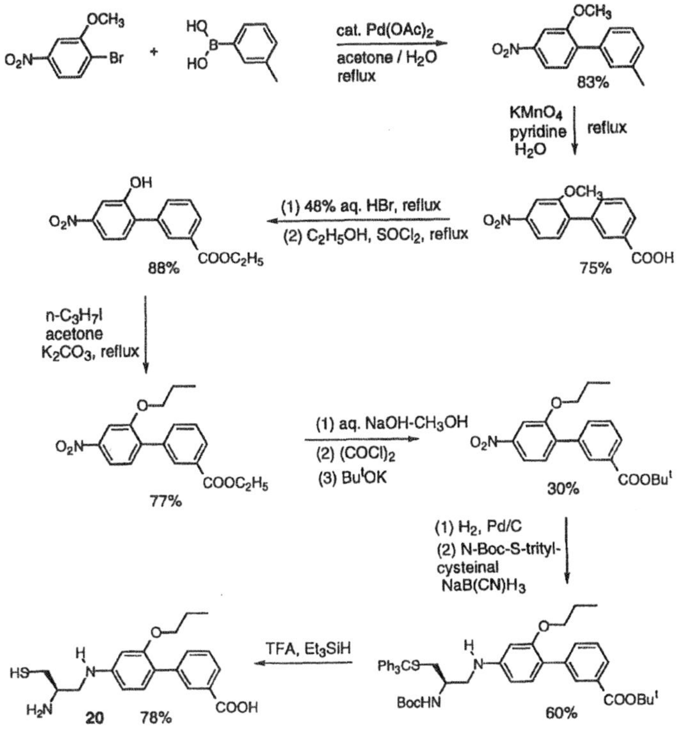

Scheme 7. Synthesis of compound **20**.

The published sequence-dependence studies of tetrapeptides as FTase inhibitors have shown that the most potent inhibitor in the CA_1A_2X series is Cys-Ile-Phe-Met, with an IC_{50} value of 30 nM.[39] Non-peptide molecules **19** and **20** (Table 6) are as potent as CIFM despite the large difference between their structures. These results confirm the success of the hydrophobic strategy in the design of non-peptide FTase inhibitors.

As seen in Table 6, a free carboxylic acid at the 3'-position of biphenyl is important for biological activity. Methylation, deletion or replacement with a methyl group all result in 5- to 10-fold loss of activity (compounds **16**, **17**, and **18** in Table 6).[38] Poor inhibition activity of compound **15** when compared with compound **18** (fourfold difference, Table 5 and Table 6) suggested that the 4'-position of the aromatic ring is close to a hydrophobic region in the enzyme.

Substitution of hydrogen by methoxy or propanoxy at the 2-position of the C-terminal aromatic ring all improved inhibition potency by three- to fourfold (compounds **19** and **20**, Table 6). This suggested there might be a hydrophobic interaction between the enzyme active site and the substituents at the 2-position of the inhibitors. This is reasonable since both valine and isoleucine in CVIM have large and hy-

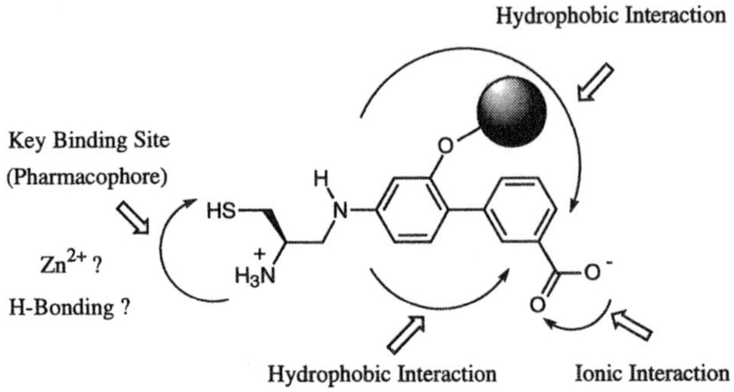

Figure 15. Proposed inhibition mechanism between non-peptide inhibitor and FTase active site.

drophobic side chains. It is also possible that the increased potency is due to the restricted flexibility of the biphenyl.

One interesting result from Table 6 comes from the inhibitors' selectivity for FTase over GGTase-I. Compounds **19** and **20** have similar affinity for FTase. However the difference in their affinity for GGTase-I is dramatic. Larger hydrophobic groups at the 2-position increased GGTase-I inhibition more than FTase inhibition (compare **19** and **20**). This result suggested GGTase-I might have a larger hydrophobic pocket than FTase.

Based on the rigidity of the biphenyl group and the inhibition activity dependence on thiol and carboxylic acid groups in non-peptide inhibitors, a model for the enzyme active site interaction is proposed as shown in Figure 15. One of the biggest uncertainties is the interaction mode at the mercaptan region.

10. KINETICS OF FARNESYLTRANSFERASE INHIBITORS

FTase utilizes two substrates, Ras protein and farnesylpyrophosphate. FTase can bind to either substrate and the binding complex of FTase-FPP can be isolated by gel filtration.[40] Kinetic analysis has demonstrated that binding of FTase to FPP to form a FTase–FPP complex is the first step in the catalysis and that the rate-limiting step is product release.[41] Since the binding of FTase to FPP, forming FTase-FPP, is a fast equilibrium, the two substrate reactions can be considered a pesudo-one substrate reaction under conditions where the FPP concentration is close to enzyme saturation.

In a kinetic study of FTase-catalyzed Ras farnesylation, the reaction rate was measured based on the amount of incorporated [^{3}H]-farnesyl group into Ras protein. Ras protein was first precipitated from the reaction mixture under acidic conditions and then filtered through fiber filter paper. The amount of farnesylated Ras was calculated from liquid scintillation counting. Negative controls were per-

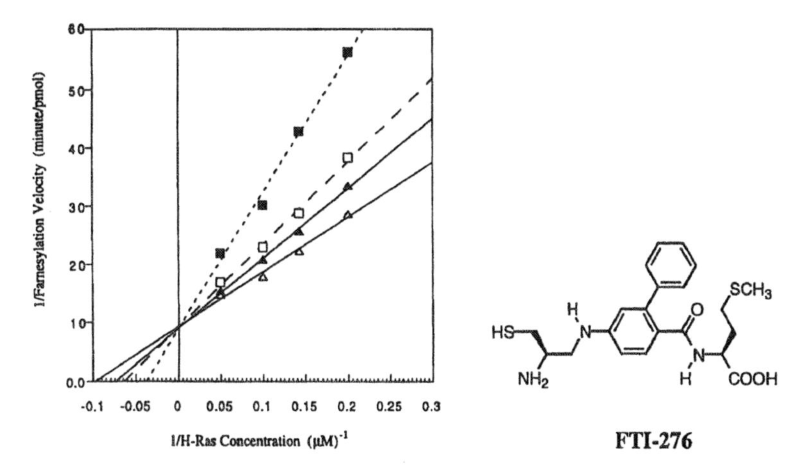

Figure 16. Lineweaver-Burk double reciprocal plot of (1/velocity) versus (1/H-Ras concentration) at 0.0 nM (Δ) 200 pM (▲), 400 pM (ÿ), and 1 nM (þ) of **FTI-276. FTI-276** is competitive to H-Ras, $K_i = 0.74$ nM.

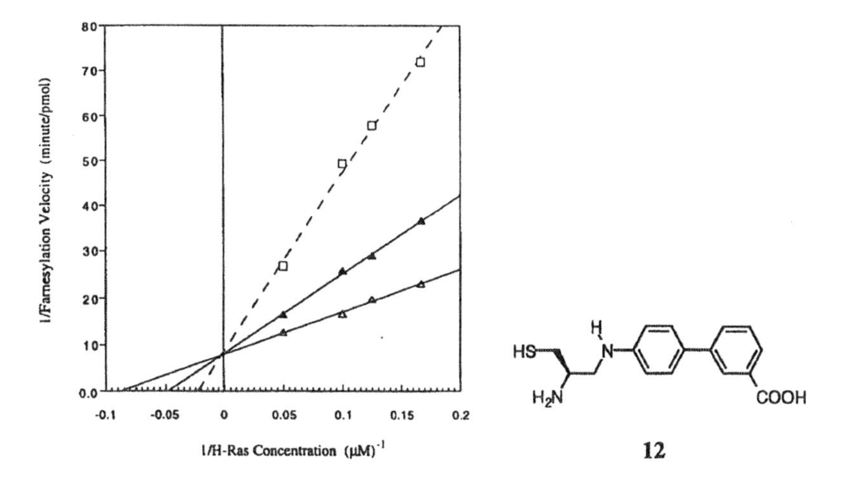

Figure 17. Lineweaver-Burk double reciprocal plot of (1/velocity) versus (1/H-Ras concentration) at 0.0 nM (Δ) 25 M (▲) and 100 nM (□) of non-peptide FTase inhibitor **12**. Compound **12** is competitive to H-Ras, $K_i = 30$ nM.

formed in the absence of one of the three components, FPP, Ras, or FTase. For the inhibition study, a positive control was performed in the absence of inhibitors. The partially purified FTase from a human Burkitt lymphoma cell line (Daudi) was used and FPP concentration was kept at 600 nM. The initial velocity was measured when less than 10% of FPP was incorporated into the H-Ras protein.

Two compounds were characterized to illucidate their inhibition mechanisms. **FTI-276**, one of the most potent FTase inhibitors reported so far, was shown to be competitive to H-Ras protein. The inhibition constant (K_i) was 0.74 nM based on Lineweaver–Burk double reciprocal plots (Figure 16). Compound **12** is an important target for kinetic investigation as it is non-peptidic and, containing a hydrophobic core linked to an anionic carboxylate substituent, could conceivably interact with the FPP-binding region. Figure 17 shows that this compound is also competitive to H-Ras protein and the measured K_i was 30 nM.

11. CONCLUSIONS

In this review we have tried to demonstrate that the development of inhibitors for farnesyltransferase represents an exciting and very active field of bioorganic chemistry. Outstanding progress has been made particularly in the design of molecules that mimic the four terminal amino acids of Ras. These peptidomimetics reproduce key features of the CAAX sequence and are effective at blocking the interaction of Ras with FTase at subnanomolar concentrations. Most importantly, these compounds show high selectivity for FTase over the closely related GGTase-1 and are able to block Ras processing in a range of Ras-transformed tumor cell lines at concentrations as low as 100 nM. Further progress has been made in demonstrating the efficacy of these peptidomimetics in animal models including their ability to block tumor growth in nude mice.

ACKNOWLEDGMENTS

We wish to thank a bright and enthusiastic group of students and research associates whose efforts have made the above work possible. In particular, Churl Min Seong, Renae Fossum, Anil Vasudevan and David Knowles in Chemistry and Andreas Vogt, Edwina Lerner, Jiazhi Sun, Terence McGuire and Michelle Blaskovich in Pharmacology have all made critical contributions. We also thank our senior collaborators, Professors Channing Der and Adrienne Cox (University of North Carolina) for their participation and support. This work was supported by the National Cancer Institute (CA67771).

NOTE ADDED IN PROOF

A crystal structure of farnesyltransferase has recently been published.[42]

REFERENCES

1. Barbacid, M. *Ann. Rev. Biochem.* **1987**, *56*, 779.
2. McCormick, F. *Nature* **1993**, *363*, 15.
3. Milburn, M. V.; Tong, L., deVos, A. M.; Brünger, A.; Yamaizumi, Z.; Nishimura, S.; Kim, S. H. *Science (Washington, DC)* **1990**, *247*, 939.

4. Jackson, J. H.; Cochrane, C. G.; Bourne, J. R.; Solski, P. A.; Buss, J. E.; Der, C. J. *Proc. Natl. Acad. Sci. USA* **1990**, *87*, 3042.
5. Vogt, A.; Sun, J.; Qian, Y.; Tan-Chiu, E.; Hamilton, A. D.; Sebti, S. M. *Biochemistry* **1995**, *34*, 12398.
6. Koblan, K. S.; Culberson, J. C.; Desolms, S. J.; Giuliani, E. A.; Mosser, S. D.; Omer, C. A.; Pitzenberger, S. M.; Bogusky, M. J. *Protein Sci.* **1995**, *4*, 681.
7. Sebti, S. M.; Tkalcevic, G. T.; Jani, J. P. *Cancer Commun.* **1991**, *3*, 141.
8. Reiss, Y.; Goldstein, J. L.; Seabra, M. C.; Casey, P. J.; Brown, M. S. *Cell* **1990**, *62*, 81. Reiss, Y.; Stradley, S. J.; Gierasch, L. M.; Brown, M. S.; Goldstein, J. L. *Proc. Natl. Acad. Sci. USA* **1991**, *88*, 732.
9. Sebti, S. M.; Hamilton, A. D. *Pharmacol. Ther.* **1997**, *74*, 103. Bishop, W. R.; Bond, R.; Petrin, J.; Wang, L.; Patton, R.; Doll, R.; Njoroge, G.; Catino, J.; Schwartz, J.; Windsor, W.; Syto, R.; Schwartz, J.; Carr, D.; James, L.; Kirschmeier, P. *J. Biol. Chem.* **1996**, *270*, 30611.
10. See for example, Gibbs, J. B.; Pompliano, D. L.; Mosser, S. D.; Rands, E.; Lingham, R. B.; Singh, S. B.; Scolnick, E. M.; Kohl, N. E.; Oliff, A. *J. Biol. Chem.* **1993**, *268*, 7617.
11. Marciano, D.; Baruch, G.; Marom, M.; Egozi, Y.; Haklai, R.; Kloog, Y. *J. Med. Chem.* **1995**, *38*, 1267.
12. Patel, D. V.; Gordon, E. M.; Schmidt, R. J.; Weller; H. N.; Young, M. G.; Zahler, R.; Barbacid, M.; Carboni, J. M.; Gullo-Brown, J. L.; Hunihan, L.; Ricca, C.; Robinson, S.; Seizinger, B. R.; Tuomari, A. V.; Manne, V. *J. Med. Chem.* **1995**, *38*, 435.
13. Graham, S. L.; Desolms, S. J.; Giuliani, E. A.; Kohl, N. E.; Mosser, S. D.; Oliff, A. I.; Pompliano, D. L.; Rands, E.; Breslin, M. J.; Deana, A. A.; Garsky, V. M.; Scholtz, T. H.; Gibbs, J. B.; Smith. R. L. *J. Med. Chem.* **1994**, *37*, 725. Garcia, A. M.; Rowell, C.; Ackermann, K.; Kowalczyk, J. J.; Lewis, M. D. *J. Biol. Chem.* **1993**, *268*, 18415.
14. Kohl, N. E.; Wilson, F. R.; Mosser, S. D.; Giuliani, E.; Desolms, S. J.; Conner, M. W.; Anthony, N. J.; Holtz, W. J.; Gomez, R. P.; Lee, T. J.; Smith, R. L.; Graham, S. L.; Hartman, G. D.; Gibbs, J. B. *Proc. Natl. Acad. Sci. USA* **1994**, *91*, 9141.
15. James, G. L.; Goldstein, J. L.; Brown, M. S.; Rawson, T. E.; Somers, T. C.; McDowell, R. S.; Crowley, C. W.; Lucas, B. K.; Levinson, A. D.; Marsters, J. C. *Science,* **1993**, *260*, 1937.
16. Clerc, F-F.; Guitton, J-D.; Fromage, N.; Lelievre, Y.; Duchesne, M.; Tocque, B.; James-Surcouf, E.; Commercon, A.; Becquart, J. *Biorg. Med. Chem. Lett.* **1995**, *5*, 1779.
17. Gibbs, J. B.; Oliff, A.; Kohl, N. E. *Cell,* **1994**, *77*, 175. Buss, J. E.; Marsters, J. C. *Chem. & Biol.* **1995**, *2*, 787. Qian, Y.; Sebti, S. M.; Hamilton, A. D. *Peptide Sci.* **1997**, *43*, 25-41.
18. Goldstein, J. L.; Brown, M. S.; Stradley, S. J.; Reiss, Y.; Gierasch, L. M. *J. Biol. Chem.* **1991**, *266*, 15575.
19. Stewart, F. *Aust. J. Chem.* **1983**, *36*, 2511.
20. Bodanszky, M.; Bodanszky, A. *The Practice of Peptide Synthesis;* Springer-Verlag, Berlin, 1994.
21. Nigam, M.; Seong, C.-M.; Qian, Y.; Hamilton, A. D.; Sebti, S. M. *J. Biol. Chem.* **1993**, *268*, 20695.
22. Qian, Y.; Blaskovich, M. A.; Saleem, M.; Seong, C. M.; Wathen, S. P.; Hamilton, A. D.; Sebti, S. M. *J. Biol. Chem.* **1994**, *269*, 12410.
23. James, G. L.; Goldstein, J. L.; Brown, M. S.; Rawson, T. E.; Somers, T. C.; McDowell, R. S.; Crowley, C. W.; Lucas, B. K.; Levinson, A. D.; Marsters, J. C. *Science (Washington, DC)* **1993**, *260*, 1937.
24. Stradley, S. J.; Rizo, J.; Gierasch, L. M. *Biochemistry* **1993**, *32*, 12586.
25. Koblan, K. S.; Culberson, J. C.; deSolms, S. J.; Giuliani, E. A.; Mosser, S. D.; Omer, C. A.; Pitzenberger, S. M.; Bogusky, M. J. *Protein Sci.* **1995**, *4*, 681.
26. Garcia, A. M.; Rowell, C.; Ackermann, K.; Kowalczyk, J. J.; Lewis, M. D. *J. Biol. Chem.* **1993**, *268*, 18415.
27. Fincham, C. I.; Higginbottom, M.; Hill, D. R.; Horwell, D. C.; O'Toole, J. C.; Ratcliffe, G. S.; Rees, D. C.; Roberts, E. *J. Med. Chem.* **1992**, *35*, 1472.

28. Fehrentz, J.-A.; Castro, B. *Synthesis* **1983**, 676.
29. Pearson, D. A.; Blanchette, M.; Baker, M. L.; Guindon, C. A. *Tetrahedron Lett.* **1989**, *30*, 2739.
30. Qian, Y.; Blaskovich, M. A.; Seong, C.-M.; Vogt, A.; Hamilton, A. D.; Sebti, S. M. *Bioorg. Med. Chem. Lett.* **1994**, *4*, 2579.
31. Wallow, T. I.; Novak, B. M. *J. Org. Chem.* **1994**, *59*, 5034.
32. Bellamy, F. D.; Ou, K. *Tetrahedron Lett.* **1984**, *25*, 839.
33. Watanabe, T.; Miyaura, N.; Suzuki, A. *Synlett* **1992**, 207.
34. Lerner, E. C.; Qian, Y.; Blaskovich, M. A.; Fossum, R. D.; Vogt, A.; Sun, J.; Cox, A. D.; Der, C. J.; Hamilton, A. D.; Sebti, S. M. *J. Biol. Chem.* **1995**, *270*, 26802.
35. Lerner, E. C.; Qian, Y.; McGuire, T. F.; Hamilton, A. D.; Sebti, S. M. *J. Biol. Chem.* **1995**, *270*, 26770. Whyte, D. B., Kirschmeier, P., Hockenberry, T.N.; Nunez-Oliva, I.; James, L.; Catino, J. J.; Bishop, W. R.; Pai, J. K. *J. Bol. Chem.* **1997**, *272*, 14459. Rowell, C. A.; Kowalczyk, J. J.; Lewis, M. D.; Garcia, A. M. *J. Biol. Chem.* **1997**, *272*, 14093.
36. Sun, J.; Qian, Y.; Hamilton, A. D.; Sebti, S. M. *Cancer Res.* **1995**, *55*, 4243.
37. Vogt, A.; Qian, Y.; Blaskovich, M. A.; Fossum, R. D.; Hamilton, A. D.; Sebti, S. M. *J. Biol. Chem.* **1995**, *270*, 660.
38. Qian, Y.; Vogt, A.; Sebti, S. M.; Hamilton, A. D. *J. Med. Chem.* **1996**, *39*, 217.
39. Graham, S. L.; deSolms, S. J.; Giuliani, E. A.; Kohl, N. E.; Mosser, S. D.; Oliff, A. I.; Pompliano, D. L.; Rands, E.; Breslin, M. J.; Deana, A. A.; Garsky, V. M.; Scholz, T. H.; Gibbs, J. B.; Smith, R. L. *J. Med. Chem.* **1995**, *37*, 725.
40. Reiss, Y.; Brown, M. S.; Goldstein, J. L. *J. Biol. Chem.* **1992**, *267*, 6403.
41. Furfine, E. S.; Leban, J. J.; Landavazo, A.; Moomaw, J. F.; Casey, P. J. *Biochemistry* **1995**, *34*, 6857.
42. Park, H. W.; Bodulari, S. R.; Moomaw, J. F.; Casey, P. J.; Beese, L. S. *Science* **1997**, *275*, 1800.

SYNTHETIC ROUTES TO LACTAM PEPTIDOMIMETICS

Jeffrey Aubé

Advances in Amino Acid Mimetics and Peptidomimetics
Volume 1, pages 193-232

ISBN: 0-7623-0200-3

ABSTRACT

Freidinger lactams are loosely defined as peptidomimetics in which the α-carbon of an amino acid residue in a chain has been cyclized onto the nitrogen of the adjacent residue in the N→C-terminal direction. The attachment may replace either the α-hydrogen of the first residue or the α-side chain, with the latter being most common. Freidinger lactams and their close relatives have been used in a variety of medicinal chemistry applications. Those discussed in this chapter include hormone agonists (such as luteinizing-hormone releasing hormone), angiotensin-converting enzyme and/or neutral endopeptidase antagonists, renin inhibitors, immunosupressants, human growth hormone and dopamine agonists, neurokinin, and thrombin antagonists. The second part of the chapter is concerned with the general synthetic routes used by scientists to prepare the various types of Freidinger lactams, with particular attention being paid to the stereochemical consequences of each route.

1. INTRODUCTION AND OVERVIEW

Some of the most important chemical messengers and their macromolecular targets used in human and animal biology are peptides and proteins. The efforts of medicinal chemists to design new drugs based on these targets have been hampered by some inherent disadvantages of peptides, including poor bioavailability, rapid metabolism, or action at more than one biochemical site. These concerns have been addressed through a number of approaches, including novel formulation or delivery strategies.[1-5] Alternatively, peptides can be thought of as lead compounds for the development of non-peptidic therapeutic agents (i.e. *peptidomimetics*). Numerous reviews have dealt with the general principles of peptidomimetic design, which can entail anything from the chemical replacement of a swatch of a peptide chain to the use of unnatural amino acids, side chain modifications, or, most germane to this review, various kinds of cyclizations.[6,7]

Cyclization strategies can involve the formation of biogenically inspired disulfide or amide bonds, N- to C-backbone connections, or the introduction of artificial rings. Among the latter, Freidinger lactams and their close relatives have proved among the most generally useful in drug design. A Freidinger lactam is informally derived from a peptide in which the α-position of an arbitrarily chosen amino acid residue i_1 has been connected to a downstream amide nitrogen (i_2 residue) through the addition of a carbon bridge (Figure 1). There are three substituents on the i_1 resi-

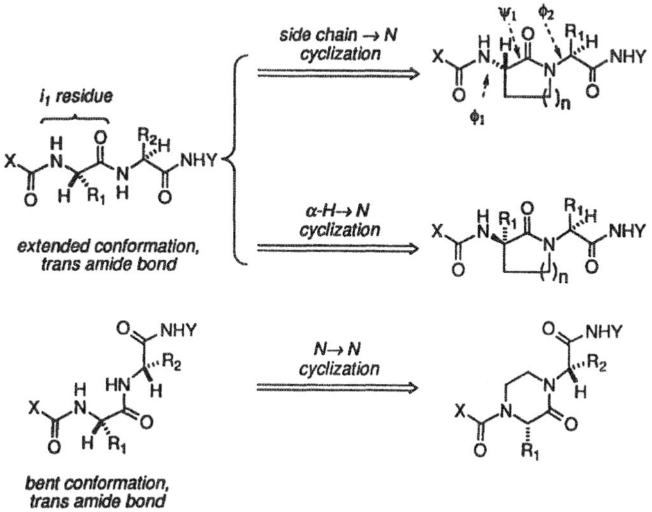

Figure 1. Cyclizations of i_n residues onto the i_{n+1} nitrogen atom, resulting in Freidinger lactams and other peptidomimetics.

due that can be attached in this way. The original Freidinger lactam[8] and the motif most commonly used involves the formation of a ring from the position normally occupied by the i_1 side chain, forming a 3-amino or 3-amido lactam. Alternatively, a C–C bond could be made by replacing the α-hydrogen of this residue with the bridging ring while retaining the normal side chain. Peptidomimetics in which the nitrogen atom of the i_1 residue is attached to the adjacent nitrogen have also been reported.[9] An excellent review of cyclization strategies toward peptide mimicry has been published by Toniolo.[10]

Both types of Freidinger lactams constrain the lactam in several important ways. First, the progenitor peptides are in extended conformations, and placing the amide bond into the lactam format ensures that it will adopt a *trans* configuration with respect to the peptide backbone (this is the same thing as saying the amide bond is *cis* from the point of view of the lactam ring). An example of a lactam in which the amide bond is constrained into a *cis*-backbone conformation is shown in Figure 2a.[11] Second, the ψ_1-bond is also constrained by virtue of its placement within the lactam ring. Although the original Freidinger contribution involved a five-membered ring, the dihedral angle of the ψ_1-bond can be usefully varied by experimenting with different ring sizes, thus extending the utility of this species of peptidomimetic. Finally, the bonds connected to the ring (corresponding to ϕ_1 and ϕ_2) should experience different nonbonded interactions relative to the original extended peptide and may be rotationally restricted as well. A theoretical study of the conformational possibilities of γ-lactam inserts has appeared.[12] All of these features can have a significant effect on the potency of a given lactam toward its target. Further tweak-

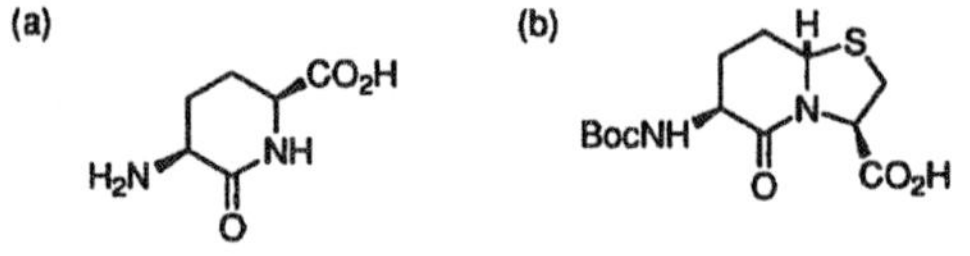

Figure 2. (**a**) A peptidomimetic containing an amide bond constrained into a *cis* geometry. (**b**) An example of a bicyclic lactam commonly used for β-turn mimicry and other types of peptide mimics.

ing is possible by adding substituents on the ring or side chains, inserting heteroatoms into the ring system, or by adding additional cyclic constraints resulting in fused- or spiro-bicyclic or even tricyclic ring systems. By far the most common of these are the fused-bicyclic lactams very commonly used as β-turn mimics (Figure 2b[13]). This important extension of the Freidinger lactam concept is the subject of a very recent review[14] and will not be covered here.

This review is designed to give the reader a sense of the possibilities inherent in using Freidinger lactams in medicinal chemistry, so the emphasis will be on selected examples of the art. The discussion will focus on the structural types presented in Figure 1, but some related compounds will be mentioned so that the reader can appreciate the impact of the Freidinger lactam concept on peptidomimetic design in general. In fact, I would argue strongly against the over-formalization of any given concept in peptidomimetic design. After all, the proof is in the pudding inasmuch as the overall goal is to devise pharmacologically active and metabolically robust new agents, irrespective of structure. In this spirit, the next section will detail different Freidinger lactams that have been tested in some pharmacological context, with an emphasis on the different structural types that have been examined. Interspersed will be some conformational analysis work that has been reported by various groups. Finally, some general synthetic methods that have been developed to provide lactam peptidomimetics will be discussed.

2. STRUCTURES AND BIOLOGICAL EVALUATION OF FREIDINGER LACTAMS

2.1. The Original Contributions: LHRH, Enkephalin, and Sheep Rumen Fermentation

Freidinger and his colleagues at Merck reported the first uses of C_α–N cyclizations for the constraint of peptides in 1980. By far the most influential paper published in this field concerned the design and synthesis of a potent analogue of the hypothalamic hormone luteinizing-hormone-releasing hormone (LHRH, also known as gonadotropin-releasing hormone or GRH) (Figure 3).[8] The impetus for the introduction of a lactam into this hormone was the putative presence of a Tyr-

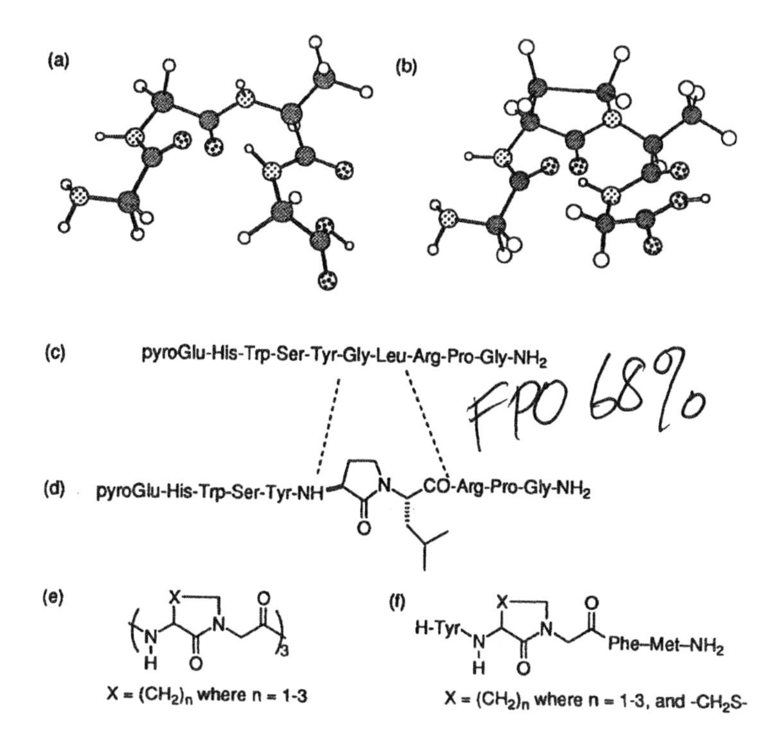

Figure 3. (**a**) An idealized type II' β-turn. (**b**) A five-membered Freidinger lactam showing how this particular bridge can accommodate the type II' β-turn structure shown in (**a**). (**c**) The primary sequence of LHRH and (**d**) an LHRH analogue containing a Freidinger lactam isostere for Gly-Leu. (**e**) Cyclic lactam-containing hexapeptide equivalents used to inhibit rumen fermentation. (**f**) General structure for a series of Met-enkephalin analogues.

Gly-Leu-Arg type II' β-turn in the bioactive conformation of the molecule.[15] It was hypothesized that the idealized type II' turn shown in Figure 3a would be stabilized by the addition of the two-carbon lactam between the pro-*S* α-hydrogen of the glycine unit and the nitrogen of the adjacent leucine because these two bonds are nearly coplanar. A rough depiction of the resulting lactam is given in Figure 3b (these ball-and-stick models were not minimized in any way; in addition, Gly-Gly-Ala-Gly was used instead of Tyr-Gly-Leu-Arg for clarity). The utilization of the pro-*S* position of the glycine residue was especially interesting because replacement of Gly6 with L-Ala decreased potency of the peptide whereas utilization of the D-isomer gave an analogue that was 3.7 times as active as LHRH; this observation had supported the notion of a type II' β-turn at this position in the first place.

The lactam-containing analogue shown in Figure 3d was found to be 8.9 times as potent as LHRH in an *in vitro* pituitary cell culture system. *In vivo*, the lactam-containing analogue was found to be 2.4 times as potent in effecting luteinizing-hormone release in adult ovariectomized female rats. Together, the results suggested that the increased activity was due to enhanced receptor binding thanks to an increased population of the β-turn-containing conformation and not decreased inactivation via proteolytic cleavage (since the *in vitro* system also showed a significant activity enhancement). Paul and coworkers, however, have cautioned against assuming that the introduction of this type of Freidinger lactam will necessarily result in a turn conformation.[12]

The Merck group also investigated the use of these lactams for several other biological targets at about the same time. One example unusual both in the structure of the lactam and in its biological target involved the use of cyclic lactam–Gly trimers for altering rumen fermentation activity in sheep, inspired by *cyclo*(Ala-Sar)$_3$ which has similar activity (Figure 3e).[16,17] These compounds were synthesized with various ring sizes and, in some cases, as mixtures of stereoisomers. Molecular mechanics calculations were carried out on model compounds for each of the three lactam sizes examined, where "slightly puckered", half-chair, and chair conformations were found to be minimum structures for the five-, six-, and seven-membered rings, respectively. However, other conformations were also found within only a few kcal/mol of these minimum structures. NMR data (H–C_α–C_β–H coupling constants) were consistent with these minimum conformations for all ring sizes except for the piperidone ring, in which half-chair and boat conformations were both permitted and were calculated to have very similar strain energies. In the end, only one compound had comparable activity to *cyclo*(Ala-Sar)$_3$ (X = $(CH_2)_2$, *SSS* isomer) in blocking formation of methane in sheep stomach fluid.

Finally, a brief series of compounds was examined for their ability to act as Met-enkephalin analogues (Figure 3f).[17] In this study, only six-membered lactams (X = $(CH_2)_2$ or, significantly, –CH_2S–) had appreciable activity in naloxone-binding or guinea pig ileum assays. In each case, the lactam of D configuration was more active than its L-isomer. However, the absolute potency of the best compounds only approached 10% that of Met-enkephalin, and the series was presumably dropped at this point.

2.2. ACE and NEP Inhibitors

The most in-depth exploration of Freidinger lactams has undoubtedly been for the control of blood pressure by interfering with the renin–angiotensin axis. In attempts to follow up on the enormous commercial success of the angiotensin-converting enzyme (ACE) inhibitors captopril and enalapril, several therapeutic targets in this field were among the most extensively studied biological areas in the 1970s. Besides new ACE inhibitors, these targets included renin inhibitors and, more recently, neutral endopeptidase (NEP) inhibitors and angiotensin II (AII) receptor antagonists.

ACE catalyzes the cleavage of angiotensin I to the potent vasopressor hormone angiotensin II.[18] ACE is a well-established target for the control of blood pressure, and several ACE inhibitors are important in the clinic. In particular, the prototypical ACE inhibitors captopril and enalapril are based on the Ala-Pro sequence present in the angiotensin substrate. The basis for the use of Freidinger lactams as potential ACE inhibitors stemmed from the conformational constraint of the Ala moiety via cyclization (Figure 4). In the simplest Freidinger lactams used for this purpose, the C-terminal constraint is conformationally freed by removing the more "natural" constraint of the proline ring that is still present in capropril and enalapril. This constraint has been restored in bicyclic inhibitors such as cilazaprilat (Figure 4). Detailed discussions of the ACE binding site as elucidated by inhibitor design have been published and are recommended to the interested reader.[19,20]

Thorsett and coworkers at Merck examined a large number of Freidinger lactams for ACE inhibitory activity and also carried out some of the most detailed structural analyses of the class.[21,22] This discussion will focus on those compounds containing a C-3 N-substituent that is most commonly outfitted with a side chain similar to that of enalaprilat. Similar compounds containing other groups at C-3 have also been reported, most notably the sulfhydryl moiety of captopril.[23] A systematic study of ring size afforded the data shown in Table 1.

These data confirm the virtues of examining a variety of ring sizes for the lactam constraint. In this series, medium-sized rings are actually more conducive to enzyme inhibition relative to five- and six-membered lactams, with the best results being obtained in the eight-membered ring series. It is not surprising that the most active compounds have relative and absolute configurations that can be directly mapped onto enalaprilat (*S,S*, in this case). The Merck group carried out detailed computational and, in two cases, X-ray crystallographic studies that focused on the conformation of the ring and particularly on the dihedral angle of the C-2–C-3 car-

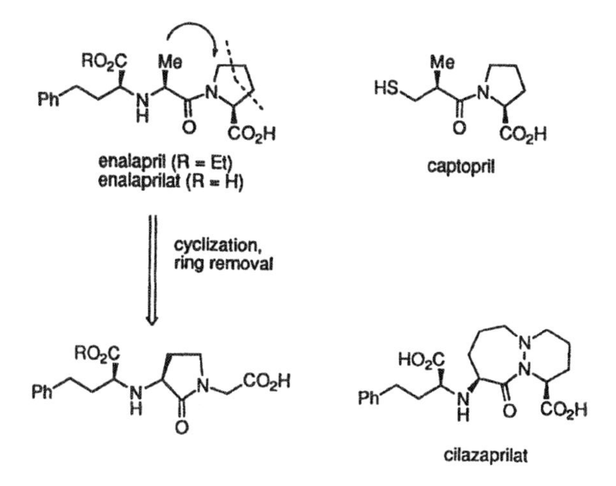

Figure 4. The structure of some important ACE inhibitors showing the formal conversion of enalaprilat to a generalized Freidinger-inspired mimic.

Table 1. Effect of Ring Size and Configuration on ACE-Inhibitory Activity of Some Simple Freidinger Lactams[a]

Isomer A

Isomer B

Ring Size	IC_{50} Values		ψ (calcd, deg)	ψ (X-ray, deg)	Stereochemistry
	Isomer A	*Isomer B*			
5	1.2 x10^{-5}	1.9 x 10^{-5}	-132		racemic
6	4.3 x 10^{-7}	1.7 x 10^{-6}	-138		racemic
7	7.0 x 10^{-7}	1.9 x 10^{-8}	166	166	racemic
8	1.7 x 10^{-6}	4.8 x 10^{-9}	145	159	racemic
8		2.0 x 10^{-9}			*S,S*
8		9.2 x 10^{-8}			*R,R*
9	1.3 x 10^{-6}	8.1 x 10^{-9}	135		racemic
enalaprilat	1.2 X 10^{-9}				

Note: [a] Data from reference 22.

bon–carbon single bond (ψ in Table 1). In this way, they defined a window of about ψ = 130–170° for optimal activity.

Two X-ray structures of Freidinger lactams taken from our own work are shown in Figure 5.[24] In both cases, the N-Boc group appears in a pseudoequatorial position. The key ψ-bond angle cited by Thorsett are shown in the drawings and agree reasonably well with the those obtained for analogous compounds in their work (± ca. 9°, Table 1).[22] Finally, the seven-membered ring adopts a well-defined chair form in the solid state, whereas the homologue sits in a hybrid boat/chair conformation. However, one must always be aware that crystal structures of small molecules do not necessarily represent the reality of solution-phase structures. Moreover, Thorsett et al. point out that their modeling studies yielded multiple low-energy conformations at room temperature for medium-sized rings, with the possibility of *cis*-amide bonds becoming an issue in nine-membered or higher rings.[22]

Such conformational flexibility is hardly unexpected for ≥7-membered rings, however, and the key issue regarding enzyme inhibition is not necessarily lactam conformation but rather how the side chains are displayed to the binding site. For enalaprilat, the key interactions are very roughly shown in Figure 6: three important hydrophobic pockets, several hydrogen bonding sites, and an essential interaction with zinc. The ability of a given inhibitor to occupy this site will depend not only on the number and kind of functional groups present, but also on a compound's ability to encounter a conformation where it can comfortably reach the necessary sites.

Several attempts to improve the potency of lactam-containing ACE inhibitors, then, have involved the addition of hydrophobic substituents onto various positions

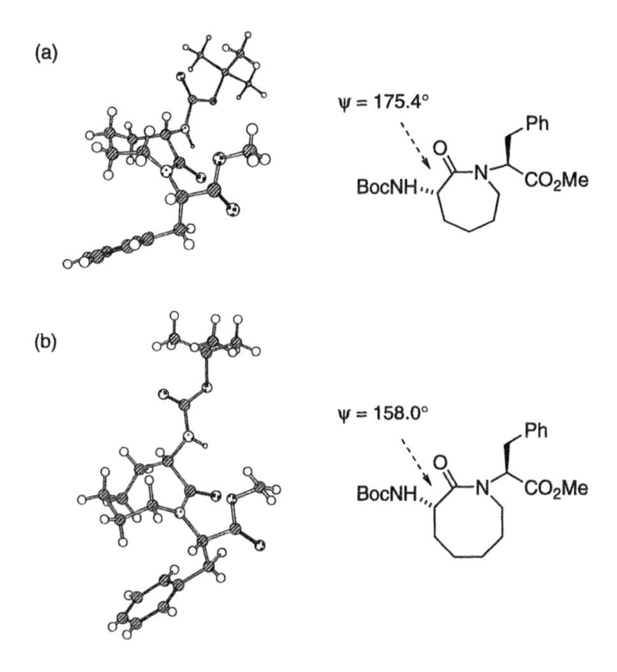

Figure 5. X-ray crystallographic structures of (**a**) seven- and (**b**) eight-membered-ring Freidinger lactam cores.

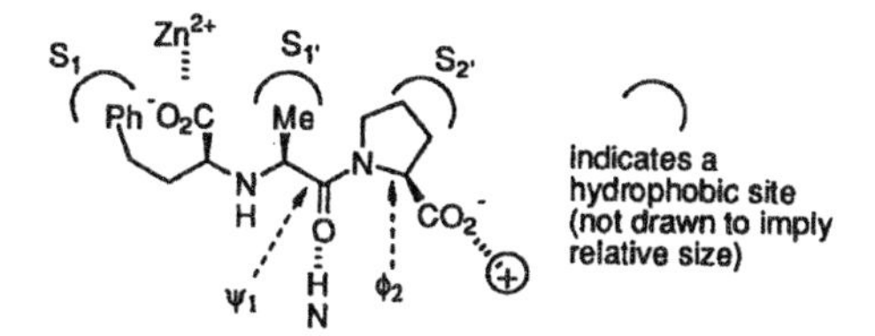

Figure 6. A cartoon depicting the important inhibitor–enzyme interactions between enalaprilat and ACE.

of the ring. In the late 1980s, Yanagisawa and coworkers at the Sankyo company published an extensive series of substituted seven-membered lactam-containing ACE inhibitors of the type shown in Figure 7a[25,26] expanding on an earlier, smaller series published by the Merck group.[21] Both groups found that aryl or alkyl substitution at these positions were consistent with single-digit nanomolar IC_{50} values due to enhancement of hydrophobic interactions with the S2' site of ACE. Expanding on this theme, benzo-fused compounds (Figure 7b, X = CH_2[27] or X = S[28]), the re-introduction of the C-terminal constraint (e.g. Figure 7c,[29] or a combination of both strategies (Figure 7d[30]) were found to result in some extremely potent com-

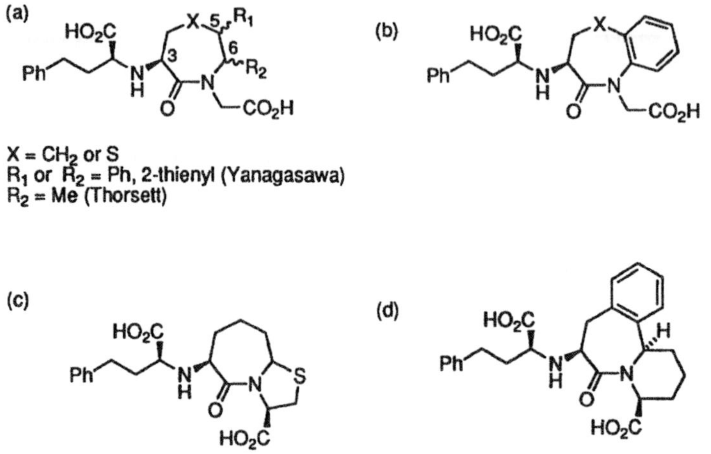

$X = CH_2$ or S
R_1 or R_2 = Ph, 2-thienyl (Yanagasawa)
R_2 = Me (Thorsett)

Figure 7. Representative examples of substituted Freidinger lactams and related compounds that act as ACE inhibitors.

pounds. For example, the compound in Figure 7d was reported to inhibit ACE with a K_i value of 1.2 x 10^{-11} M.[30] Molecular modeling analyses carried out by most of these researchers (especially Yanagisawa[25,26]) revealed low-energy chair conformations for each of the seven-membered lactams with the expected substituents, such as phenyl groups, generally occupying the pseudoequatorial positions.

Some advantages of some bicyclic inhibitors (cilazaprilat in Figure 4; also Figures 7c,d) are that the C-terminal carboxylic acid is constrained to a particular position via cyclization and that a hydrophobic bridge has an opportunity to interact favorably with the S2' binding site.[20] An intermediate approach would simply involve the use of a hydrophobic amino acid at the C-terminal end instead of the glycine used almost universally in the early ACE work described above. A substituent at this position should encounter some steric interaction with the amide carbonyl so, although not rigorously constrained, rotation would be restricted. This point was examined by several groups (Table 2).[21,24] Although it appears that some modest improvement might have occurred by the replacement of glycine with L-alanine, use of the D-amino acid or residues containing larger side chains was not productive (especially see the progression in entries 8–10). There is too little data as of this writing to make generalizations pertaining to the effect of the stereochemistry of the N-substituent on the side-chain conformation. Examination of the X-ray data presented in Figure 5 show similar side-chain conformations in each compound, but the standard caveats regarding possible differences between solid- and solution-state structures are especially apt for such a rotatable bond. Still, there are enough examples of biologically active Freidinger lactams containing a non-glycine C-terminus that effective methods for preparing these compounds are highly sought.

Table 2. Effect of C-Terminal Residue on ACE Inhibition by Some Monocyclic Lactams

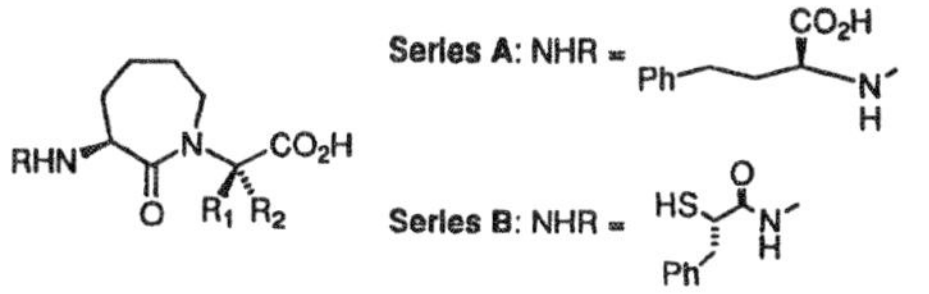

Entry	*Series*	R_1	R_2	IC_{50} *(nM)*	*Reference*
1	A	H	H	19	21
2	A	H	Me	8	21
3		Me	H	57	21
4	A	H	Bn	>200	24
5	A	Bn	H	>200	24
6	A	H	*i*-Bu	165	24
7	A	*i*-Bu	H	>200	24
8	B	H	Me	26	38
9	B	H	Bn	479	38

More recently, attention has been focused on the development of compounds that can simultaneously inhibit both ACE and neutral endopeptidase (NEP). The latter is an enzyme found in the brush border cells of the kidney, where it is responsible for the cleavage of atrial natriuretic peptide (ANP). Since ANP exhibits a variety of effects that lower blood pressure (vasodilation, inhibition of aldosterone formation, sodium excretion/diuresis), the inhibition of NEP becomes a reasonable mechanism for blood pressure control. Fortuitously, both ACE and NEP are zinc metalloproteases and have remarkably similar active sites, raising the hope that one could inhibit both enzymes with a single agent and thus have a two-pronged attack on hypertension. A number of groups have examined Freidinger lactams or their cousins as combined ACE/NEP inhibitors.[20,31-39]

A direct comparison of substituted lactam peptidomimetics was done by Robl and coworkers at Bristol-Myers Squibb; their results and those of some other groups are collected in Figure 8 and Table 3. As was the case for ACE inhibition alone, a number of structural types of these lactams are accommodated in the NEP active site. The series shown in Figures 8b–h[36,37] established that benzo-fused Freidinger lactams worked best with both enzymes, that the position of the aromatic ring and the ring size did have some effect on the potency, and that the (*S*)-configured side chain was preferred. Of course, some types of substitution—including the use of a Leu moiety for the C-terminus of the peptidomimetic—were not handled well at all (Figure 8a[31]), whereas improved results were obtained by resorting to multicyclic lactams of the type shown in Figures 8j[38] and k.[32] This short excerpt does not do justice to the tremendous amount of work done in this field, but fairly

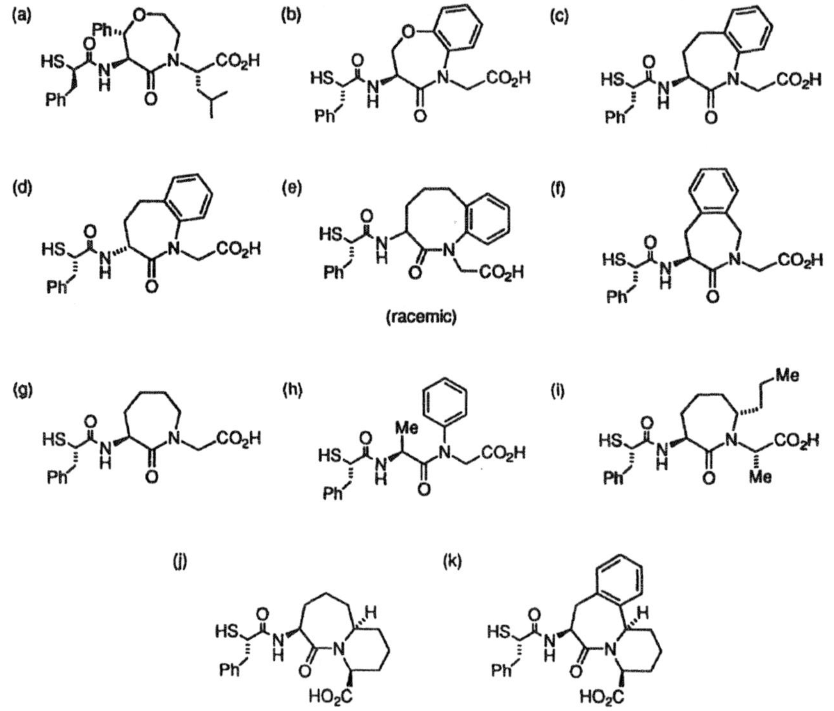

Figure 8. Compounds examined as potential combined ACE/NEP inhibitors.

Table 3. Comparison of Some Combination ACE/NEP Enzyme Inhibitors

	IC_{50} values (nM)		
Structure[a]	*ACE*	*NEP*	*Reference*
a	50,000	330	31
b	28	3.1	36
c	12	6	37
d	306	560	37
e	25	1124	37
f	33	175	37
g	22	82	37
h	14	137	37
i	5	17	38
j	11	25	38
k	0.11	0.18	32

Note: [a] See Figure 8 for structures of inhibitors.

detailed analyses of the binding sites of ACE and NEP, respectively, are given by Bohacek[20] along with a good discussion of the likely conformations of the bioactive combination inhibitors of several classes.

2.3. Other Approaches to Antihypertensives: Renin and Angiotensin II Antagonists

The great success of ACE inhibition for the control of blood pressure spurred scientists to find other macromolecular targets for antihypertensive agents. As is the case with many drug discovery programs, the goals were to find efficacious drugs that might have fewer side effects. Since ACE has biological targets other than angiotensin I (such as bradykinin), renin was identified as a possible alternative target enzyme.[40] Renin hydrolyzes angiotensinogen to afford angiotensin I, and as such is upstream in this vasoconstriction pathway. Alternatively, the ultimate action of ACE is to catalyze the formation of angiotensin II, which binds to blood vessel receptors (leading to vasoconstriction) or adrenal gland receptors (leading to the release of aldosterone, a steroid hormone that causes sodium retention by the kidney).

Despite a tremendous effort in renin inhibitor design in the 1980s, no renin inhibitor has yet reached the market. Still, much has been learned from the basic science perspective of inhibitor design. For example, the fact that both renin and the HIV protease are aspartyl proteases suggests that the strong effort toward renin inhibition may well have primed the medicinal chemistry community for the design of HIV-protease inhibitors some years later. In general, renin inhibitors are pentapeptide-derived transition state inhibitors containing a hydroxymethylene isostere of some type at the active site.

The progression of substrate to inhibitor is depicted for one class of renin inhibitors in Figure 9. The replacement of scissile bond in angiotensinogen with a nonhydrolyzable amino acid insert such as statine or ACHPA is the main event, but Williams et al. have published a rather extensive survey of lactam-containing substrates that attain very high levels of ACE inhibition.[41] Although not "technically" Freidinger lactams, compounds of the general structure shown (Figure 9b) proved effective in this area and demonstrate how lactam constraints can improve potency over strictly peptidyl agents [cf. data for the all-methyl derivative of the lactam in Figure 9b with the unconstrained amide (Figure 9c)]. A similar peptidomimetic insert used in a series of HIV- protease inhibitors published by Merck serves as an example of the renin–HIV protease connection mentioned in the previous paragraph (Figure 9d).[42,43]

However, cyclization is not always profitable, such as when the constraint introduces unfavorable steric interactions with its macromolecular target or when it establishes a conformation unsuitable for binding. An interesting comparison of some renin inhibitors was made by workers at Upjohn and Merck. Given the generic N-terminus of a renin inhibitor shown at the top of Figure 10, three different sites of cyclization were tried. Thaisrivongs et al. investigated a series of inhibitors

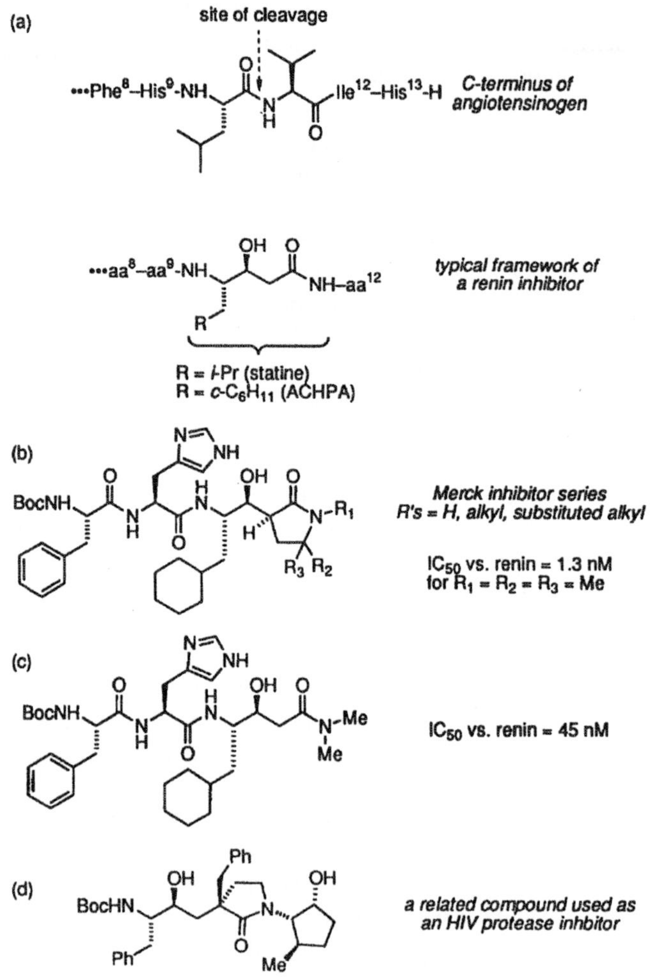

(a)

site of cleavage

•••Phe[8]–His[9]-NH … Ile[12]–His[13]-H

C-terminus of angiotensinogen

•••aa[8]–aa[9]-NH … NH–aa[12]

typical framework of a renin inhibitor

R = *i*-Pr (statine)
R = *c*-C_6H_{11} (ACHPA)

(b)

Merck inhibitor series
R's = H, alkyl, substituted alkyl

IC_{50} vs. renin = 1.3 nM
for $R_1 = R_2 = R_3$ = Me

(c)

IC_{50} vs. renin = 45 nM

(d)

a related compound used as
an HIV protease inhbitor

Figure 9. (**a**) Mapping of the C-terminus of angiotensinogen with several typical renin inhibitors. Renin inhibitors (**b**) with and (**c**) without lactam constraints. (**d**) An HIV protease inhibitor containing a similar lactam-derived transition state analogue.

in which the α-H of the *i* residue was connected to NH_{i+1}, affording a butyrolactam that exhibited a modest increase in potency over its analogue lacking the two-carbon bridge (not shown, $IC_{50} = 26$ nM).[44] In another series of compounds, cyclizations were made to each diastereomeric position on the benzylic carbon (cyclization **b**) and to the ortho position of the side chain phenyl group (cyclization **c**; note the similarity of the resulting compound to some shown in Figures 7 and 8).[45] In the former case, both the stereochemistry of the phenyl substituent on the resulting

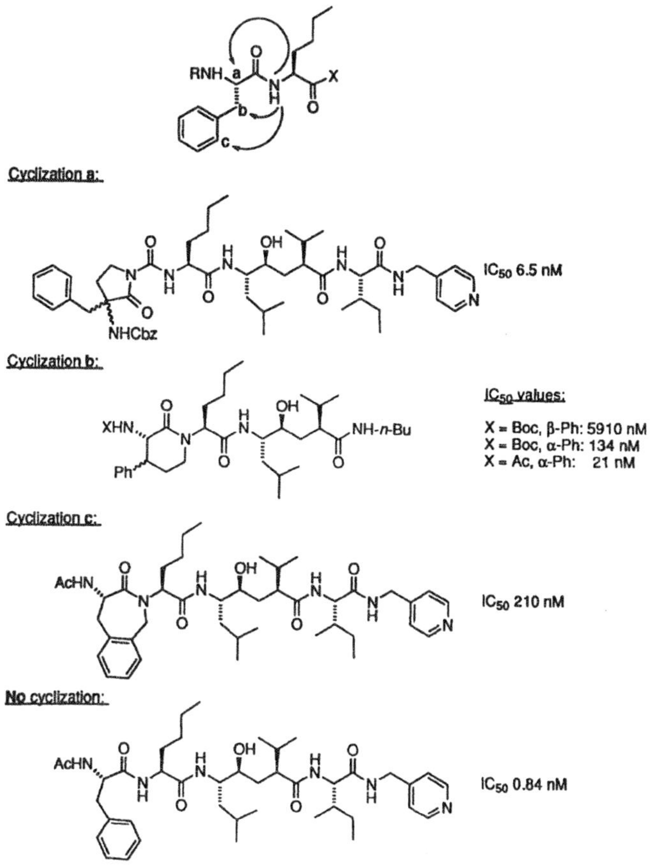
RNH
a
b
c
X
Cyclization a:
NHCbz
IC50 6.5 nM
Cyclization b:
XHN
Ph
NH-n-Bu
IC50 values:
X = Boc, β-Ph: 5910 nM
X = Boc, α-Ph: 134 nM
X = Ac, α-Ph: 21 nM
Cyclization c:
AcHN
IC50 210 nM
No cyclization:
AcHN
IC50 0.84 nM

Figure 10. Various cyclizations used in the design of renin inhibitors, along with a member of the parent series.

piperidone ring and the nature of the N-terminal capping group proved important, as expected. For the fused lactam, the best compound was also that which contained an *N*-acetyl group. In none of these cases, however, were results better than the acyclic parent compound obtained. The authors suggested that the problem may be due to the inability of the inhibitors to attain an extended conformation when cyclizations **b** and **c** (but not **a**) are imposed. Other types of interesting cyclizations, including some rather dramatic macrocyclizations, have been reported in the renin area but will not be discussed here.[46-49]

Another recent approach to blood pressure control involves the development of angiotensin II receptor antagonists (AII antagonists), such as the non-peptidic inhibitor losartin. Samanen and coworkers reported the use of the standard Freidinger constraint into a hexapeptide precursor, [Sar1, Ile8]-angiotensin II (Figure 11). Use

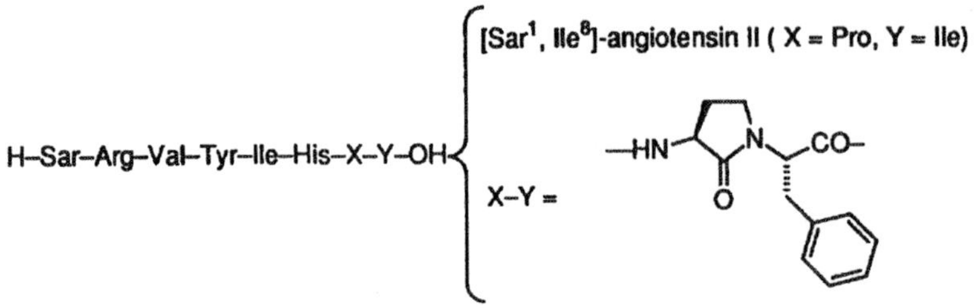

H–Sar–Arg–Val–Tyr–Ile–His–X–Y–OH
[Sar[1], Ile[8]]-angiotensin II (X = Pro, Y = Ile)
X–Y = —HN–(lactam)–N–CO–

Figure 11. An angiotensin II analogue with a lactam-containing receptor inhibitor.

of the five-membered lactam along with replacement of Ile8→Phe8 gave a compound that lost substrate activity but was instead an antagonist of moderate potency.[50,51]

2.4. Cyclosporin and a Farnesylated Yeast Pheromone

Since their introduction, lactam constraints have become standard tools for the medicinal chemist interested in modifying the pharmacological profile of a peptide. This section begins a survey of some other classes of compounds that have been subjected to "lactamization" that will end with peptidomimetics which, although they are not Freidinger lactams per se, are related in some way.

Rich and coworkers, working at the University of Wisconsin, looked at the effect of adding a lactam constraint to the immunosupressive agent cyclosporin A, which is itself a cyclic compound (Figure 12a).[52,53] The use of a Freidinger constraint was suggested by conformational analyses of cyclosporin A that indicated the presence of a type II' β-turn centered around residues 3 and 4. Adding a bridge between the Sar3, MeLeu4 residues afforded γ- and δ-lactams, respectively (Figure 12b, $n = 0$ or 1). Unfortunately, this gambit did not afford very active derivatives—in no case was potency better than 17% of the parent compound. Comparative NMR studies of these compounds and cyclosporin A did not reveal large conformational differences across the series, which led the authors to conclude that the poor activity was due to steric crowding in the receptor which binds this portion of the molecule. Still, the installation of the lactam constraint into such a complex setting has to be considered a noteworthy achievement on its own merits, biological activity aside.

A better outcome was realized more recently by a group studying a farnesylated dodecapeptide yeast pheromone.[54] The eukaryotic yeast *Saccharomyces cerevisiae* utilizes the farnesylated pheromone YIIKGVFWDPAC(*S*-farnesyl)-OMe (termed the **a**-factor) in its sexual reproduction cycle. As was the case for LHRH (see Figure 3 and associated text above[8]), replacement of a Gly residue with D-Ala led to an active analogue, whereas use of the L-antipode caused a diminution of activity, suggesting the possibility for a type II' β-turn in the active conformation at that position. Once again, replacement of the Lys4-Gly5 portion with Freidinger's γ-la-

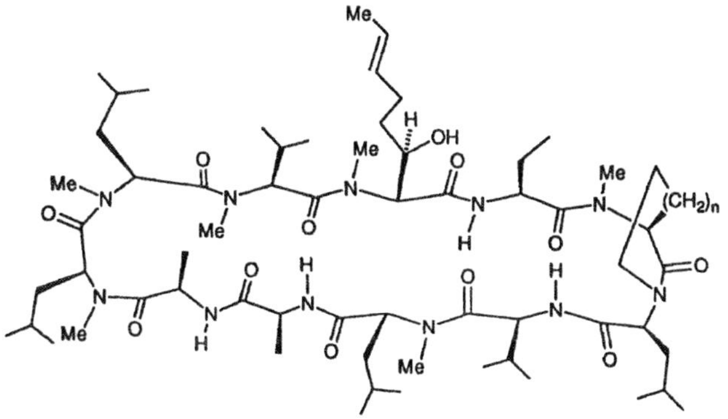

Figure 12. (a) Cyclosporin A. (b) Lactam analogues of cyclosporin A.

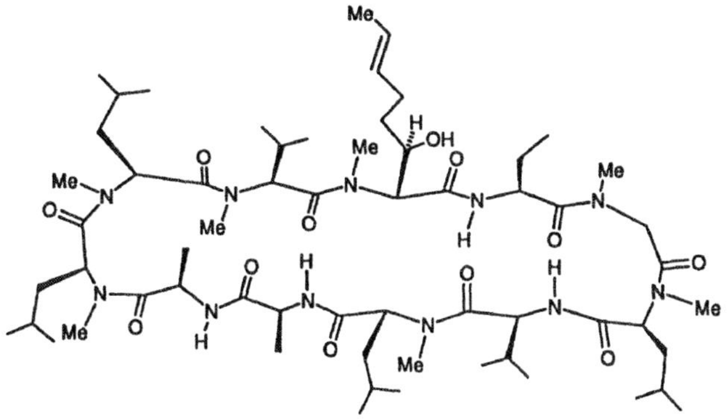

ctams afforded very potent agonists having up to 32-fold greater activity when the (*R*)-lactam was used (not shown). The authors concluded that the type II' β-turn is found in the active conformation of the **a**-factor, but the results of Paul and coworkers mentioned earlier[12] suggest that such a suggestion be made (and taken) prudently in the absence of confirming information (for an instructive case, see the next section).

2.5. Dopamine Agonists

The tripeptide Pro-Lys-Gly-NH_2 (PLG) is a modulator of dopamine receptors in the central nervous system, leading to such pharmacological effects as potentiating the effects of L-DOPA and apomorphine, ameliorating tremors caused by oxotremorine, and others. Johnson's group at Minnesota examined a variety of lactamized analogues of this very simple peptide, entailing the use of either antipode of the γ- to ε-lactams in place of the Lys-Gly end of PLG (Figure 13a).[55,56] An interesting oxo analogue was also examined (Figure 13b). The activity of each com-

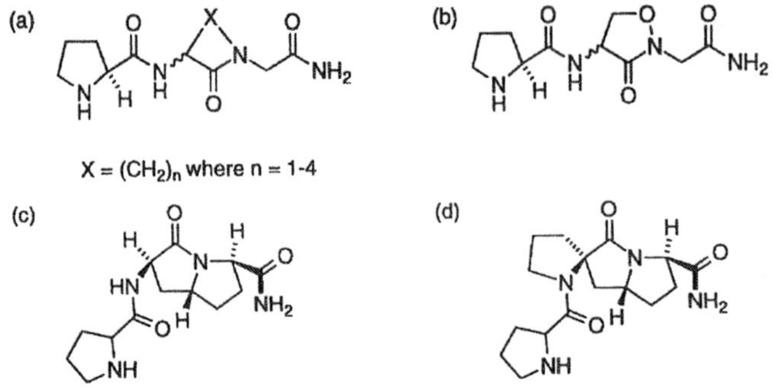

Figure 13. PLG analogues used as dopamine receptor modulators.

pound was assayed by the *in vitro* enhancement of the dopamine agonist 2-amino-6,7-dihydroxy-1,2,3,4-tetrahydronaphthalene to striatal dopamine receptors. Several of the analogues investigated were active. Indeed, the tripeptide mimic containing the (*R*)-configured-5-membered lactam was found to be extremely potent, having 4 orders of magnitude greater activity than PLG itself.

Curiously, X-ray studies published of a few of these analogues showed that, although PLG was found to crystallize in a turn conformation (carrying a hydrogen bond between the Pro C–O and the acetyl NH), only the inactive (*S*)-γ-lactam had a similar conformation, whereas the highly potent (*R*)-γ-lactam was found in an extended conformation.[57,58] On the surface, one might take this as evidence against a β-turn conformation being essential for activity in this series. However, it is the view of the present author that *any* conclusions about active site conformation be addressed by any and all means available and that hypotheses are shaky when based on any single piece of evidence (including small-molecule X-ray). Indeed, later work by the same group entailed the synthesis of still more highly constrained bicyclic (Figure 13c)[59] and tricyclic (Figure 13d)[60] analogues, both of which were found to have dopamine receptor modulation activity of the same general magnitude of PLG. Careful analysis of these structures leaves little doubt as to their conformational similarity with type II β-turns[61] and so the preponderance of the evidence is consistent with a type II structure for PLG (the use of spirocyclic lactams will be discussed more thoroughly below in the context of neurokinin antagonists). The unusually high potency of the (*R*)-γ-lactam has not yet been explained, but may be attributable to metabolism or other effects.

2.6. Human Growth Hormone Analogues

In the examination of a C-terminal hexapeptide derived from human growth hormone (hGH[6-13]) as a potentiator of insulin, it was found that the bioactivity did not result from this peptide itself, but rather the imide resulting from intramolecular

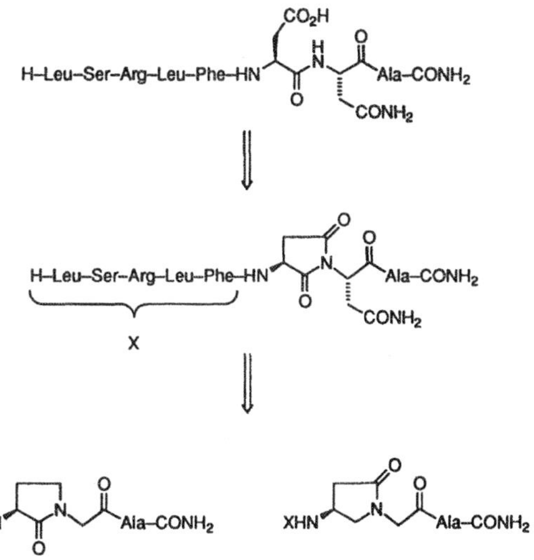

Figure 14. Discovery and development of human growth hormone analogues.

cyclization of the Asp residue (Figure 14). Imides of this type (Asu) have been identified as intermediates in peptide decomposition pathways resulting in the transformation of arginine to aspartic acid or its β-linked isomer, depending on which carbonyl group of the imide undergoes hydrolysis. The lability of the Asu residue prevents the use of this moiety as a peptidomimetic unit, but the activity of the peptide containing it suggested other possibilities.[62,63] Thus, removal of each carbonyl leads to a lactam freed of the extreme lability of the imide. One of these, of course, is the five-membered Freidinger lactam, whereas an isomeric lactam results from the omission of the α-carbonyl group. Bioassay demonstrated that all of these compounds had some activity, but the "iso-Freidinger" lactam was somewhat less potent as an insulin potentiator. Nuclear Overhauser effect studies on all three classes of compounds suggested type II' β-turn conformations in each. Although this particular application of the new 4-amido lactam isostere was not successful, this structural unit may well prove useful in other applications. In addition, alternative imide-derived isosteres have begun to find use as peptide isosteres.[64]

2.7. Neurokinin Antagonists

The many pharmacological effects of substance P (SP), neurkokinin A, and neurokinin B (collectively the tachykinins) may be matched only by the intense interest that medicinal chemists and pharmacologists have lavished on these compounds over the past decade. The effects of the tachykinins are mediated through three different receptors (NK_1, NK_2, and NK_3) and can include nociception, water homeostasis, anxiolytic effects, histamine release, GABA release, and others.

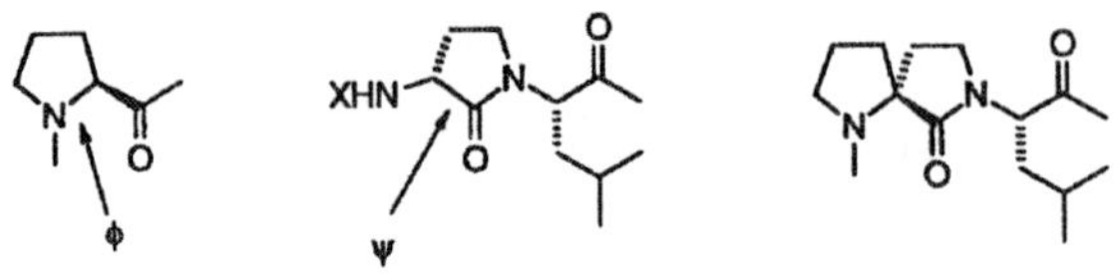

Figure 15. Comparison of constraints arising from proline vs. Freidinger lactam utilization with the corresponding spirolactam.

Scientists at Glaxo have investigated Freidinger lactams, and their spirocyclic cousins, as specific antagonists of the NK_1 receptor. The decision to use the spirocyclic lactam isostere was made on the basis of a straightforward conformational argument (Figure 15).[65] The exchange of glycine for proline in a peptide automatically constrains the corresponding ϕ angle to a range of -55° to -95°. On the other hand, the Freidinger lactam insert, (*R*)-γ-lactam, constrains the adjacent ψ bond to ca. 130° to 150°. Since the spirocyclic lactam of *R* absolute configuration is simply the combination of these two constraints, it follows that both of the above ranges will pertain. And, from parity, the (*S*)-spirolactam will carve out a conformational space of $\phi = 75° \pm 20°$ and $\psi = -140° \pm 10°$. However, the two antipodal spirolactams lead to diastereomers when installed in a peptide, and therefore a different backbone conformation (i.e. extended vs. turn) is likely to result. Other researchers have also investigated the conformational properties of similar spirolactams.[66,67]

A hexapeptide derived from the C-terminal end of SP provided the platform for assessing neurokinin agonist potency of peptides in which positions 9 and 10 (SP numbering) were systematically varied (Table 4, with Ava = δ-aminovaleryl). Thus, the replacement of Gly9 with L- or D-Pro (entries 4 and 5) demonstrated that the former (occupying negative values for ϕ) was preferred by the NK_1 receptor and the latter by NK_2 (ϕ being positive). The preferred conformation was further refined by noting that only the use of the (*R*)-γ-lactam in the Gly9,Leu10 positions was tolerated by either receptor (entries 5 and 6), which seem to require positive values for the positive range. Accordingly, the (*R*)-spirolactam was a full agonist, albeit one with about 10-fold less activity than the Gly-Lys-containing parent compound. The loss of activity could be due to untoward steric interactions of the bridge with the receptor target or possibly to overly constraining the compound into an adequate, but less than ideal conformation. Still, the utility of this approach was verified through further refinement, leading to highly active antagonists[68,69] containing a spirocyclic linkage as well as a γ-lactam-containing agonist.[70]

2.8. Thrombin Antagonists

Returning to the general field of protease inhibition brings us to thrombin, a serine protease that plays several important roles in blood coagulation and hemostasis

Table 4. Potency of [Ava6]-SP(6-11) Analogue[a]

Entry	Substitution	EC_{50}, nM NK1	EC_{50}, nM NK2	Comments
1	Gly9	34	1190	
2	Ala9	5.9	1860	
3	D-Ala9	638	452	
4	Pro9	17.7	31,500	
5	D-Pro9	1960	67	
6	(*R*)-γ-lactam	75	67	Freidinger lactam
7	(*S*)-γ-lactam	>49,000	9990	Freidinger lactam
8	(*R*)-spirolactam	307	2390	Extended conformation
9	(*S*)-spirolactam	>98,000	>80,000	Type II' β-turn
10	Substance P	4.91	167.4	

Note: [a] Data form reference 65.

processes. Thrombin inhibitors may be useful in the treatment of cardiovascular diseases in which abnormal clotting plays a role, such as thromboses, pulmonary embolism, and angina.[71] A class of previously identified thrombin inhibitors (Figure 16a, left) was subjected to lactamization through attachment of the phenylalanine side chain to the adjacent proline residue with concomitant phenyl removal and proline ring opening.[72] This maneuver was made in close analogy to that carried out in the conversion of enalaprilat to Freidinger lactam-containing ACE inhibitors (see Figure 4 and accompanying text). The C-terminal residue of the resulting molecules comprise a unit that resembles a kind of reduced Freidinger lactam. A closer inspection, however, shows that this residue is merely the lactol resulting from attack of the argininal side chain onto its aldehyde; reversal of this process reveals the electrophilic aldehyde, which is ostensibly the reactive portion of this class of transition-state inhibitor (Figure 16b). It turns out that the introduction of the lactam unit leads to active compounds, of which the most potent are those containing six- and seven-membered lactam rings.

Another group carried out an α-carbon-to-amide constraint in another class of thrombin inhibitors (Figure 16c).[73] The resulting compounds, each containing a quaternary center, were also potent inhibitors of thrombin. In contrast to the previous example, the most active member of the series was the γ-lactam version.

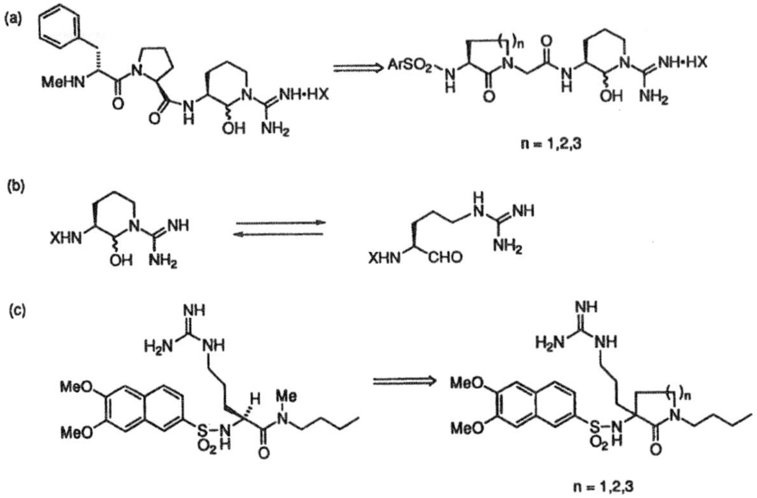

Figure 16. (**a**) Lactamization of a class of thrombin inhibitors. (**b**) Equilibrium between closed and open forms of an arginine-derived aldehyde. (**c**) Another class of thrombin inhibitors resulting from α-H lactamization.

2.9. Relationship to Other Peptidomimetic Classes

Several classes of peptidomimetics related to Freidinger lactams have been briefly mentioned. Chief among these are the very useful and diverse bicyclic lactams reviewed by Hanessian and Lubell (for examples, see Figures 2b, 4, 7c, 7d, 8j, 8k, 13c, and 13d),[14] although any heterocycle containing a nitrogen atom derived from a natural or unnatural amino acid should be considered to some extent. As bicyclic and spirocyclic lactams are often considered to be mimics of β-turns, Huffman and coworkers have published extensively on the use of various seven-membered lactams derived from amino acids as γ-turn mimics. One example, an HIV protease inhibitor, is shown in Figure 17a.[74,75]

The utility of lactams containing a carbon adjacent to the lactam carbonyl instead of the nitrogen found in Freidinger lactams was mentioned in the context of renin inhibitor design (see Figure 9 and accompanying text). Other examples of useful compounds that fall into this category are a potent and orally active fibrinogen receptor antagonist (Figure 17b)[76] and a bradykinin analogue (Figure 17c) that was also cited as a possible γ-turn peptidomimetic.[77] No rule outlaws Freidinger-style lactams containing additional unsaturation like the pyridone in Figure 17d, and such compounds may be useful in cases where some of the stereocenters near the lactam constraint are superfluous.[78,79] Finally, benzodiazepines that contain nitrogen substitution are ubiquitous in medicinal chemistry and share many design features with Freidinger lactams. The practicing medicinal chemist would be well-

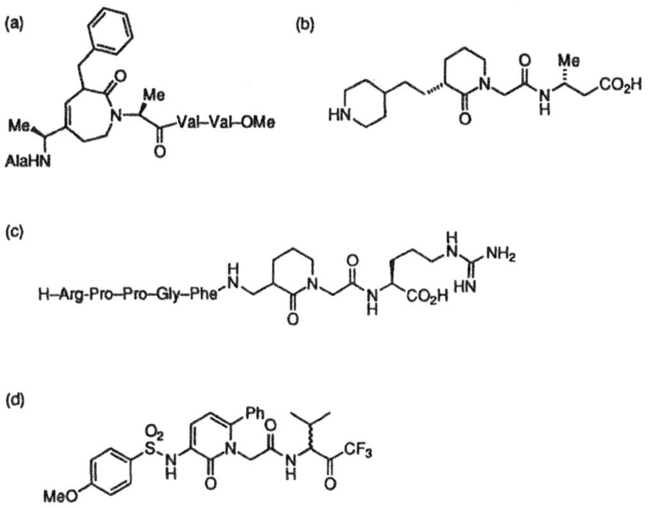

(a) (b) (c) (d)

Figure 17. Some peptidomimetics related to Freidinger lactams: (**a**) a γ-turn mimic, (**b**) fibrinogen receptor antagonist, (**c**) a bradykinin analogue, and (**d**) a pyridone synthesized as a non-peptidic human leukocyte elastase inhibitor.

advised to look carefully at all of these classes of compounds when considering a lactam constraint in her or his work.

3. SYNTHETIC ROUTES TO LACTAM PEPTIDOMIMETICS

An ideal method for the synthesis of lactam peptidomimetics would address several issues. It would (1) be applicable to lactams of various sizes, (2) allow the installation of potentially expensive and/or scarce amino acid moieties late in the scheme, permitting divergence to a number of amino acid substitutions, (3) incorporate protection schemes consonant with standard peptide chemistry, and (4) allow control over relative and absolute stereochemistry. As of this writing, there is no single method that incorporates all of these wishes, but a number of approaches with complementary strengths and deficiencies.

The goal of this section is to summarize some of the important routes to the various classes of Freidinger lactams and their relatives finding application in modern medicinal chemistry. Although no attempt will be made to recount the synthesis of every compound discussed in the previous section, the merits of the most practically important and chemically interesting strategies will be discussed. The interested reader is referred to the original literature as cited in the previous section for detailed information regarding the preparation of an individual compound. This section will begin with a discussion of basic cyclization strategies and functional group transformations without particular regard for stereocontrol. A discussion of newer methods that either afford a high degree of stereocontrol or have the inherent potential for doing so will follow, and we will finish by commenting on methods that afford heterocyclic analogues of Freidinger lactams.

3.1. Standard Cyclization and Functionalization Strategies Leading to Freidinger Lactams

The original route[8] to Freidinger lactams is still one of the simplest and most useful methods for the preparation of γ-lactam peptidomimetics (Figure 18).[80] The sulfide moiety of a dipeptide of the type Boc–Met–Xyz–OR can be converted to a leaving group by treatment with methyl iodide, and subsequently cyclized under basic conditions. This route allows for easy incorporation into peptide chains using solution[8] or solid-phase[54,81] techniques because typical peptide protecting groups are used at both ends. The syntheses of large-ring cyclosporin-containing lactams under solution-phase conditions are some of the more impressive applications of this technology.[52,53]

Another useful aspect of this chemistry is its tolerance of a wide variety of residues for the C-terminal peptide residue. Obviously, methionine or cysteine would cause regiochemistry problems as would residues containing other nucleophilic side chains. Overall, though, the direct incorporation of chiral amino acids into the sequence must be considered a real plus. Of several simple cases examined by Freidinger et al., only the use of phenylalanine was found to result in partial epimerization in the cyclization process.[8] The equally obvious limitation of this method is that it is limited to γ-lactam synthesis in view of the relative inaccessibility of homomethionine in enantiomerically pure form. One could, of course, pursue such a route if willing to procure the appropriate starting materials, but it is unlikely that any direct cyclization unit would be much use for lactams containing more than seven members.

Another class of cyclization reactions that has seen a great deal of action in this field utilize reductive amination processes for precursor synthesis. Of many such examples, two from the spirolactam literature will serve as illustrations (Figure 19).

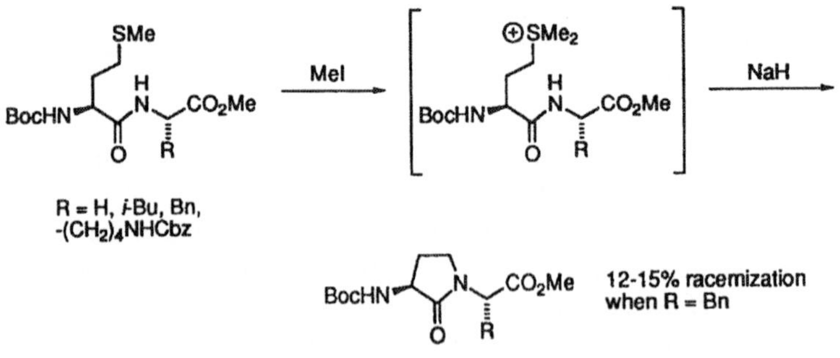

Figure 18. Methionine cyclization method of γ-lactam synthesis.

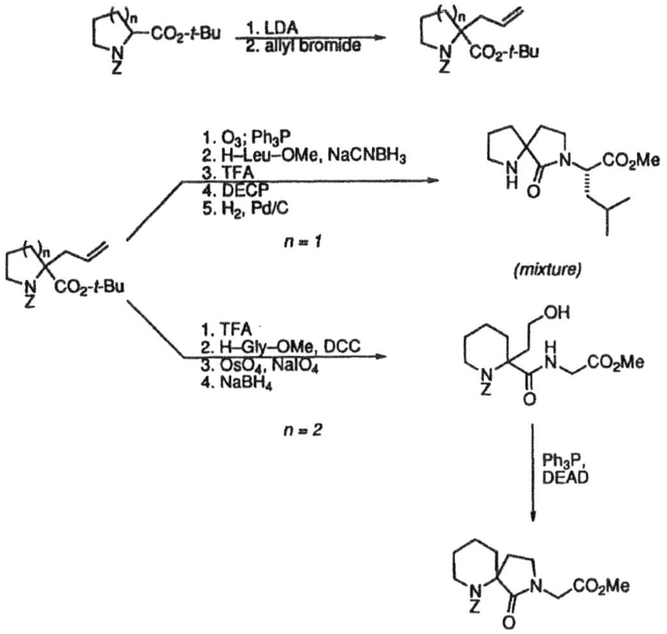

Figure 19. α-Allylation as a tool in Freidinger lactam synthesis, including several methods for ring closure.

Allyl groups are particularly useful in organic chemistry because of their excellent reactivity in alkylation chemistry and the easy manipulation of the olefin linkage via ozonolysis to the corresponding aldehyde. Thus, α-alkylation of Cbz-protected amino esters with allyl bromide affords a quaternary carbon-containing intermediate. For the synthesis of the [5.5]spirocyclic lactam shown in Figure 19 ($n = 1$),[65] direct ozonolysis of the double bond is followed by reductive amination with a leucine ester. After protecting group removal, amide bond formation follows readily under normal peptide-coupling conditions (in the case shown, diethylcyanophosphonate was used). This method is probably the most commonly employed technique for Freidinger lactam synthesis today. An interesting alternative was used in the synthesis of the [6.5]spiro system indicated, wherein the allyl group was used in the formation of the C–N single bond.[67] Thus, ester deprotection with acid was followed by amide bond formation and alkene cleavage. Cyclization was completed using Mitsunobu conditions.

An alternate cyclization strategy that also used a reductive amination step was reported by Freidinger (Figure 20).[80] Condensation of protected ornithine ($n = 1$) or lysine ($n = 2$) with glyoxylic acid under hydrogen afforded amino acids or esters that were cyclized thermally in DMF or acetonitrile. The pyrimidone was readily cyclized from the carboxylic acid, but the seven-membered lactam required conver-

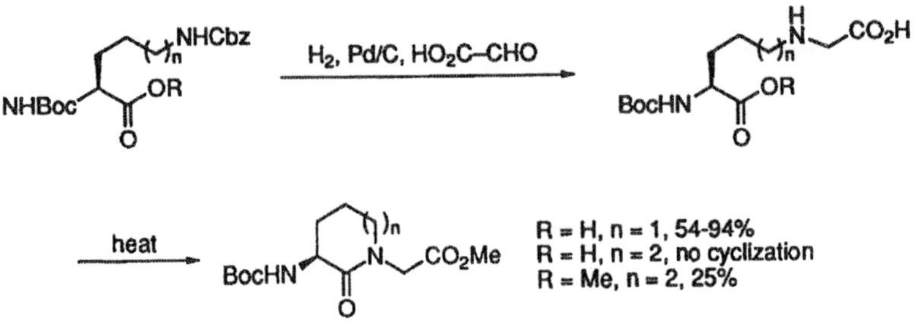

Figure 20. Ring closures leading to six- and seven-membered lactams.

sion to the ester. In this case, the reductive amination specifically provided C-terminal glycine dipeptide isosteres only; higher amino acids would necessitate the use of pyruvic acid conditions in the reductive amination step, in which a mixture of diastereomeric compounds would have resulted anyway, unless asymmetric hydrogenation conditions could be employed. As of this writing, asymmetric catalysts for reductive amination exist but are limited to particular substrate types.

These strategies were used in the synthesis of quaternary compounds as shown in Figure 21.[82] In this case, the α-quaternary center was prepared by allylation of phenylalanine disguised as an oxazolidinone. The oxazolidinone could be turned into the corresponding allylated amino ester using the unremarkable conditions shown. Either ozonolysis/reductive amination as shown earlier (route **a**) or extension of the allyl group via a hydroboration reaction (route **b**) led to amino esters that were subjected to thermal cyclization. γ-Lactams were obtained through room temperature cyclization in DMF/toluene mixtures, but the reactions leading to δ-lactams were much slower, requiring extended heating times and the addition of base to obtain decent yields (55–66%). Improved cyclization results were obtained in the latter cases through the direct use of the oxazolidinones in the cyclization step (route **c**).

An alternative approach to a quaternary-substituted γ-lactam featured the amination of a doubly activated cyclopropane and a Curtius rearrangement to install the carbamoyl group (Figure 22).[83] The amination reaction led directly to the lactam containing a 3-carboxylic acid group. Direct reaction of the amido acid with an electrophile under basic conditions gave the quaternary-center-containing product in modest yield and as a mixture of diastereoisomers. Curtius rearrangement using diphenylphosphoryl azide in a sea of *tert*-butanol afforded the 3-BocNH-substituted γ-lactam directly.

Additional complexity was introduced using methodology published by Garvey et al. (Figure 23).[84] Formylation occurred specifically at the β-carbon of a differentially protected derivative of aspartic acid, allowing either direct reductive amination or O-alkylation with allyl iodide. In the latter example, Claisen rearrangement and utilization of the reductive amination/cyclization sequence gave the 4-propyl-γ-lactam product as a mixture of diastereomers.

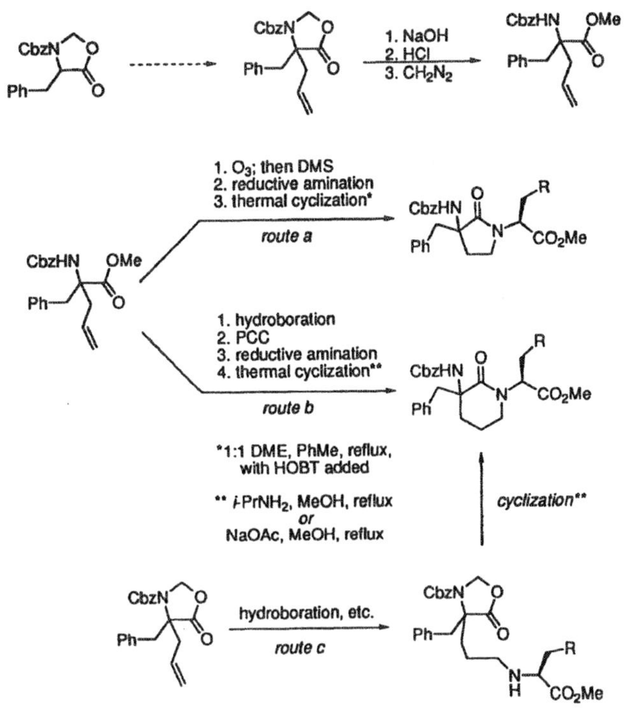

Figure 21. More cyclization techniques.

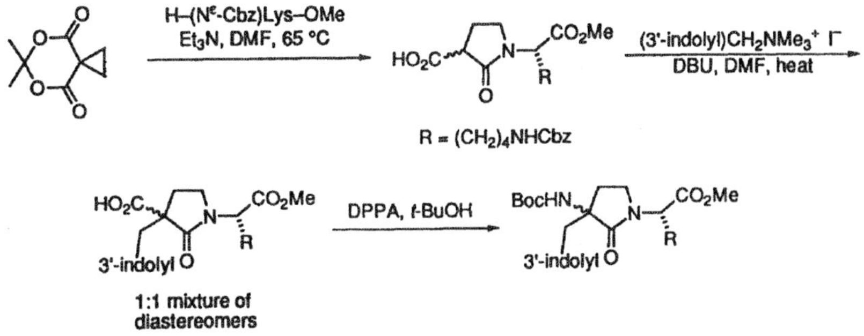

Figure 22. Use of a cyclopropane ring-opening/Curtius rearrangement approach to peptidomimetic synthesis.

With the exception of Figure 22, all of the above schemes involved cyclization reactions in which all of the elements of the Freidinger lactam were in place prior to cyclization. Some important routes take a more indirect approach, and specifically involve the introduction of the 3-amido nitrogen after lactam formation or require an N-alkylation step to install the C-terminal amino acid residue.

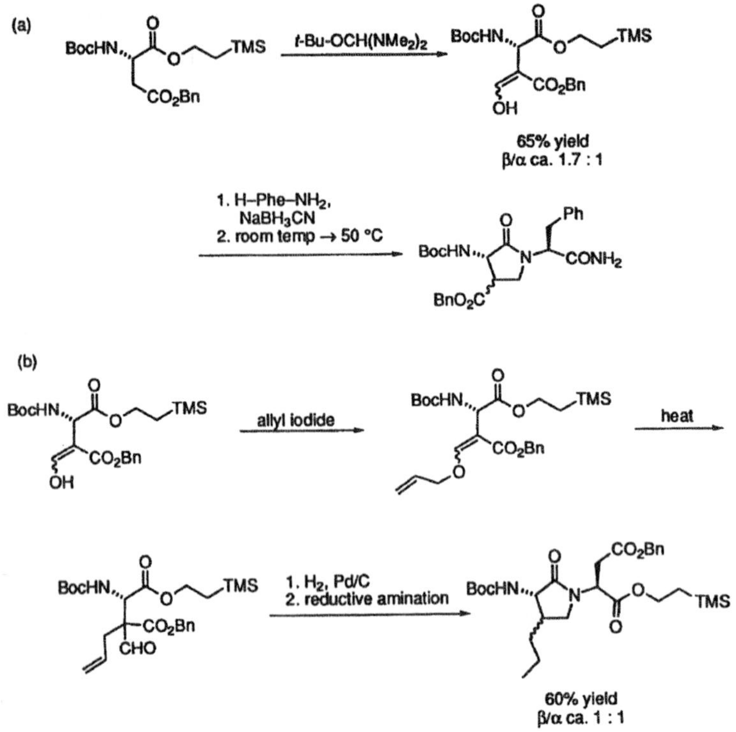

Figure 23. Formylation routes to more complex γ-lactams, involving (**a**) a simple reductive amination and (**b**) a Claisen rearrangement.

In general, direct cyclization reactions are awkward when the ring size of the final lactam exceeds seven. In contrast, simple lactams of all of the usual ring sizes are available, often using a Schmidt reaction to effect a ring expansion of a ketone or by more involved routes leading to simply substituted lactams (e.g. those in Figure 7a).[26] Accordingly, 3-amidation may be carried out on a preformed lactam as shown in Figure 24a.[22] α-Bromination is carried out using phosphorous pentabromide[22] or Br_2/PBr_5[26] and the resulting halide displaced via treatment with sodium azide. Occasionally, more involved α-halogenation procedures may be required.[27] Eventually, the resulting 3-azido lactam is N-alkylated with an iodoacetate ester; azide reduction followed by protecting group removal affords the desired isostere. It is worth noting that some of the 3-amino-lactams synthesized by Thorsett et al. could be resolved to enantiomeric purity using classical methods.[22]

The chemoselective alkylation of an amide nitrogen can also be carried out at the stage of a 3-*tert*-butoxycarbonylamido lactam (Figure 24b); these lactams are readily available for the smaller ring sizes shown (and the seven-membered version is α-Boc-cyclolysine).[22] Finally, an interesting turn of events occurred in the particu-

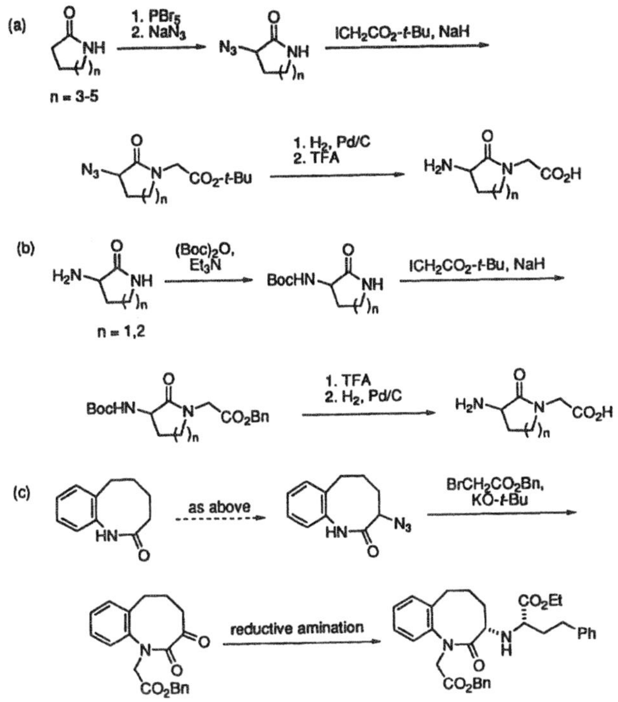

Figure 24. Alkylation reactions of (**a**) 3-azido and (**b**) 3-butoxycarbamoyl lactams. (**c**) An azide decomposition reaction that nonetheless led to a useful ACE inhibitor.

lar example shown in Figure 24c. In an attempt to alkylate the amide nitrogen of an eight-membered 3-azido lactam, the azide underwent an unusual base-promoted decomposition under long reaction times, affording the keto lactam shown.[27] The ketone thus formed was then exploited in a reductive amination step that diastereoselectively afforded a side chain reminiscent of enalapril.

3.2. Stereochemical Considerations

Increasingly, a key consideration in the preparation of peptidomimetics is stereochemistry. Although racemic compounds or mixtures of diastereoisomers may be acceptable in the early going, a detailed view of an active site or the identification of the "best" isomer for further investigation will eventually be necessary for an ongoing program. In addition, the development of more selective lactam peptidomimetics implies an ever-increasing level of structural complexity that needs to be addressed through effective asymmetric syntheses.

Of course, some compounds can be readily prepared in diastereo- or enantiomerically form using the techniques already developed. However, two specific problems often arise. First, a dipeptide isostere immediately implies the possibility of two asymmetric centers. The stereoselective synthesis of a given diastereomer is easy if both centers can be prepared separately and then coupled, as in Freidinger's original route involving methionine cyclization (Figure 18). The limitations of this route with respect to ring size have already been noted. The alternative technique of preparing an enantiomerically pure 3-amido lactam and then subjecting it to N-alkylation works well for C-terminal glycine, but is practically limited to that residue (Figure 25a). For example, the attempted alkylation of an enantiomerically pure lactam derived from L-lysine with the chiral triflate reagent shown gave isomeric mixtures regardless of conditions.[85] On the other hand, racemization and epimerization are concerns that permeate all of peptide-related synthesis, even when enantiomerically pure building blocks can be assembled. In one case, the thionium-ion cyclization shown in Figure 25b suffered significant racemization en route to product.[80]

A key requirement is the availability of enantiomerically pure starting materials. Obviously, this is not a problem when one can rely on the naturally occurring amino acids and their simple derivatives, but more complex building blocks may be more problematic. The virtues of using α-allylated amino acid derivatives in lactam synthesis were previously stated (Figure 19). Several groups have utilized some of the significant recent advances in the synthesis of α-alkylated amino acids[86] for Freidinger lactam synthesis. The most useful of these has been the concept of "self-regeneration of chirality" as developed by Seebach[87] and Karady.[88] The concept as demonstrated by Acton[89] involves the cyclization of an amino acid with an aldehyde, which generates a new stereocenter (Figure 26a). The original α-center is lost upon deprotonation, but alkylation is still directed by the aminal center and the quaternary carbon is formed with generally excellent stereoselectivity. In this case, the use of methionine in the alkylation routine allows the activation and displacement

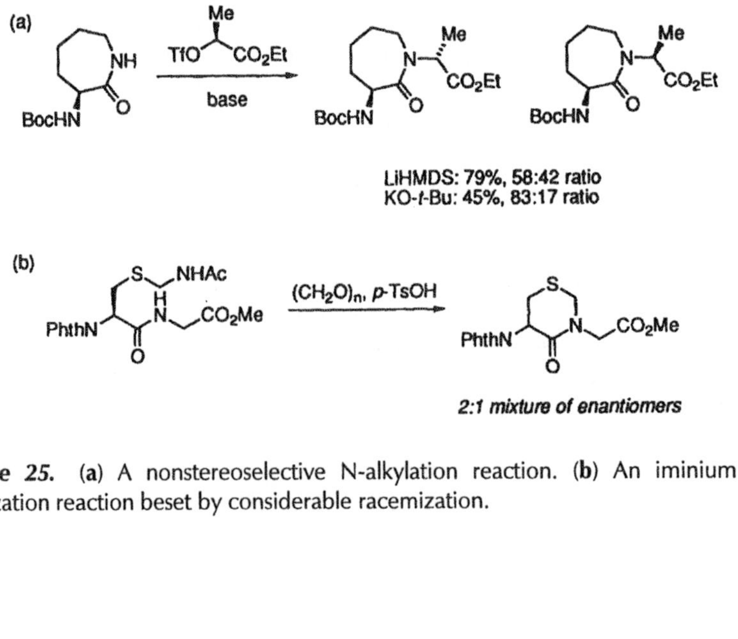

(a)

BocHN NH O — TfO Me CO_2Et / base → BocHN N Me CO_2Et O + BocHN N Me CO_2Et O

LiHMDS: 79%, 58:42 ratio
KO-*t*-Bu: 45%, 83:17 ratio

(b)

PhthN S NHAc H N CO_2Me O — $(CH_2O)_n$, *p*-TsOH → PhthN S N CO_2Me O

2:1 mixture of enantiomers

Figure 25. (**a**) A nonstereoselective N-alkylation reaction. (**b**) An iminium ion cyclization reaction beset by considerable racemization.

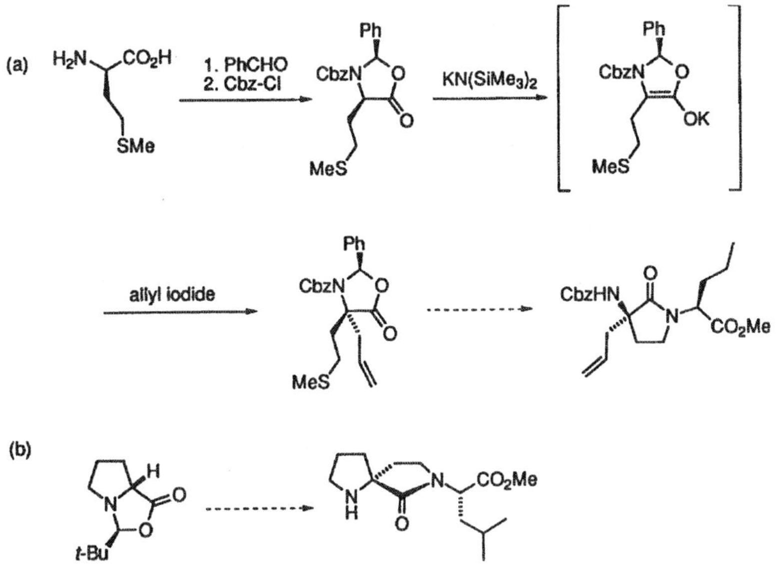

Figure 26. Application of the "self-regeneration of chirality" approach to synthesis of (**a**) a γ-lactam containing a quaternary center at C-3 and (**b**) a spirocyclic lactam.

of the thiomethyl group in the Freidinger manner. Functional group manipulation permits a reductive alkylation step and affords, stereoselectively, a very useful dipeptide mimic containing an allyl group suitable for further elaboration. Another example, from Ward and coworkers, demonstrates the utility of this strategy for spirocyclic lactam synthesis (Figure 26b); in this case, the allyl group was used for lactam formation in direct analogy to the scheme in Figure 19.[65]

A new method for lactam synthesis has a great deal of promise for the stereocontrolled synthesis of Freidinger lactams (Figure 27).[90] The Grubbs technique for lactam synthesis uses a ring-closing metathesis procedure that allows for carbon–carbon bond formation under mild conditions that are compatible with a number of functional groups. Although the single example published by this group used racemic building blocks and therefore led to a mixture of isomers, the success

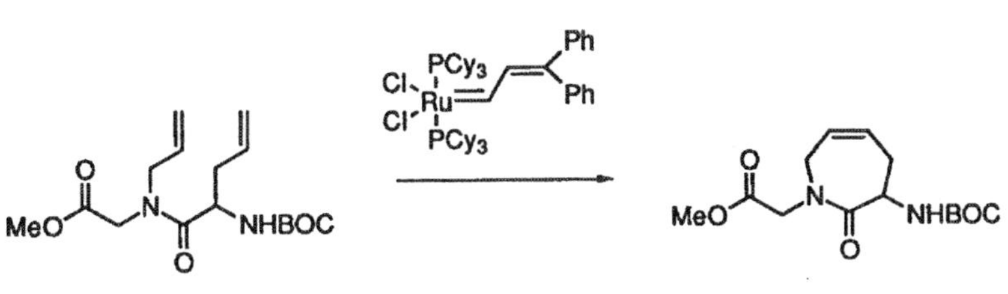

Figure 27. Ring-closing metathesis route to a caprolactam.

of a number of researchers in using this reaction in demanding total synthesis applications[91-94] bodes well for its extension to diastero- and enantioselective syntheses.

Some particularly versatile methods for the stereocontrolled synthesis of lactam peptidomimetics of all types use acyl iminium ion cyclizations as a key step. Much of this may have been inspired by a 1987 paper by Flynn and coworkers in which they reported the use of an arene–acyl iminium cyclization to obtain a tricyclic inhibitor of ACE.[30] A direct descendant of this method was used by de Laszlo and coworkers in their renin work shown in Figure 28a; approximately 10% epimerization (of an unidentified stereocenter) was encountered during their key step.[45] An alter-

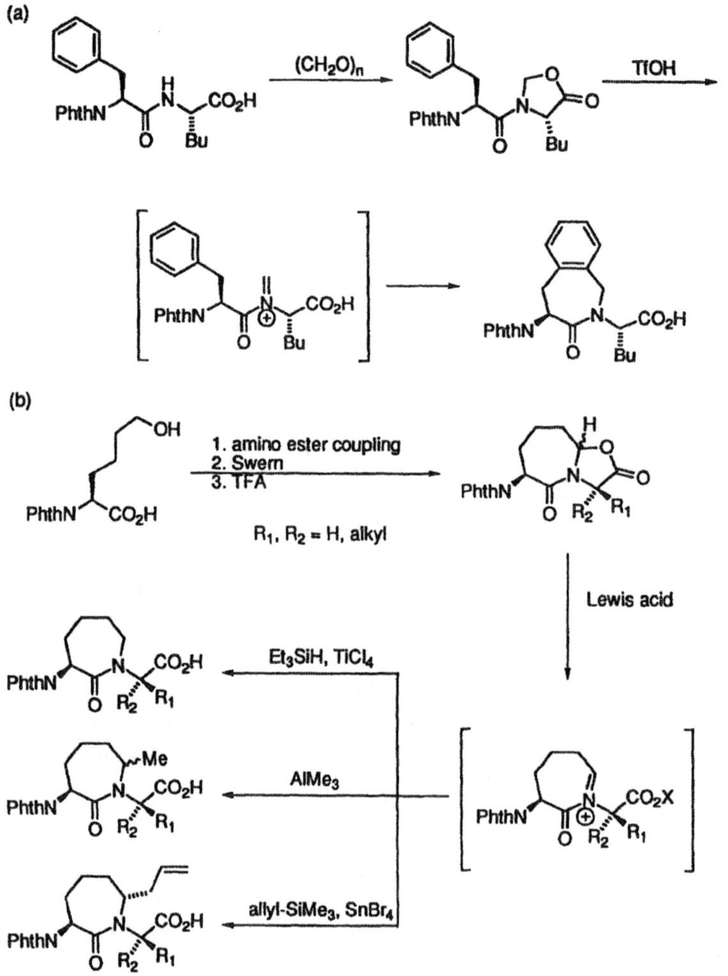

(a) PhthN, H, N, CO_2H, O, Bu, $(CH_2O)_n$, TfOH, CO_2H

(b) OH, PhthN, CO_2H, 1. amino ester coupling, 2. Swern, 3. TFA, R_1, R_2 = H, alkyl, Lewis acid, Et_3SiH, $TiCl_4$, $AlMe_3$, allyl-$SiMe_3$, $SnBr_4$, Me, CO_2X

Figure 28. Iminium ions as intermediates in stereoselective lactam synthesis.

native use of iminium intermediates has been used extensively by Robl and coworkers (Figure 28b). Instead of using the acyl iminium species in a ring-forming step, these investigators chose a seven-membered ring via aminal formation (although irrelevant to the rest of this route, the aminal stereocenter is generally formed with very good stereoselectivity (≥92% ds).[85] The iminium ion could be revealed under a variety of Lewis acid conditions and trapped by various nucleophiles ranging from hydride to allyl silanes. The stereoselectivity of this sequence depends on both the Lewis acid and substituents used.

A few stereoselective methods for the introduction of nitrogen adjacent to a lactam carbonyl have been reported. Most generally applicable would be a number of methods for the electrophilic amination of enolates that have appeared in the literature.[95-98] A specific example that uses an Evans oxazolidinone chiral auxiliary has been reported by de Laszlo (Figure 29a).[45] In this case, azidation of the amide enolate with trisyl azide occurred with high selectivity (≥95% ds) and a subsequent reductive amination was followed by spontaneous cyclization. Although not applied to Freidinger lactam synthesis per se, the stereoselective azidation (following silica gel equilibrium) of the enolate of the lactam shown in Figure 29b is interesting and might find an application to peptide mimicry, particularly if the bromination reactions are selective when other amino alcohols are used.[99] The stereoselective introduction of a halogen α to the lactam carbonyl has also been attempted using a catalytic hydrogenation of a α,α-dihalolactam, but high enantioselectivities have not yet been achieved in this process.[100]

We have reported a new and potentially general synthesis of Freidinger lactams and applied it to the preparation of some inhibitors of angiotensin-converting enzyme.[24,101] Accordingly, 2-*N*-Boc-aminocycloalkanones varying from five-seven-membered rings were synthesized via the corresponding amino alcohols, which

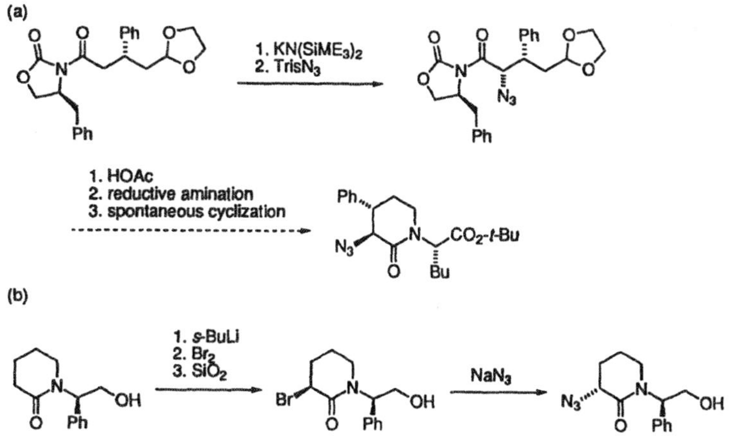

Figure 29. Stereoselective introduction of nitrogen adjacent to the lactam carbonyl.

were in turn prepared by opening a cycloalkene oxide with a chiral amine (Figure 30).[102] A more modern approach would probably involve one of several new methods for the synthesis of chiral azido alcohols in enantiomerically pure form.[103-106]

The stereoselective synthesis of oxaziridines via the corresponding imines had been a subject of great interest in our laboratory.[107,108] The α-carbamoyl ketones were condensed with a variety of amino esters, such as L-phenylalanine; use of the tin catalyst[109] was necessary to effect the condensation reaction at room temperature, lest racemization occurs. Imine oxidation with *m*-CPBA then led to an oxaziridine in which the group on nitrogen is *trans* to the more highly substituted carbon substituent. The stereoselectivity of this step was found to depend on the nature of the amino ester used; those bearing hydrophobic side chains gave high selectivity, whereas others (Asp, Glu) led to mixtures of isomers at the C-2 position of the cycloalkanone ring. Finally, it had been established that the photolyses of spirocyclic oxaziridines led to the preferential migration of the group syn to the nitrogen substituent.[107,108] In contrast, typical Beckmann and Schmidt rearrangements occur so that the more substituted carbon migrates to nitrogen, therefore the more common methods for nitrogen insertion would be unsuitable for the preparation of this class of peptidomimetics.[110] In the cases examined—including the synthesis of an eight-membered lactam—exclusive formation of the desired Freidinger lactams and none of their regioisomers resulted. This rearrangement chemistry has also been used for the synthesis of a series of β-turn peptidomimetics.[111]

3.3. Synthesis of Heteroatom-Containing "Lactam" Peptidomimetics

The construction of Freidinger-like lactams containing additional heteroatoms will be briefly mentioned here. Often, the introduction of an oxygen or sulfur sim-

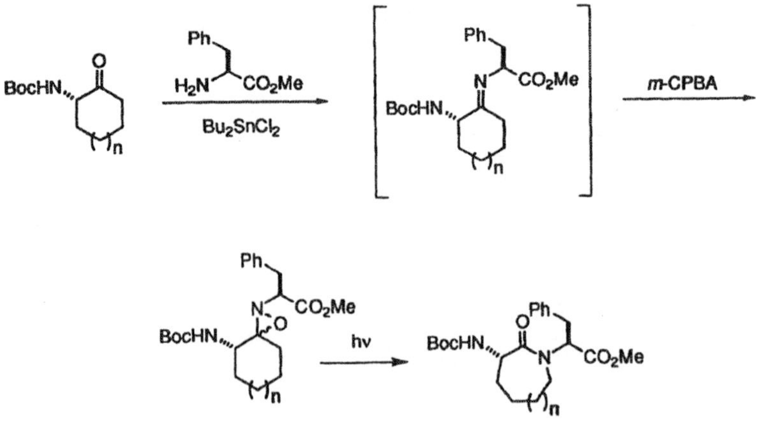

Figure 30. Oxazirindine route to seven- and eight-membered lactams.

plifies the synthetic procedures necessary for the synthesis of a series of compounds, thanks to the very wide variety of methods available for the synthesis of carbon–heteroatom bonds. A few of such syntheses will be outlined here as examples.

Two interesting bond constructions are the basis for the two syntheses shown in Figure 31. Yanagisawa has extensively utilized the conjugate addition of sulfides to vinyl nitro species as shown in Figure 31a; this approach allowed the use of enantiomerically pure cysteine as the chiral educt for this synthesis.[25] Simple techniques (reduction, amide bond coupling) then permitted this group to make a very large number of ACE inhibitor precursors with the general structure shown.

An oxazolidinone-mediated aldol reaction of a bromoacetate equivalent gave the 1,2-bromohydrin product shown in Figure 31b with high enantio- and diastereoselectivity.[31] Displacement of the bromide, allylation, and miscellaneous functional group manipulation afforded the allylic ether. Ozonolytic cleavage of the olefin, reductive amination, and cyclization as previously discussed lead to an N,O-containing dipeptide isostere. The use of benzaldehyde as the aldol partner probably simplifies the aldol reaction considerably, as it would be presumably nontrivial to replace it with a formaldehyde equivalent.

A reaction useful in the synthesis of benzo-fused dipeptide mimics is the nucleophilic aromatic substitution of nitro compounds. As shown in Figure 32, this process allows one to attach a fairly complex alcoholic acid onto the aromatic ring.[36] Reduction of the nitro group and peptide coupling complete the synthesis of the

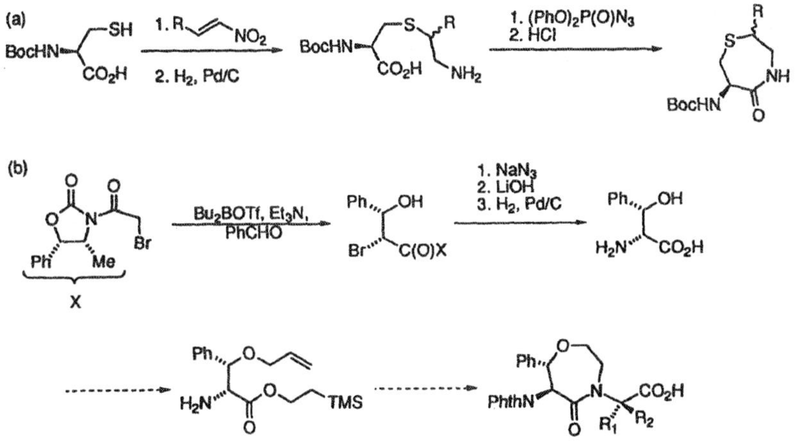

(a) BocHN, SH, CO2H; 1. R–CH=CH–NO2; 2. H2, Pd/C; BocHN, S, R, CO2H, NH2; 1. (PhO)2P(O)N3; 2. HCl; S, R, NH, BocHN, O

(b) O, O, O, N, Br, Ph, Me, X; Bu2BOTf, Et3N, PhCHO; Ph, OH, Br, C(O)X; 1. NaN3; 2. LiOH; 3. H2, Pd/C; Ph, OH, H2N, CO2H

Ph, O, H2N, O, O, TMS; Ph, O, PhthN, N, CO2H, O, R1, R2

Figure 31. Heteroatom containing dipeptide mimics prepared by routes featuring (a) conjugate addition and (b) aldol reaction methodology.

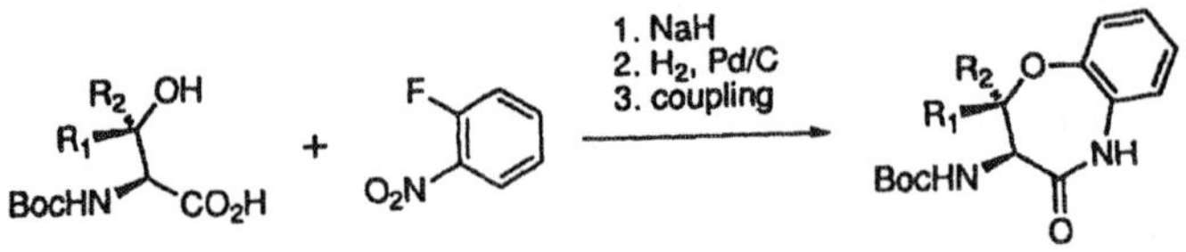

Figure 32. Nucleophilic aromatic substitution routes leading to heteroatam-containing dipeptide mimics.

peptidomimetic core, with similar processes being available for thiol aromatic additions as well.[28]

4. FINAL COMMENTS

Compounds in the 1980 vintage of classic Freidinger lactams, especially those directed toward LHRH agonism, have to be considered some of the earliest deliberate attempts to influence the conformation of a bioactive peptide through synthetic modification. Of course, the field of peptidomimetics has virtually exploded since that time, with the publication of this series of books being proof positive of this fact. The concept of short-range cyclization has proven an extremely valuable and versatile tool for such constraints and it seems likely that medicinal chemists of the future will continue to use this concept in their work. It is equally clear that there has been a two-way exchange of progress with respect to the design of these compounds and their synthesis. Indeed, the activity of peptidomimetic synthesis has slowly begun to receive the attention and respect it deserves among those who would consider themselves to be strictly synthetic organic chemists (even if it has not quite caught up with natural products synthesis). As of this writing, the development of an inherently synthesis-driven mode of drug discovery—combinatorial chemistry[112]—has become entrenched and is still gaining steam. No generalizable routes to Freidinger libraries yet exist (at least not in the public domain), but their development will constitute the next immediate challenge for chemists interested in working with this fascinating and useful class of organic molecules.

ACKNOWLEDGMENTS

First and foremost, I would like to thank my colleagues who have experimentally contributed to our work on Freidinger lactams, especially Professor Michael Wolfe and Dr. Dinah Dutta. I would also like to acknowledge Dr. Shankar Venkatraman's contributions to this chapter, which included the writing of a preliminary draft. I thank Mary MacDonald and Klaas Schildknegt, who helped with research and editorial details. Finally, our own work in this area was funded by the National Institutes of Health (GM-49093) and the American Cancer Society – Kansas Affiliate, and their continuing support is greatly appreciated.

REFERENCES

1. Veber, D. F.; Freidinger, R. M. *TINS* **1985**, 392.
2. Plattner, J. J.; Norbeck, D. J., In *Drug Discovery Technologies*; Clark, C. R.; Moos, W. H., Eds.; John Wiley & Sons, New York, 1990, pp 92-126.
3. Hruby, V. J.; Al-obeidi, F.; Kazmierski, W. *Biochem. J.* **1990**, *268*, 249.
4. Sawyer, T. K., In *Peptide-Based Drug Design*; Taylor, M. D.; Amidon, G. L., Eds.; American Chemical Society, Washington, DC, 1995, pp 387-422.
5. Nestor, J. J., In *Peptide-Based Drug Design*; Taylor, M. D.; Amidon, G. L., Eds.; American Chemical Society, Washington, DC, 1995, pp 449-471.
6. Giannis, A.; Kolter, T. *Angew. Chem. Int. Ed. Engl.* **1993**, *32*, 1244.
7. Olson, G. L.; Bolin, D. R.; Bonner, M. P.; Bös, M.; Cook, C. M.; Fry, D. C.; Graves, B. J.; Hatada, M.; Hill, D. E.; Kahn, M.; Madison, V. S.; Rusiecki, V. K.; Sarabu, R.; Sepinwall, J.; Vincent, G. P.; Voss, M. E. *J. Med. Chem.* **1993**, *36*, 3039.
8. Freidinger, R. M.; Veber, D. F.; Perlow, D. S.; Brooks, J. R.; Saperstein, R. *Science* **1980**, *210*, 656.
9. Di Maio, J.; Bellau, B. *J. Chem. Soc., Perkin Trans. 1* **1989**, 1687.
10. Toniolo, C. *Int. J. Peptide Protein Res.* **1990**, *35*, 287.
11. Kemp, D. S.; Sun, E. T. *Tetrahedron Lett.* **1982**, 3759.
12. Paul, K. P. C.; Burney, P. A.; Campbell, M. M.; Osguthorpe, D. J. *J. Computer-Aided Mol. Design* **1990**, *4*, 239.
13. Nagai, U.; Sato, K.; Nakamura, R.; Kato, R. *Tetrahedron* **1993**, *49*, 3577.
14. Hanessian, S.; McNaughton-Smith, G.; Lombart, H.-G.; Lubell, W. D. *Tetrahedron* **1997**, in press.
15. Rose, G. D.; Gierasch, L. M.; Smith, J. A., In *Advances in Protein Chemistry*; Anfinsen, C. B., Edsall, J. T.; Richards, F. M., Eds.; Academic, Orlando, 1985, Vol. 37, pp 1-109.
16. Freidinger, R. M.; Veber, D. F.; Hirschmann, R.; Paege, L. M. *Int. J. Peptide Protein Res.* **1980**, *16*, 464.
17. Freidinger, R. M., In *Peptides: Synthesis, Structure, Function*; Rich, D. H.; Gross, E., Eds.; Pierce Chemical Company, Rockford, IL, 1981, pp 673-683.
18. Kostis, J. B.; DeFelice, E. A. *Angiotensin Converting Enzyme Inhibitors*; Alan R. Liss, New York, 1987.
19. Pascard, C.; Guilhem, J.; Vincent, M.; Rémond, G.; Portevin, B.; Laubie, M. *J. Med. Chem.* **1991**, *34*, 663.
20. Bohacek, R.; De Lombaert, S.; McMartin, C.; Priestle, J.; Grütter, M. *J. Am. Chem. Soc.* **1996**, *118*, 8231.
21. Thorsett, E. D.; Harris, E. E.; Aster, S. D.; Peterson, E. R.; Tristram, E. W.; Snyder, J. P.; Springer, J. P.; Patchett, A. A.; Ulm, E. H., In *Peptides: Structure and Function*; Hruby, V. J.; Rich, D. H., Eds.; Pierce Chemical Company, Rockford, IL, 1983, pp 555-558.
22. Thorsett, E. D.; Harris, E. E.; Aster, S. D.; Peterson, E. R.; Snyder, J. P.; Springer, J. P.; Hirshfield, J.; Tristram, E. W.; Patchett, A. A.; Ulm, E. H.; Vassil, T. C. *J. Med. Chem.* **1986**, *29*, 251.
23. Thorsett, E. D.; Harris, E. E.; Aster, S.; Peterson, E. R.; Taub, D.; Patchett, A. A.; Ulm, E. H.; Vassil, T. C. *Biochem. Biophys. Res. Commun.* **1983**, *111*, 166.
24. Wolfe, M. S.; Dutta, D.; Aubé, J. *J. Org. Chem.* **1997**, *62*, 654.
25. Yanagisawa, H.; Ishihara, S.; Ando, A.; Kanazaki, T.; Miyamoto, S.; Koike, H.; Iijima, Y.; Oizumi, K.; Matsushita, Y.; Hata, T. *J. Med. Chem.* **1987**, *30*, 1984.
26. Yanagisawa, H.; Ishihara, S.; Ando, A.; Kanazaki, T.; Miyamoto, S.; Koike, H.; Iijima, Y.; Oizumi, K.; Matsushita, Y.; Hata, T. *J. Med. Chem.* **1988**, *31*, 422.
27. Watthey, J. W. H.; Stanton, J. L.; Desai, M.; Babiarz, J. E.; Finn, B. M. *J. Med. Chem.* **1985**, *28*, 1511.
28. Slade, J.; Stanton, J. L.; Deb-David, D.; Mazzenga, G. C. *J. Med. Chem.* **1985**, *28*, 1517.

29. Wyvratt, M. J.; Tischler, M. H.; Ikeler, T. J.; Springer, J. P.; Tristram, E. W.; Patchett, A. A., In *Peptides: Structure and Function*; Hruby, V. J.; Rich, D. H., Eds.; Pierce Chemical Company, Rockford, IL, 1983, pp 551-554.
30. Flynn, G. A.; Giroux, E. L.; Dage, R. C. *J. Am. Chem. Soc.* **1987**, *109*, 7914.
31. Burkholder, T. P.; Huber, E. W.; Flynn, G. A. *Bioorg. Med. Chem. Lett.* **1992**, *3*, 231.
32. Flynn, G. A.; Beight, D. W.; Mehdi, S.; Koehl, J. R.; Giroux, E. L.; French, J. F.; Hake, P. W.; Dage, R. C. *J. Med. Chem.* **1993**, *36*, 2420.
33. MacPherson, L. J.; Bayburt, E. K.; Capparelli, M. P.; Bohacek, R. S.; Clarke, F. H.; Ghai, R. D.; Sakane, Y.; Berry, C. J.; Peppard, J. V.; Trapani, A. J. *J. Med. Chem.* **1993**, *36*, 3821.
34. Fournié-Zaluski, M.-C.; Copric, P.; Turcaud, S.; Rousselet, N.; Gonzalez, W.; Barbe, B.; Pham, I.; Jullian, N.; Michel, J.-B.; Roques, B. P. *J. Med. Chem.* **1994**, *37*, 1070.
35. Delaney, N. G.; Barrish, J. C.; Neubeck, R.; Natarajan, S.; Cohen, M.; Rovnyak, G. C.; Huber, G.; Murugesan, N.; Ravindar, G.; Sieber-McMaster, E.; Robl, J. A.; Asaad, M. M.; Cheung, H. S.; Bird, J. E.; Waldron, T.; Petrillo, E. W. *Bioorg. Med. Chem. Lett.* **1994**, *4*, 1783.
36. Robl, J. A.; Simpkins, L. M.; Stevenson, J.; Kelly, Y. F.; Sun, C.-Q.; Murugesan, N.; Barrish, J. C.; Asaad, M. M.; Bird, J. E.; Schaeffer, T. R.; Trippodo, N. C.; Petrillo, E. W.; Karanewsky, D. S. *Bioorg. Med. Chem. Lett.* **1994**, *4*, 1789.
37. Robl, J. A.; Simpkins, L. M.; Sulsky, R.; Sieber-McMaster, E.; Stevenson, J.; Kelly, Y. F.; Sun, C.-Q.; Misra, R. N.; Ryono, D. E.; Asaad, M. M.; Bird, J. E.; Trippodo, N. C.; Karanewsky, D. S. *Bioorg. Med. Chem. Lett.* **1994**, *4*, 1795.
38. Robl, J. A.; Cimarusti, M. P.; Simpkins, L. M.; Barown, B.; Ryono, D. E.; Bird, J. E.; Asaad, M. M.; Schaeffer, T. R.; Trippodo, N. C. *J. Med. Chem.* **1996**, *39*, 494.
39. Fournié-Zaluski, M.-C.; Coric, P.; Thery, V.; Gonzalez, W.; Meudal, H.; Turcaud, S.; Michel, J.-B.; Roques, B. P. *J. Med. Chem.* **1996**, *39*, 2594.
40. Greenlee, W. J. *Med. Res. Rev.* **1990**, *10*, 173.
41. Williams, P. D.; Perlow, D. S.; Payne, L. S.; Holloway, M. K.; Siegl, P. K. S.; Schorn, T. W.; Lynch, R. J.; Doyle, J. J.; Strouse, J. F.; Vlasuk, G. P.; Hoogsteen, K.; Springer, J. P.; Bush, B. L.; Halgren, T. A.; Richards, A. D.; Kay, J.; Veber, D. F. *J. Med. Chem.* **1991**, *34*, 887.
42. Vacca, J. P.; Fitzgerald, P. M. D.; Holloway, M. K.; Hungate, R. W.; Starbuck, K. E.; Chen, L. J.; Darke, P. L.; Anderson, P. S.; Huff, J. R. *Bioorg. Med. Chem. Lett.* **1994**, *4*, 499.
43. Hungate, R. W.; Chen, J. L.; Starbuck, K. E.; Vacca, J. P.; McDaniel, S. L.; Levin, R. B.; Dorsey, B. D.; Guare, J. P.; Holloway, M. K.; Whittier, W.; Darke, P. L.; Zugay, J. A.; Schleif, W. A.; Emini, E. A.; Quintero, J. C.; Lin, J. H.; Chen, I.-W.; Anderson, P. S.; Huff, J. R. *Bioorg. Med. Chem.* **1994**, *2*, 859.
44. Thaisrivongs, S.; Pals, D. T.; Turner, S. R.; Kroll, L. T. *J. Med. Chem.* **1988**, *31*, 1369.
45. de Laszlo, S. E.; Bush, B. L.; Doyle, J. J.; Greenlee, W. J.; Hangauer, D. G.; Halgren, T. A.; Lynch, R. J.; Schorn, T. W.; Siegl, P. K. S. *J. Med. Chem.* **1992**, *35*, 833.
46. Sham, H. L.; Blois, G.; Stein, H. H.; Fesik, S. W.; Marcotte, P. A.; Plattner, J. J.; Rempel, C. A.; Greer, J. *J. Med. Chem.* **1988**, *31*, 284.
47. Rivero, R. A.; Greenlee, W. J. *Tetrahedron Lett.* **1991**, *32*, 2453.
48. Thaisrivongs, S.; Blinn, J. R.; Pals, D. T.; Turner, S. R. *J. Med. Chem.* **1991**, *34*, 1276.
49. Yang, L.; Weber, A. E.; Greenlee, W. J.; Patchett, A. A. *Tetrahedron Lett.* **1993**, *34*, 7035.
50. Samanen, J.; Hempel, J. C.; Narindray, D.; Regoli, D., In *Chemistry and Biology: Proceedings of the Tenth American Peptide Symposium*; Marshall, G. R., Ed.; ESCOM, Leiden, 1988, pp 123-125.
51. Samanen, J.; Cash, T.; Narindray, D.; Brandeis, E.; Adams, W., Jr.; Weideman, H.; Yellin, T.; Regoli, D. *J. Med. Chem.* **1991**, *34*, 3036.
52. Aebi, J. D.; Guillaume, D.; Dunlap, B. E.; Rich, D. H. *J. Med. Chem.* **1988**, *31*, 1805.
53. Lee, J. P.; Dunlap, B.; Rich, D. H. *Int. J. Peptide Protein Res.* **1990**, *35*, 481.
54. Zhang, Y. L.; Dawe, A. L.; Jiang, Y.; Becker, J. M.; Naider, F. *Biochem. Biophys. Res. Commun.* **1996**, *224*, 327.

55. Yu, K.-L.; Rajakumar, G.; Srivastava, L. K.; Mishra, R. K.; Johnson, R. L. *J. Med. Chem.* **1988**, *31*, 1430.
56. Sreenivasan, U.; Mishra, R. K.; Johnson, R. L. *J. Med. Chem.* **1993**, *36*, 256.
57. Valle, G.; Crisma, M.; Toniolo, C.; Yu, K.-L.; Johnson, R. L. *J. Chem. Soc., Perkin Trans. 2* **1989**, 83.
58. Valle, G.; Crisma, M.; Toniolo, C.; Yu, K.-L.; Johnson, R. L. *Int. J. Peptide Protein Res.* **1989**, *33*, 181.
59. Subsinghe, N. L.; Bontems, R. J.; McIntee, E.; Mishra, R. K.; Johnson, R. L. *J. Med. Chem.* **1993**, *36*, 2356.
60. Genin, M. J.; Mishra, R. K.; Johnson, R. L. *J. Med. Chem.* **1993**, *36*, 3481.
61. Genin, M. J.; Johnson, R. L. *J. Am. Chem. Soc.* **1992**, *114*, 8778.
62. Ede, N. J.; Lim, N.; Rae, I. D.; Ng, F. M.; Hearn, M. T. W. *Peptide Res.* **1991**, *4*, 171.
63. Ede, N. J.; Rae, I. D.; Hearn, M. T. W. *Int. J. Peptide Protein Res.* **1994**, *44*, 568.
64. Abell, A. D.; Oldham, M. D. *J. Org. Chem.* **1997**, *62*, 1509.
65. Ward, P.; Ewan, G. B.; Jordan, C. C.; Ireland, S. J.; Hagan, R. M.; Brown, J. R. *J. Med. Chem.* **1990**, *33*, 1848.
66. Hinds, M. G.; Richards, N. G. J.; Robinson, J. A. *J. Chem. Soc., Chem. Commun.* **1988**, 1447.
67. Genin, M. J.; Gleason, W. B.; Johnson, R. L. *J. Org. Chem.* **1993**, *58*, 860.
68. Hagan, R. M.; Ireland, S. J.; Jordan, C. C.; Beresford, I. J. M.; Stephens-Smith, M. L.; Ewan, G.; Ward, P. *Br. J. Pharmacol.* **1990**, *99*, 62P.
69. Hagan, R. M.; Ireland, S. J.; Bailey, F.; McBride, C.; Jordan, C. C.; Ward, P. *Br. J. Pharmacol.* **1991**, *102*, 168P.
70. Hagan, R. M.; Ireland, S. J.; Jordan, C. C.; Beresford, I. J. M.; Deal, M. J.; Ward, P. *Neuropeptides* **1991**, *19*, 127.
71. Edmunds, J. J.; Rapundalo, S. T. *Ann. Rep. Med. Chem.* **1996**, *31*, 51.
72. Semple, J. E.; Rowley, D. C.; Brunck, T. K.; Ha-Uong, T.; Minami, N. K.; Owens, T. D.; Tamura, S. Y.; Goldman, E. A.; Siev, D. V.; Ardecky, R. J.; Carpenter, S. H.; Ge, Y.; Richard, B. M.; Nolan, T. G.; Håkanson, K.; Tulinsky, A.; Nutt, R. F.; Ripka, W. C. *J. Med. Chem.* **1996**, *39*, 4531.
73. Okayama, T.; Seki, S.; Ito, H.; Takeshima, T.; Hagiwara, M.; Morikawa, T. *Chem. Pharm. Bull.* **1995**, *43*, 1683.
74. Newlander, K. A.; Callahan, J. F.; Moore, M. L.; Tomaszek, T. A., Jr.; Huffman, W. F. *J. Med. Chem.* **1993**, *36*, 2321.
75. Hoog, S. S.; Zhao, B.; Winborne, E.; Fisher, S.; Green, D. W.; DesJarlais, R. L.; Newlander, K. A.; Callahan, J. F.; Moore, M. L.; Huffman, W. F.; Abdel-Meguid, S. S. *J. Med. Chem.* **1995**, *38*, 3246.
76. Duggan, M. E.; Naylor-Olsen, A. M.; Perkins, J. J.; Anderson, P. S.; Chang, C. T.-C.; Cook, J. J.; Gould, R. J.; Ihle, N. C.; Hartman, G. D.; Lynch, J. J.; Lynch, R. J.; Manno, P. D.; Schaffer, L. W.; Smith, R. L. *J. Med. Chem.* **1995**, *38*, 3332.
77. Sato, M.; Lee, J. Y. H.; Nakanishi, H.; Johnson, M. E.; Chrisciel, R. A.; Kahn, M. *Biochem. Biophys. Res. Commun.* **1992**, *187*, 999.
78. Bernstein, P. R.; Gomes, B. C.; Kosmider, B. J.; Vacek, E. P.; Williams, J. C. *J. Med. Chem.* **1995**, *38*, 212.
79. Beholz, L. G.; Benovsky, P.; Ward, D. L.; Barta, N. S.; Stille, J. R. *J. Org. Chem.* **1997**, *62*, 1033.
80. Freidinger, R. M.; Perlow, D. S.; Veber, D. F. *J. Org. Chem.* **1982**, *47*, 104.
81. Ede, N. J.; Rae, I. D.; Hearn, M. T. W. *Tetrahedron Lett.* **1990**, *31*, 6071.
82. Zydowsky, T. M.; Dellaria, J. F., Jr.; Nellans, H. N. *J. Org. Chem.* **1988**, *53*, 5607.
83. Freidinger, R. M. *J. Org. Chem.* **1985**, *50*, 3631.
84. Garvey, D. S.; May, P. D.; Nadzan, A. M. *J. Org. Chem.* **1990**, *55*, 936.
85. Robl, J. A.; Cimarusti, M. P.; Simpkins, L. M.; Weller, H. N.; Pan, Y. Y.; Malley, M.; DiMarco, J. D. *J. Am. Chem. Soc.* **1994**, *116*, 2348.

86. Wirth, T. *Angew. Chem., Int. Ed. Engl.* **1997**, *36*, 225.
87. Seebach, D.; Boes, M.; Naef, R.; Schweizer, W. B. *J. Am. Chem. Soc.* **1983**, *105*, 5390.
88. Karady, S.; Amato, J. S.; Weinstock, L. M. *Tetrahedron Lett.* **1984**, *25*, 4337.
89. Acton, J. J., III; Jones, A. B. *Tetrahedron Lett.* **1996**, *37*, 4319.
90. Miller, S. J.; Grubbs, R. H. *J. Am. Chem. Soc.* **1995**, *117*, 5855.
91. Martin, S. F.; Liao, Y.; Chen, H.-J.; Pätzel, M.; Ramser, M. N. *Tetrahedron Lett.* **1994**, *35*, 6005.
92. Kinoshita, A.; Mori, M. *J. Org. Chem.* **1996**, *61*, 8356.
93. Nishikawa, N.; Komazawa, H.; Orisaka, A.; Yoshikane, M.; Yamaguchi, J.; Kojima, M.; Ono, M.; Itoh, I.; Azuma, I.; Fujii, H.; Murata, J.; Saiki, I. *Bioorg. Med. Chem. Lett.* **1996**, *6*, 2725.
94. Overkleeft, H. S.; Pandit, U. K. *Tetrahedron Lett.* **1996**, *37*, 547.
95. Gennari, C.; Colombo, L.; Bertolini, G. *J. Am. Chem. Soc.* **1986**, *108*, 6394.
96. Evans, D. A.; Britton, T. C.; Dorow, R. L.; Dellaria, J. F. *Tetrahedron* **1988**, *44*, 5525.
97. Oppolzer, W.; Moretti, R. *Tetrahedron* **1988**, *44*, 5541.
98. Oppolzer, W.; Tamura, O.; Sundarababu, H.; Signer, M. *J. Am. Chem. Soc.* **1992**, *114*, 5900.
99. Rodríguez, R.; Estiarte, M. A.; Diez, A.; Rubiralta, M.; Colell, A.; García-Ruiz, C.; Fernández-Checa, J. C. *Tetrahedron* **1996**, *52*, 7727.
100. Blaser, H.-U.; Boyer, S. K.; Pittelkow, U. *Tetrahedron: Asymm.* **1991**, *2*, 721.
101. Aubé, J.; Wolfe, M. S. *Bioorg. Med. Chem. Lett.* **1992**, *2*, 925.
102. Aubé, J.; Wolfe, M. S.; Yantiss, R. K.; Cook, S. M.; Takusagawa, F. *Synth. Commun.* **1992**, *22*, 3003.
103. Durst, T.; Koh, K. *Tetrahedron Lett.* **1992**, *33*, 6799.
104. Besse, P.; Veshambre; Dickman, M.; Chênevert, R. *J. Org. Chem.* **1994**, *59*, 8288.
105. Martinez, L. E.; Leighton, J. L.; Carsten, D. H.; Jacobsen, E. N. *J. Am. Chem. Soc.* **1995**, *117*, 5897. See this paper for references to earlier approaches to this problem.
106. Chang, H.-T.; Sharpless, K. B. *Tetrahedron Lett.* **1996**, *37*, 3219.
107. Aubé, J.; Wang, Y.; Hammond, M.; Tanol, M.; Takusagawa, F.; Vander Velde, D. *J. Am. Chem. Soc.* **1990**, *112*, 4879.
108. Aubé, J. *Chem. Soc. Rev.* **1997**, in press.
109. Stetin, C.; de Jeso, B.; Pommier, J. C. *Synth. Commun.* **1982**, *12*, 495.
110. Gawley, R. E. *Org. React.* **1988**, *35*, 1.
111. Kitagawa, O.; Vander Velde, D.; Dutta, D.; Morton, M.; Takusagawa, F.; Aubé, J. *J. Am. Chem. Soc.* **1995**, *117*, 5169.
112. Gallop, M. A.; Barrett, R. W.; Dower, W. J.; Fodor, S. P. A.; Gordon, E. M. *J. Med. Chem.* **1994**, *37*, 1233.

TOWARD RATIONALLY DESIGNED PEPTIDYL–PROLYL ISOMERASE INHIBITORS

Juris Paul Germanas, Kyonghee Kim, and Jean-Philippe Dumas

Advances in Amino Acid Mimetics and Peptidomimetics
Volume 1, pages 233-250

ISBN: 0-7623-0200-3

ABSTRACT

The design, synthesis, and structural and biological evaluation of novel peptidomimetic inhibitors of peptidyl–prolyl isomerases (PPIases) is described. The lactams were designed to mimic the solid-state enzyme-bound conformation of peptide substrates of the PPIase cyclophilin. A series of lactams, differing in the identity of functional groups representing the side chains of amino acids, were prepared in a stereo- and enantioselective fashion. Comparison of the crystallographically determined structures of the lactams with those of the enzyme-bound substrates revealed exceptional conformational similarity. A number of lactams bound to cyclophilin with micromolar affinity; lactams with aryl side chains displayed the greatest affinity. These results suggest that the *s-cis* isomer of the substrate selectively binds to the enzyme in solution, and that potent inhibitors of PPIases can be prepared using rational design methods.

1. INTRODUCTION

1.1. Proline Isomerism in Biology

The chemical bond which connects individual amino acid residues in peptides and proteins is the amide, or peptide, bond, which is formed between the carboxyl group of one residue and the amino group of the subsequent residue. Amide bonds are planar structures as a result of delocalization of the lone pair of electrons of the sp^2-hybridized nitrogen atom onto the π-system of the carbonyl group; consequently, the C(O)–N bond can exhibit isomerism and exist in *s-cis* or *s-trans* configurations.[1] In peptides the *s-trans* isomer of an amide bond involving a primary amino group is substantially more stable than the *s-cis* isomer (ΔG^o = -2.6 kcal/mol).[2] In contrast the difference in free energy between *s-cis* and *s-trans* amides involving secondary amino acids such as proline is small (ΔG^o = -0.5 kcal/mol). As a result readily detectable quantities of both isomers are often present at equilibrium (Figure 1).[2]

Interconversion of the *s-cis* and *s-trans* forms of proline-containing peptides occurs by rotation about the C(O)–N bond. This process has been determined to be a difficult one, with free energy barriers in the 18–21 kcal/mol range.[3] The substantial hindrance to rotation has been postulated to arise from loss of the resonance stabilization of the amide linkage in the transition state, which is characterized by an orthogonal orientation of the lone pair of the nitrogen atom and the carbonyl π-bond, as in structure **I** (Figure 1)

Proline isomerism has been implicated in a number of important biochemical processes.[4] For example, isomerization of XaaPro bonds from "non-native" to "native" configurations has been shown to be the rate-determining step of folding of many proteins.[5] Specific cases of receptor-mediated transmembrane signaling have been demonstrated to be mediated by Pro peptide bond isomerization.[6] Also the isomerization of specific proline residues of viral coat proteins has been postulated to control the assembly of the human immunodeficiency virus (HIV).[7]

1.2. Peptidyl–Proline Isomerases: Structure and Function

In living organisms, XaaPro bond isomerization is catalyzed by a class of enzymes called peptidyl–prolyl isomerases, or PPIases.[8] PPIases are divided into two distinct families: the cyclophilins[9] and the FKBPs.[10] Although these enzymes share a common activity, they display no primary or tertiary structural homology. Analysis of the primary structures of PPIases from prokaryotic and eukaryotic sources has shown that the amino acid sequences are strongly conserved within each enzyme family.[8] Additionally the intracellular concentrations of the PPIases have been demonstrated to be substantial, further signifying a critical role for these enzymes in maintaining cellular viability.[9]

It is now known that the cyclophilins and FKBPs occur as multiple isoforms which are localized in different subcellular compartments (Table 1).[11-14] In humans, for example, cyclophilin A, the most abundant isoform, is located in the cytoplasm[11] along with another variant, cyclophilin 40.[12] Cyclophilins B and C are targeted to the endoplasmic reticulum (ER),[13] while cyclophilin D locates itself within mitochondria.[14] Although homology among the isoforms is generally high, examination of the sequences of human cyclophilin isoforms reveals that, besides extensions at the termini for targeting to specific organelles, considerable sequence variability also exists within each gene.

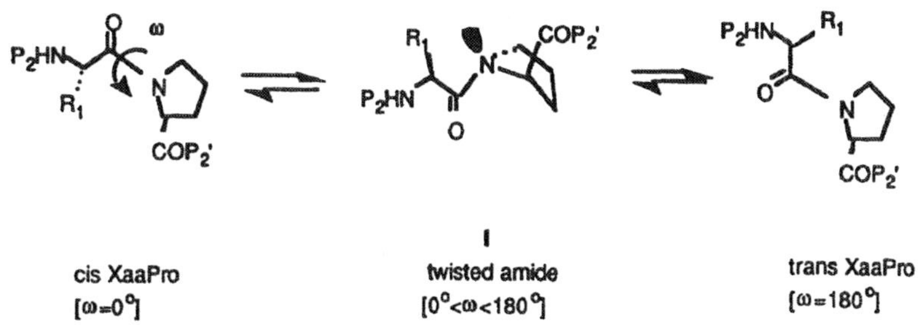

Figure 1. Ground and transition states for isomerization of XaaPro dipeptides.

Table 1. Properties and Locations of Cyclophilin Isoforms[11-15]

Name	*Source*	MW (*kDa*)	*Relative Activity (AAPFpNA)*	*Cyclosporin A Affinity (nM)*	*Cellular Location*
Cyclophilin A	*H. sapiens*	18	1.0	6	cytoplasm
Cyclophilin B	*H. sapiens*	21	0.5	9	ER
Cyclophilin D	*H. sapiens*	18	0.5	10	mitochondria
Cyclophilin 40	*H. sapiens*	40	0.2	300	cytoplasm
Cyclophilin	*E. coli*	18	4.0	3000	cytoplasm

Two recent medically important discoveries have considerably increased interest in PPIase structure and function. The first finding was that the cyclophilins and the FKBPs were the cellular receptors for the immunosuppressive agents, cyclosporin A and FK 506, respectively.[9,10] These compounds formed tight complexes with their respective receptors, binding to their active sites and inhibiting their PPIase activities. Further studies revealed that inhibition of isomerase activity was not sufficient for immune system suppression. The immunosuppressant–PPIase complex in fact mediated the immunosuppressive activity of the agents by inhibiting the phosphatase function of calcineurin, an enzyme involved in signal transduction. Phosphatase binding involved an effector domain on the suppressants that did not contact the PPIase active site.[16]

A second finding was the discovery that HIV-1 virions contain high concentrations of cyclophilin A, which formed a tight complex with the Gag polyprotein of the virus.[17,18] Mutagenesis studies established that cyclophilin A interacted with the HIV-1 Gag polyprotein at a specific, highly conserved section of the protein in the vicinity of Pro222. The PPIase has been postulated to promote correct assembly of the mature virion by facilitating folding of the p24 capsid protein, a structural component of the viral coat.[19] The presence of the cyclophilin A–Gag complex was essential for viral pathogenicity, since disruption of cyclophilin binding to Gag with cyclosporin A eliminated viral replication.

1.3. Mechanism of PPIase-Mediated XaaPro Bond Isomerization

As a result of their ubiquitous nature and the importance of their function, a great deal of attention has focused on the mechanism by which PPIases accelerate the rate of amide bond isomerization. Kinetic studies of peptide bond isomerization in the presence of PPIases have revealed that these enzymes lower the free energy of activation for XaaPro isomerization from 15 to 20 kcal/mol to about 8 kcal/mol.[20] Initial mechanistic hypotheses speculated that the carbonyl carbon of the amide bond underwent nucleophilic attack, affording a "tetrahedral intermediate," which had a dramatically lowered activation barrier for C–N bond rotation.[21] Analyses of mutant enzymes[22] and determinations of kinetic isotope effects on substrate isomerization,[23] however, discounted the intermediacy of such species.

Presently one particular mechanistic pathway that is consistent with the considerable body of experimental evidence appears to have gained wide acceptance.[20,23,24] This proposal has been termed the "twisted amide" mechanism, in reference to the transition state through which the substrate passes. According to this proposal, as the C(O)–N bond of the bound XaaPro substrate undergoes rotation, the planar amide bond is distorted until a transition state is achieved in which the proline nitrogen lone pair is no longer coplanar with the carbonyl π-orbital, as in **I** (Figure 1). In this conformation the overlap between the π-system of the carbonyl group and the lone pair on the nitrogen atom is abolished, and localization of electron density onto the proline nitrogen results in rehybridization of the nitrogen center from sp^2 to sp^3.

One of the main assertions of enzymology is that enzyme active sites are preorganized to bind transition states of reactions rather than ground states.[25] Using models constructed from the high-resolution X-ray structure of FKBP, two groups have investigated the pathway for isomerization of the peptide substrate AlaAlaProPhe by this enzyme using *ab initio* calculations and molecular dynamics simulation.[26,27] The main features of the two proposed mechanisms are similar. According to this model, the peptide first binds to the active site in a ground state conformation with a planar AlaPro amide bond exclusively in the *s-cis* configuration. Rotation about the C-N bond results in a transition-state structure that is characterized by a 40° to 90° twist angle ω of the amide bond out of coplanarity with the nitrogen lone pair, and by a proline nitrogen atom that is significantly pyramidalized (Figure 1). According to the model, this twisted amide structure is stabilized by hydrogen bonds from the side chain of Tyr82 to the substrate Ala carbonyl oxygen and from the backbone amide of Phe4 to the substrate Pro nitrogen, as well as by specific van der Waals interactions.[26,27] A similar model has been proposed for the mechanism of cyclophilin, although the precise enzyme residues involved in transition-state stabilization are different.[28]

1.4. Structural Studies of Cyclophilin–Peptide Complexes

Mechanistic speculations regarding cyclophilin-mediated peptide isomerization have been greatly aided by the availability of high-resolution structural information on noncovalent enzyme–substrate complexes. For example crystal structures have been determined for complexes of cyclophilin A with four different XaaPro dipeptides,[28,29] and for complexes of cyclophilin A with two different tetrapeptide substrates.[30,31] In all cases the substrates were found to occupy the same groove on the enzyme's surface and adopt similar conformations. Additionally the conformation of the tetrapeptide substrate *N*-succinylAlaAlaProPhe *p*-nitroanilide (sucAAPF-pNA) in the solution phase bound to cyclophilin A has been examined by NMR spectroscopy.[32,33] A consistent observation from these structural studies is that the XaaPro peptide bond of the bound substrate is decidedly planar. A further observation from the crystal structures is that the XaaPro linkage of the bound peptide exists entirely in the *s-cis* configuration.

The structural details of the cyclophilin–substrate complexes provide insight into mechanism of PPIase-mediated peptide isomerization. For example, it is clear from the studies that the free energy of formation of the complex of cyclophilin with the ground state of the substrate in the *s-cis* configuration is sizable, since only the *s-cis* isomer is detected in the complex. In fact the stability of the *s-cis* peptide-enzyme complexes has evoked the suggestion that cyclophilin has evolved to preferentially catalyze *s-trans* to *s-cis* peptide interconversions.[29] Secondly the cyclophilin–peptide structures have made possible the identification of enzyme amino acids that are in contact with atoms of the substrate and that would likely play roles in catalysis of amide isomerization (Figure 2). Consistent with these conjec-

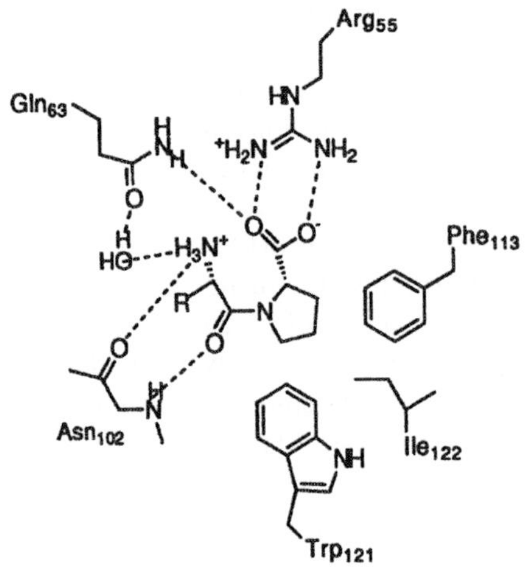

Figure 2. X-ray-derived model of the active site of cyclophilin A-dipeptide complexes.

Table 2. Catalytic Activity and Cyclosporin A Inhibition of Mutant Human Cyclophilin A Proteins[34]

Enzyme	k_{cat}/K_M ($mM^{-1}\ s^{-1}$)	K_i *Cyclosporin A* (nM)
WT	16.0	17 ±2
H54Q	2.4	40 ±10
Q111A	2.4	130 ±20
W121A	1.4	290 ±20
F113A	0.48	190 ±30
H126Q	0.084	ND[a]
F60A	0.051	ND
R55A	0.016	ND

Note: [a] Not determined.

tures are results from site-directed mutagenesis studies, in which enzyme variants with alterations at several of these conserved residues displayed substantially diminished activity (Table 2).[34]

2. TOWARD NEW CYCLOPHILIN INHIBITORS

2.1. Probing the Catalytic Mechanism of Cyclophilin with XaaPro Peptidomimetics

Despite the considerable amount of information on PPIase-mediated peptide isomerization, significant questions regarding the catalytic mechanism remain unanswered. For example, it is not clear whether the enzymes bind one configurational isomer of the substrate more strongly than the other in solution. Thus while the *s-cis* isomer of the peptide substrate sucAAPF*p*NA appears to be preferentially bound to cyclophilin in the solid state, NMR studies have indicated that both *s-cis* and *s-trans* isomers bind to the enzyme in the solution phase.[33] Additionally the precise roles of individual amino acids of cyclophilin in stabilizing the ground state or transition state of the substrate have not been established.

According to the X-ray structural studies, the active site of cyclophilin is preorganized to bind substrates in a well-defined backbone conformation characterized by a *s-cis* peptide bond between the Xaa and Pro residues. More specifically the conformation of the XaaPro segment bound to cyclophilin A corresponds to the type VI reverse or β-turn conformation, a common structural feature of peptides and proteins (structure **II** of Figure 3).[35] This backbone conformation is identical to the conformation of the LeuPro segment of a peptide substrate bound to the PPIase FKBP, according to the computer-generated model.[26,27] Altogether these findings suggest that the type-VI turn conformation is a molecular recognition feature of PPIases.

A novel method of investigating the structural features of the receptor-bound state of a flexible ligand is to prepare a conformationally constrained analogue of the ligand and determine its affinity to the receptor. In the case of the PPIases, valuable insight into the catalytic mechanism could be elucidated by the synthesis and evaluation of XaaPro analogues constrained to either *s-cis* or *s-trans* configura-

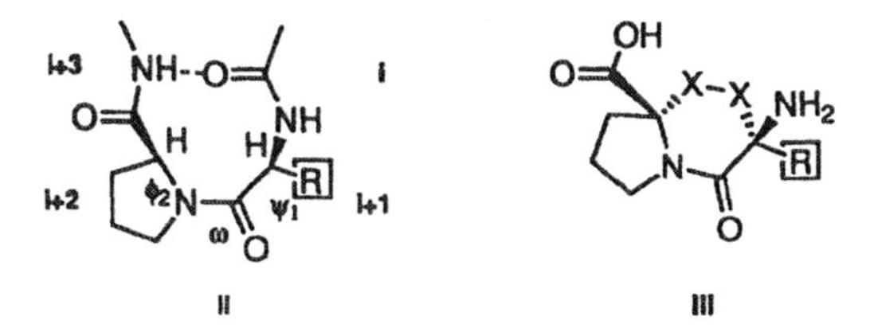

Figure 3. Structures of the type VI β-turn (**II**) and the dipeptide mimic (**III**).

tions. The well-defined conformation of cyclophilin-bound XaaPro peptides in the solid state inspired us to design and synthesize rigid structural mimics of the solid-state conformation. The enzyme-binding properties of such compounds would establish whether the *s-cis* conformation is also adopted in the solution phase.

Our approach to the design of a conformationally constrained analogue of the type VI turn was initiated by a structural database search using the program CAVEAT.[36] Based on the information obtained through this query, a series of compounds of the general structure **III** were chosen as first generation type VI turn mimics (Figure 3). The lactam **III** ($X = CH_2$) was selected for the first synthetic forays. The incorporation of a two atom bridge joining the α-carbons of the Xaa and Pro residues, as in bicycle **III**, removed the possibility of rotation about the $\psi 1$, ω, and $\phi 2$ bonds. Functional groups R in structure **III** corresponded to side chains of amino acids Xaa in constrained XaaPro dipeptides.

The suitability of the lactam **III** as a general type VI turn mimetic was first probed by molecular modeling. Structure **III** ($X = CH_2$, R = H) was built and its minimum energy conformation determined using molecular mechanics with the CHARMM force field. A conformational search was applied to the six-membered ring of the lactam, and different conformations were used as starting points for the calculation. One specific conformation of lactam **III** corresponded to the minimum energy structure. According to the model, the six-membered ring of lactam **III** adopted a pseudo-chair conformation, with the pendant amino group located in the pseudo-equatorial position. Next an overlay algorithm that minimizes the deviation between the positions of analogous atoms of two or more structures was used to assess the structural similarity of the lowest energy conformation of lactam **III** ($X = CH_2$, R = H) with the type VI turn. Using the main-chain and side-chain atoms of the type VI turn of RNase (Tyr92-Pro93) as a template, a superimposition of the two structures revealed high conformational similarity, with an RMS deviation of less than 0.15 Å for the main chain atoms (Figure 4).

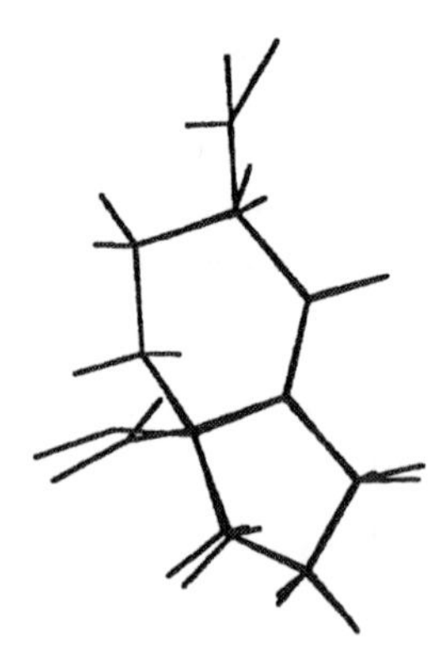

Figure 4. Superimposition of minimum energy conformation of compound **III** with analogous atoms of Tyr92-Pro93 of RNase.

2.2. Synthesis of Novel XaaPro Type VI Turn Mimics

The starting material for the preparation of the mimics was (*R*)-2-(2-propenyl)proline (**1**), which was readily available as a single enantiomer from (*S*)-proline following the elegant protocol of Seebach.[38] Protection of the amino and carboxyl functionalities as benzyloxycarbamates and methyl esters, respectively, furnished the ester **2** (Scheme 1). Oxidative cleavage of the double bond with periodate and osmium tetroxide afforded the aldehyde **3** in good yield. Introduction of a glycyl anion equivalent was achieved by coupling aldehyde **3** with the anion of *N*-(*tert*-butyloxycarbonyl)-α -phosphonoglycine dimethyl ethyl ester;[39] an equimolar mixture of alkenes **4*E*** and **4*Z*** was obtained after silica gel chromatography. Exposure of the isomeric mixture of alkenes to hydrogen in the presence of a palladium catalyst effected concomitant reduction of the double bond and hydrogenolysis of the benzyloxycarbonyl group to provide the amino ester **5** as a mixture of isomers at the chiral center of the pendant alkyl chain. Thermolysis of the amino ester **5** in refluxing toluene afforded the isomeric bicyclic lactams **6** and **7**.

The thermolysis products could be separated by silica gel chromatography and individually characterized. The similarities of their NMR spectral characteristics supported the hypothesis that these compounds were the isomeric lactams **6** and **7**.

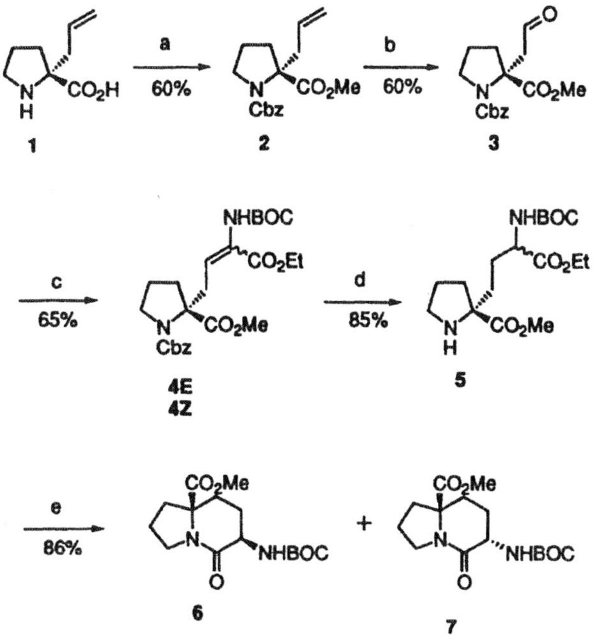

Scheme 1. (**a**) (i) CbzCl, Et_3N, RT, 3 d; (ii) CH_2N_2, Et_2O; (**b**) cat. OsO_4; $NaIO_4$; (**c**) $BOCHNCH[PO(OCH_3)_2]CO_2Et$, LDA, -78 to 0° C; (**d**) H_2, Pd-C; (**e**) Et_3N, $PhCH_3$, relfux.

The relative stereochemistry of the substituents on the six-membered ring and the conformation of the bicyclic skeleton of each lactam isomer were determined by NMR spectroscopy, and confirmed by X-ray crystallography (*vida infra*). Thus the less polar isomer was found to have the amino and carbomethoxy groups *trans* to each other on the six-membered ring, while the two functionalities were situated *cis* in the more polar isomer. Lactam **6** represented a GlyPro dipeptide constrained to the type VI-β-turn conformation.

Modification of lactam **6** by the introduction of alkyl groups at the amino-bearing carbon atom (as in structure **III**, R = alkyl) would afford mimics of XaaPro type VI β-turn dipeptides, where the alkyl group R represented the side chain of amino acid Xaa. Such a transformation was achieved with exceptionally high stereoselectivity by alkylation of the enolate of amidine **10** (Scheme 2). For example alkylation of the enolate of amidine **10** with benzyl bromide proceeded from the α-face to afford the benzyl derivative **11a** as the exclusive product; the stereochemical outcome of the alkylation was confirmed by 2-D NMR spectroscopy as well as by X-ray crystallography. The stereoselectivity of the alkylation could be rational-

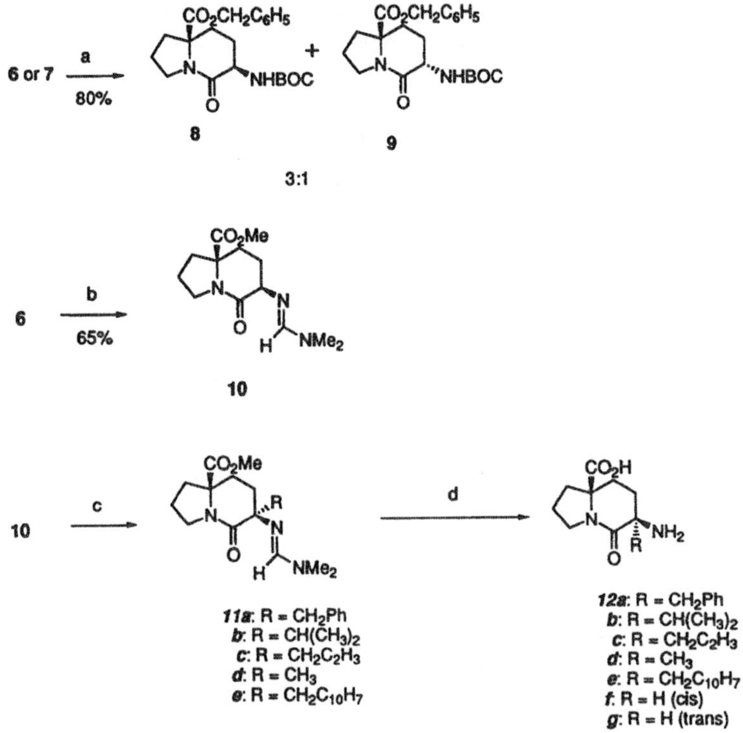

6 or 7 —a, 80%→ **8** + **9** (3:1)

6 —b, 65%→ **10**

10 —c→ **11** —d→ **12**

11a: R = CH_2Ph
b: R = $CH(CH_3)_2$
c: R = $CH_2C_2H_3$
d: R = CH_3
e: R = $CH_2C_{10}H_7$

12a: R = CH_2Ph
b: R = $CH(CH_3)_2$
c: R = $CH_2C_2H_3$
d: R = CH_3
e: R = $CH_2C_{10}H_7$
f: R = H (cis)
g: R = H (trans)

Scheme 2. (**a**) $C_6H_5CH_2OH$, Na, RT, 24 h; (**b**) (i) CF_3CO_2H, RT (ii) $Me_2NCH(OMe)_2$, reflux, 6 h. (**c**) (i) $KN(TMS)_2$, -78 °C; (ii) RX (**d**) KOH, H_2O.

ized by stereoelectronically favored pseudoaxial attack of the electrophile onto the planar enolate of formamidine **10**, as well as by sterically favored electrophilic attack away from the angular carbomethoxy group. Structurally benzyl lactam **11a** possessed the absolute stereochemistry and conformation of L-Phe-L-Pro in a type VI β-turn.

Stereoselective alkylation of amidine **10** turned out to be a general procedure for formation of XaaPro mimics. Different alkyl halides could be combined with the enolate of **10** to provide a wide variety of mimics (Table 3). In all cases only the product with *cis* stereochemistry was detected. Hydrolysis of the amidines **11** afforded the corresponding amino acids **12** in good overall yield after ion-exchange chromatography.

2.3. Structural Characterization of Type VI Turn Mimics

Insight into the structural characteristics of the type VI β-turn mimics was obtained by X-ray crystallography. The crystal structure of lactam **6** confirmed the stereochemical assignment from NMR experiments (Figure 5). In the crystalline state the six-membered ring of the bicycle indeed adopted a pseudo-chair conformation with the amido substituent oriented pseudo-equatorially, as suggested by NMR spectroscopy. The conformational similarity of lactam **6** to the type-VI turn was assessed by superimposition of the main-chain and side chain atoms of **6** with the corresponding atoms of the type-VI turn from RNase (Tyr92-Pro93). The two structures displayed exceptional similarity, with 0.15 Å RMS overall deviation in backbone and five-membered ring atom positions.[40]

Table 3. Structures and Yields of α-Alkyl Amidines **12** from **10**

Compound	*R*	*Yield (%)*
12a	$CH_2C_6H_5$	43
12b	$CH(CH_3)_2$	48
12c	CH_2CHCCH_2	65
12d	CH_3	75
12e	β-$CH_2C_{10}H_7$	46

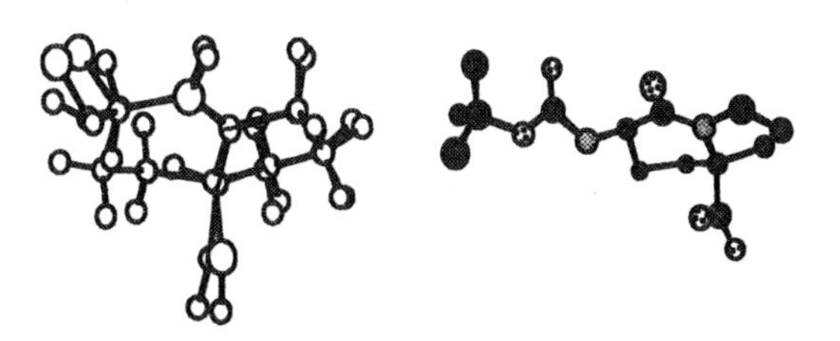

Figure 5. X-ray crystal structure of lactam **6**, and overlay of **6** with analogous atoms of Tyr92-Pro93 of RNase.

Although amides generally adopt conformations in which the six atoms of the functional group are coplanar,[1] in unusual cases, such as in strained ring systems, amide bonds may adopt conformations where the amide bond is distorted from planarity.[41] In the case of the lactam **6**, the value of ω for the amide bond was found to be $-14.6 \pm 0.5°$. In contrast the value of ω for the crystallographically characterized, unstrained amide caprolactam, which also contains a *s-cis* lactam bond, was $-4.2 \pm 0.4°$.[42] A more precise description of amide bond distortion is obtained from an analysis of atom positions by the method of Dunitz and Winkler.[43] The Dunitz parameters χN and τ' reflect the N-atom pyramidalization and rotation about the N–C(O) bond, respectively. Deviations of the values for these parameters from 0° signify distortion of a *s-cis* amide function. For the lactam bond of compound **6**, the values of χN and τ' are $6.8 \pm 0.5°$ and $-5.9 \pm 0.5°$, respectively. The values for χN and τ' for the *s-cis* amide bond of caprolactam are -1.6° and 2.6°, respectively.[42] Thus the amide bond of lactam **6** displays a distorted conformation relative to analogous monocyclic lactam bonds; the degree of nonplanarity of the amide function of lactam **6** is similar in magnitude to that seen for the amide bond in *s-cis* caprylolactam ($\chi N = -6.4°$ and τ' 4.3°),[43] oneof the more strained monocyclic amides. The distortion of the lactam bond of **6** is believed to arise from strain introduced into the piperidinone ring in order to accommodate fusion with the five-membered ring.

The crystal structure of amidine **11a** affirmed the stereochemical outcome of the alkylation reaction (Figure 6). Similarly to lactam **6**, the six-membered ring of amidine **11a** adopted a pseudo-chair conformation. Like lactam **6**, the amide bond

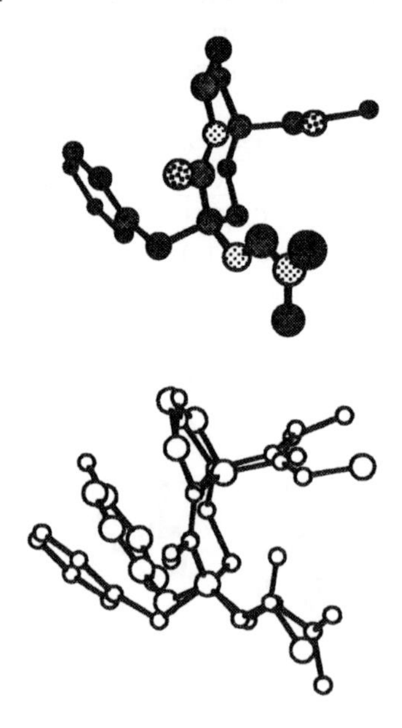

Figure 6. X-ray crystal structure of amidine **11a** and overlay of **11a** with analogous atoms of Tyr92-Pro93 of RNase.

of compound **11a** was distorted from planarity ($\omega = -12°$). In the crystalline state, the benzyl side chain of compound **11a** assumed a conformation in which the phenyl ring was directed underneath and coplanar with the pyrrolidine ring (Figure 6). Interestingly an identical orientation was found for the side chains of Phe or Tyr in PhePro and TyrPro type-VI turns in protein crystal structures[35] as well as in Xaa(aryl)Pro peptides in solution (Figure 6).[44] Though the conformation of the side chain of amidine **11a** in solution is still under investigation, the crystal structure of **11a** and the structures of Xaa(aryl)Pro peptides suggests that the stacking of the aryl and pyrollidine rings seen in these structures may be a stereoelectronically favored arrangement of these groups.

2.4. Biological Evaluation of XaaPro Type VI Turn Mimics

X-ray crystal structures of complexes of cyclophilin A with dipeptides XaaPro (Xaa = Ala, Gly, His, and Ser) have been determined to high resolution.[28,29] In all cases the backbone of the dipeptide adopted the type VI β-turn conformation and atoms of the dipeptide made similar contacts with atoms of the enzyme. Comparison of the enzyme–dipeptide structures with the enzyme–tetrapeptide structures, however, revealed differences in the contacts made between the enzyme and peptides. These differences have led to the suggestion that dipeptides are not true substrates, but rather competitive inhibitors of the enzyme.[29] The exceptional conformational similarity of mimics **12** to the enzyme-bound XaaPro peptides suggested that the mimics would form tight complexes with cyclophilin.

Two different methods were used to investigate the interaction of the bicyclic lactam dipeptide mimics **12** with cyclophilin. In the first method, the change in the intrinsic fluorescence of the enzyme upon addition of the lactams was monitored. Cyclophilin A possesses a unique Trp residue at position 121, which is adjacent to the active site. Binding of ligands such as cyclosporin A results in an increase in the fluorescence intensity of the indole ring of Trp121.[22] This effect has been attributed to an increase in the hydrophobicity of the environment surrounding the Trp residue. In the second method, the inhibition of the PPIase activity of the enzyme by lactams **12** was determined using the peptidase-coupled isomerase assay.[20] Data from both methods could be analyzed to calculate values for dissociation constants (K_D) or inhibition constants (K_i), respectively.

Human cyclophilin A was obtained from a recombinant expression system in *E. coli*.[34] The recombinant enzyme possessed specific activity similar to that of the natural enzyme with the substrate sucAAPF-pNA. The recombinant cyclophilin also bound cyclosporin A according to the fluorescence method, although the value for K_D determined by us (1 μM) was somewhat higher than the literature values (20 to 400 nM).

Titration of cyclophilin with lactams **12** resulted in changes in the fluorescence spectrum of the enzyme for several lactams. For example, addition of mimics **12f** and **12g** to the enzyme gave no detectable fluorescence change. Titration with mimics **12a-12e**, however, did result in an increase in the enzyme fluorescence intensity. Figure 7 shows the change in fluorescence intensity of cyclophilin A as a function of the concentration of lactam **12a**. The degree of fluorescence intensity enhancement by mimics **12** was found to be dependent on the nature of the side chain of **12**. The enhancement was greatest for those mimics with aryl side chains. The wavelength of the fluorescence emission maximum did not vary significantly during the titration (350 nm).

The titration data could be fit to an expression for a two-state equilibrium model, which assumed the formation of an enzyme–mimic complex with 1:1 stoichiometry (Eq. 1). In this expression, F is defined as the fluorescence of the sample, F_T is the total molar fluorescence change, $[E_T]$ is the total enzyme concentration, $[S]$ is the lactam concentration, and K_D is the dissociation constant. The data fit the model with a high correlation, signifying the formation of a simple enzyme–mimic bimolecular complex. Extraction of values for K_D was achieved by nonlinear least-squares regression (Table 4).

$$F = F_T/2\ ([E_T] + [S] + K_D) - \{([E_T] + [S] + K_D)^2 - 4[E_T][S]\}^{1/2} \qquad (1)$$

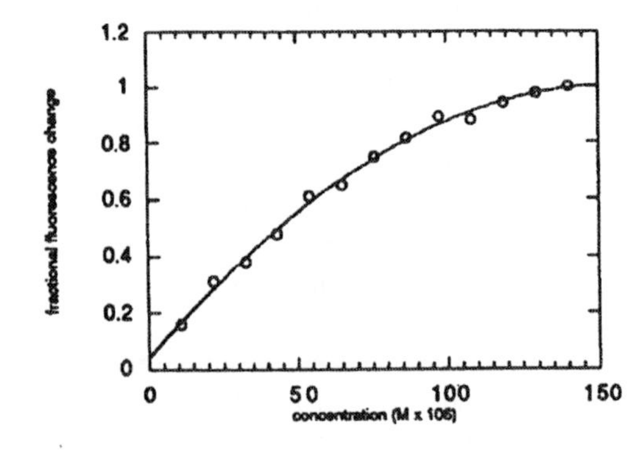

Figure 7. Fluorescence titration of cyclophilin with lactam **12a**.

Table 4. Dissociation Constants for Cyclophilin A–Lactam Complexes at RT Determined by Fluorescence Enhancement

Compound	*R*	K_D *(μM)*
12a	$CH_2C_6H_5$	40
12b	$CH(CH_3)_2$	180
12c	$CH_2CH{\subseteq}CH_2$	160
12d	CH_3	300
12e	β-$CH_2C_{10}H_7$	5
12f	*cis*-NH_2	>500
12g	*trans*-NH_2	>500

In the protease-coupled isomerase assay, the mimics **12a-e** inhibited the activity of cyclophilin in a concentration dependent manner. Analyses of the concentration dependence of the inhibition indicated that the lactams acted as competitive inhibitors. Inhibition constants were similar in magnitude to the K_D values obtained by fluorescence.

3. INSIGHT INTO THE MECHANISM OF CYCLOPHILIN GAINED FROM XAAPRO PEPTIDOMIMETIC BINDING STUDIES

3.1. Binding Mode of XaaPro Peptidomimetics to Cyclophilin

The increase in fluorescence intensity of cyclophilin as a function of XaaPro mimic concentration strongly suggested that mimics **12a-e** bound to the active site of cyclophilin. As shown in Table 4, the binding affinity of mimics **12** depended on the nature of the alkyl group R adjacent to the lactam carbonyl carbon. For example the K_D value for peptidomimetic **12f**, without any side chain, was too high to be measured accurately by the fluorescence assay. The binding affinities of compounds **12c** and **12d**, with pendant allyl and methyl groups, respectively, were relatively low, but substantially higher than that of mimic **12f**. Compounds **12a** and **12e**, with pendant benzyl and β-naphthylmethyl substituents, respectively, displayed K_D values of 40 and 5 μM, respectively. The exceptionally high affinity of these compounds approached that of cyclosporin A determined under the same conditions (1 μM).

In contrast to the substantial variation in binding affinity of mimics **12**, the value of the specificity constant k_{cat}/K_M for cyclophilin-catalyzed isomerization of peptide substrates of the form sucAlaXaaProPhe*p*NA varied only slightly with the identity of the Xaa residue.[23] Unfortunately neither the individual Michaelis-Menten constants k_{cat} and K_M nor the specificity constant have been determined for the dipeptides XaaPro. Comparison of the binding affinities of the lactams **12** to the K_M values for the corresponding dipeptides would have provided insight into the affinity of the dipeptides to the enzyme. The dissociation constants of the cyclophilin–lactam complexes determined by fluorescence were lower than the K_M values of typical tetrapeptide substrates (0.1 to 1 mM), but were not of the same order as the *k*cat values.[20] These results suggest that the lactams **12** bind to the active site of cyclophilin in a manner identical to the analogous dipeptides, and therefore represent constrained ground state analogues.[45]

The origin of the enhanced affinity of the mimics **12a** and **12b** could be rationalized by examination of the X-ray-derived models of the XaaPro dipeptide–cyclophilin complexes. According to these structures, the $\chi 1$ torsion angle of the Xaa residue assumed a value that projected the side chain partially toward the solvent. This model explains the relatively low substrate specificity of cyclophilin A toward

XaaPro sequences as a function of the Xaa residue.[23] Were the $\chi 1$ angle of the Xaa residue to adopt a conformation that would place the side chain underneath the pyrrolidine ring of the proline residue, as seen in the crystal structure of lactam **11a**, then the aryl ring would be situated in a hydrophobic pocket formed by the side chains of Phe113, Ile122, His126, and Trp121. The favorable formation of a hydrophobic cluster between the aryl rings of lactams **12a** and **12e** and the enzyme would then lead to enhanced affinity of the mimic toward the enzyme. In support of this conjecture is the observation that lactams **12a** and **12e** showed the greatest degree of fluorescence enhancement of all the mimics. The substantial fluorescence enhancement of the aryl-substituted lactams could be explained by increased hydrophobicity and by stacking interactions between the lactam aryl ring and the indole ring of Trp121.

3.2. XaaPro Dipeptide Mimics as Cyclophilin Inhibitors

In summary we have shown that rational design methods are successful in preparing inhibitors of cyclophilin. The affinity of peptidomimetics **12a** and **12e** is impressive in view of the small size of these compounds. The high affinity of the peptidomimetics is presumably due to the close correspondence between the structures of the lactams and the backbone conformations of the enzyme-bound dipeptides, to the restriction of rotation imposed by the conformational constraints, and to the specific enzyme–mimic contacts formed as a result of the unique conformational preference of the aryl side chains of mimics **12a** and **12e**.

The utility of peptidomimetics of PPIase substrates as probes of cellular biology and as therapeutic agents would be enhanced by increasing the affinity of these molecules to the enzyme. To this end, future goals are directed to improving the binding affinity of the peptidomimetics, as well as preparing derivatives with selectivity for particular cyclophilin isoforms. Currently combinatorial libraries of mimic **12** are being constructed to screen for potent inhibitors. A second generation of lactams are also being synthesized that should behave as transition state analogues of PPIases. Results from these studies will clarify structure–activity relationships in this important enzyme family.

ACKNOWLEDGMENTS

We are grateful to the R. A. Welch Foundation, the American Heart Association-Texas Affiliate (Grant 95G-448), American Cyanamid, and the University of Houston for financial support. We especially thank Dr. Gregory Tucker-Kellogg and Prof. Christopher T. Walsh (Harvard Medical School) for the expression system for human cyclophilin.

REFERENCES

1. *The Chemistry of Amides*; Zabicky, J., Ed.; Wiley, London, 1970.
2. Stewart, D. E.; Sarkar, A.; Wampler, J. E. *J. Mol. Biol.* **1990**, *214*, 253.

3. Eberhardt, E. S.; Loh, S. N.; Hinck, A. P.; Raines, R. T. *J. Am. Chem. Soc.* **1992**, *114*, 5437.
4. Yaron, A.; Naider, F. *Crit. Rev. Biochem. Mol. Biol.* **1993**, *28*, 31.
5. Kern, G.; Kern, D.; Schmid, F. X.; Fischer, G. *FEBS Lett.* **1994**, *348*, 145.
6. Brandl, C. J.; Deber, C. M. *Proc. Natl. Acad. Sci. USA* **1986**, *83*, 917.
7. Dorfman, T.; Bukovsky, A.; Ohagen, A, Hoglund, S.; Gottlinger, H. G. *J. Virol.* **1994**, *68*, 8180.
8. Schreiber, S. L. *Science* **1991**, *251*, 283.
9. Walsh, C. T.; Zydowsky, L. D.; McKeon, F. D. *J. Biol. Chem.* **1992**, *267*, 13115.
10. Rosen, M. K.; Standert, R. F. Schreiber, S. L. *Science*, **1990**, *248*, 863.
11. Handschumacher, R. E.; Harding, M. W.; Rice, J.; Drugge, R. J. *Science*, **1984**, *226*, 544.
12. Kieffer, L. J.; Thalhammer, T.; Handschumacher, R. E. *J. Biol. Chem.* **1992**, *267*, 5503.
13. Price, E. R.; Zydowsky, L. D.; Jin, M.; Baker, C. H. Walsh, C. T. *Proc. Natl. Acad. Sci. USA* **1991**, *88*, 1903.
14. Bergsma, D. J. et al. *J. Biol. Chem.* **1991**, *266*, 23204.
15. Liu, J. L.; Walsh, C. T. *Biochemistry* **1991**, *30*, 2306.
16. Liu, J.; Farmer, J. D.; Lane, W. S.; Freidman, J.; Weissman, I.; Schreiber, S. L. *Cell* **1991**, *66*, 807.
17. Franke, E. K.; Yuan, H.; Luban, J. *Nature* **1994**, *372*, 359.
18. Thali, M. et al. *Nature.* **1994**, *372*, 363.
19. Gitti, R; Lee, B. M.; Walker, J.; Summers, M. F.; Yoo, S.; Sundquist, W. I. *Science* **1996**, *273*, 231.
20. Kofron, J. L.; Kuzmic, P.; Kishore, V.; Bonilla, E.; Rich, D. H. *Biochemistry* **1991**, *30*, 6127.
21. Fischer, G.; Wittman-Liebold, B.; Lang, K.; Kiefhaber, T.; Schmid, F. X. *Nature* **1989**, *337*, 476.
22. Liu, J.; Albers, M. W.; Chen, C. M.; Schreiber, S. L.; Walsch, C. T. *Proc. Natl. Acad. Sci. USA* **1990**, *87*, 2304.
23. Harrison, R. K.; Stein, R. L. *J. Am. Chem. Soc.* **1992**, *114*, 3464.
24. Park, S. T.; Aldape, R. A.; Futer, O.; DeCenzo, M. T.; Livingston, D. J. *J. Biol. Chem.* **1992**, *267*, 3316.
25. Hackney, D. D. *The Enzymes*, Vol 19, 1991: pp 1-37.
26. Orozco, M.; Tirado-Rives, J.; Jorgensen, W. L. *Biochemistry* **1993**, *32*, 12864.
27. Fischer, S.; Michnick, S.; Karplus, M. *Biochemistry* **1993**, *32*, 13830.
28. Ke, H.; Mayrose, D.; Cao, W. *Proc. Natl. Acad. Sci. USA* **1993**, *90*, 3324.
29. Zhao, Y.; Ke, H. *Biochemistry* **1996**, *35*, 7362.
30. Kallen, J.; Wlakinshaw, M. D. *FEBS Lett.* **1992**, *300*, 286.
31. Zhao, Y.; Ke, H. *Biochemistry* **1996**, *35*, 7356.
32. Kakalis, L.; Armitage, I. M. *Biochemistry* **1994**, *33*, 1495.
33. Kern, D.; Kern, G.; Scherer, G.; Fisher, G.; Drakenberg, T. *Biochemistry* **1995**, *34*, 13594.
34. Zydowsky, L. D.; Etzkorn, F. A.; Chang, H. Y.; Ferguson, S. B.; Stolz, L. A.; Ho, S. I.; Walsh, C. T. *Protein Sci.* **1992**, *1*, 1092.
35. Richardson, J. S.; Richardson, D. C. In *Prediction of Protein Structure and the Principles of Protein Conformation;* Fasman, G. D., Ed.; Plenum Press, New York, 1989, pp 48-53.
36. Bartlett, P. A., Shea, G. T.; Tefler, S. J.; Waterman, S. In *Molecular Recognition; Roberts; S., Ed.; Royal Society: London; 1989, pp 182-196.*
37. Kim, K.; Dumas, J.-P.; Germanas, J. P. *J. Org. Chem.* **1996**, *61*, 3138.
38. Seebach, D.; Boes, M.; Naef, R.; Schweizer, W. B. *J. Am. Chem. Soc.* **1983**, *105*, 5390.
39. Schmidt, U.; Lieberknecht, A.; Wild, J. *Synthesis* **1984**, 53.
40. Dumas, J.-P.; Germanas, J. P. *Tetrahedron Lett.* **1994**, *35*, 1493.
41. (a) Dunitz, J. D.; Winkler, F. K. *Acta Cryst.* **1975**, *B31*, 251. (b) Shao, H.; Jiang, X.; Gantzel, P.; Goodman, M. *Chem. Biol.* **1994**, *1*, 231. (c) Wang, Q.-P.; Bennet, A. J.; Brown, R. S.; Santarsiero, B. D. *J. Am. Chem. Soc.* **1991**, *113*, 5757.
42. Winkler, F. K.; Dunitz, J. D. *Acta Cryst.* **1975**, *B31*, 268.
43. Winkler, F. K.; Dunitz, J. D. *J. Mol. Biol.* **1971**, *59*, 169.

44. Yao, J.; Bruschweiler, R.; Dyson, H. J.; Wright, P. E. *J. Am. Chem. Soc.* **1994**, *116*, 12051.
45. (a) Bartlett, P. A.; Marlowe, C. K. *Biochemistry* **1983**, *22*, 4618. (b) Bartlett, P. A.; Otake, A. *J. Org. Chem.* **1995**, *60*, 3107.

NATURAL PRODUCT PEPTIDOMIMETICS

Colin James Barrow and Philip Evan Thompson

Advances in Amino Acid Mimetics and Peptidomimetics
Volume 1, pages 251-293

ISBN: 0-7623-0200-3

1. INTRODUCTION

There has been considerable recent interest in the use of peptides and peptidomimetics as drug leads. Peptides themselves can be used as drugs. For example, insulin is used for the treatment of diabetes, cyclosporin is used as an immunosuppressant, and interleukins are used for the treatment of certain types of cancer. However, the presence of hydrolyzable amide bonds means that peptide drugs have a short half-life, poor oral bioavailability, and poor pharmacokinetics. In addition, conformational flexibility often causes peptides to have multiple actions and toxic side effects. Considerable effort has therefore been expended in finding non-peptide surrogates for biologically active peptides. These small molecule surrogates, or peptidomimetics, can be obtained from a random screening strategy, from *de novo* design, or from a combination of screening and design.

Peptidomimetic research has been closely tied to natural product research since the first demonstration that the endogenous ligands for opioid receptors were peptides. That morphine, the principal alkaloid of opium, is a non-peptide ligand for a peptide receptor has provided much of the impetus for peptidomimetic discovery in general, but also for the utilization of natural product sources. In this chapter, we examine four applications of natural products in the field of peptidomimetics.

The most obvious natural product leads for peptidomimetic drug discovery are the endogenous peptides themselves.[1] However, peptide ligands are more complex leads than non-peptide hormones in that they display greater conformational freedom and limited bioavailability. In the case where endogenous ligands are non-peptides—for example, epinephrine, histamine, and 5-hydroxytryptamine—drugs have been designed based on the structure of the ligand. For example, the antihypertensive β-blocker drug, propranolol, and the antiasthmatic β-agonists, terbutyline and salbutamol, were designed based on the structure of epinephrine. In fact, no totally non-peptidic drugs have been designed based solely on the structure of an endogenous peptide ligand. However, partially modified peptide analogues may prove suitable for certain therapeutic uses and the peptide ligand contains all the structural information necessary for the design of a non-peptide surrogate, thus a large amount of effort has gone into extracting this information. In Section 2 of this review we discuss selected examples of the design of peptidomimetics starting from the isolation of the endogenous peptide ligands.

Random screening for small molecule natural products has led to the discovery of a number of antagonists to endogenous peptide ligands. These antagonists have been referred to as peptidomimetics or limetics[2] and in some cases have become marketed drugs. For example, the natural product, paclitaxel, which was recently approved for the treatment of certain types of cancer, appears to act by displacing an endogenous peptide ligand. Selected examples of small molecule natural products that displace endogenous peptide ligands are discussed in Section 3 of this review.

When a peptide ligand interacts with its receptor it binds in a fixed conformation. A major obstacle to designing non-peptide drugs from peptide ligands is determining the spatial relationship among the groups involved in binding with the receptor. Similar difficulties exist for peptide venom toxins, which are often very specific and potent, but are inadequate as drugs due to their peptidic nature. Recently, small structurally rigid natural product peptides have been isolated from a number of sources, including *Conus* venoms. These peptides have generally been found to be more receptor-specific than more flexible peptides. They also offer the advantage that their conformation is more readily determined and therefore they provide a better starting point for the design of small molecule surrogates. In Section 4 we discuss some peptide venom toxins that are potential leads for drug design, either due to their relatively rigid structures or because they have a small defined active pharmacophoric region.

Design of a small drug-like molecules from a peptide lead requires replacement of the amide backbone with an appropriate surrogate. These surrogates include mimetics for a conformational portion of a peptide structure, such as β-turn or strand mimetics. Alternatively the entire peptide backbone may be relaced with a mimetic scaffold, which provides similar spatial relationships among binding groups to that provided by the peptide lead. Natural products that are conformationally rigid, stereochemically defined, and synthetically accessible are a valuable source of these scaffolds. In some cases these scaffolds have been combined with a combinatorial chemistry approach to aid with lead optimization. In Section 5, we discuss some selected examples of natural product scaffolds that have been used as surrogates for biologically active peptide leads.

2. PEPTIDOMIMETICS OF ENDOGENOUS PEPTIDE LIGANDS

Endogenous peptides and proteins represent the logical starting point for rational design of peptidomimetics. As natural products they are "purpose built" for the job they do and generally have an optimal interaction with their receptor, selective mode of action and an established mechanism of clearance from the body. However, these properties are generally offset by an inability to deliver them as drugs due to their susceptibility to biological degradation and low oral bioavailability. The challenges to the medicinal chemist upon initial identification, isolation, and characterization of the active peptide are to modify the structure so as to retain the desirable therapeutic properties of the peptide or protein while removing the impediments to delivery and longevity. Strategies to achieve this vary from relatively minor "local" modifications in structure to complete replacement of the peptide backbone and side chains, and the design process is generally an iterative one, at once reliant on but contributing to the generation of pharmacophores. This approach to peptidomimetic design is probably the most widely used strategy, since it represents the lowest risk with respect to synthetic feasibility and gaining structure-activity data which may identify important parameters relating to drug action, such as duration of action and receptor subtype selectivity. A plethora of other reviews have appeared in recent years and a number of authors have attempted to ascribe guidelines to the rational design of peptidomimetics in this way.[3-15] The intention of this section is to assess the success of this strategy from the discovery and isolation of selected peptides to their application as therapeutics or useful pharmacological probes.

2.1. From Peptide to Peptidomimetic

Peptide Isolation and Characterization

Historically, the discovery of an endogenous peptide has related to the identification of material abundant in the pathogenesis of disease, or via a biological effect induced by the experimental removal of tissues or organs, or by the presence of an exogenous substance. The discovery of insulin for which Banting and Best received the Nobel prize shows the importance of this approach. Von Mering and Minowski demonstrated that diabetic symptoms of humans could be induced in laboratory animals by removal of the pancreas in 1889.[16] In 1921, Banting and Best were able to treat these symptoms using crude pancreatic extracts which were later found to contain insulin, isolated in crystalline form in 1926 by Abel.[17] It wasn't until the 1950s that the structure of insulin was determined.[18]

Advances in analytical techniques led the expansion of the field in the early 1970s with the isolation and characterization of important peptide hormones GnRH (LHRH)—somatostatin, substance P, and enkephalin among others. In particular, advancements in chromatographic separations, receptor preparation, amino acid analysis and sequencing, peptide synthesis, and the development of mass spectrometry provided access to many new peptide hormones which had been identified but not previously characterized.

To date, molecular biology has added its considerable weight to the process, such that endogenous ligands are sought for receptors with no known biological activity. The so called "orphan receptors" which have been identified as having homologous cDNA sequences to a particular supergene family may have similarly "orphan" ligands. Transfection of cells expressing the orphan receptor provides an exquisite means of assaying extracts for endogenous ligands. High-resolution chromatographic separation techniques can be used to purify the active component to homogeneity and characterization achieved by mass spectrometry and sequencing. This material or its synthetic equivalent can then be evaluated in functional assays to determine *in vivo* biological activity. Molecular biology techniques are also breaking new ground in the discernment of receptor subtypes for which variation in affinity, tissue distribution, pharmacological and biochemical properties, and endogenous function are providing new opportunities in the design of highly selective peptidomimetic therapeutics.

Amino Acid Deletion and Substitution

An important first step in developing structure–activity relationships of an active peptide involves determining the minimum active sequence and the amino acid residues critical to activity. Truncation of peptide sequences and residue replacement studies tend to closely follow the first synthesis of the endogenous peptide. The application of these strategies has expanded markedly since the advent of peptide libraries, such that peptide structure–activity studies are now based upon up to millions of peptide analogues.[19,20] Substitution of non-ribosomal amino acids further broadens the scope of structure–function studies while introducing elements which may render the peptide more resistant to enzymic degradation.

Cyclization

The stabilization of the biologically active conformation of a peptide is important for the retention of potency, and often truncation or residue replacement merely serves to destabilize the bioactive conformer. Cyclization may have a dramatic impact upon peptide conformation, reducing the conformational freedom of a linear sequence and possibly inducing the appropriate conformation for biological activity. Cyclization, either short-range (between adjacent residues) or long-range, has been particularly successful with mimicry of reverse-turn conformers and is being more generally applied, for example, to α-helices.[21-23]

Amide Bond Surrogates

The replacement of the amide bond renders peptides resistant to proteases and thus increases the stability of the resultant pseudopeptide. However, such changes in general have a major effect on peptide conformation and as such have found only limited application in the design of receptor-binding ligands. However, the combination of these characteristics lends the amide bond surrogates to application as transition-state analogue enzyme inhibitors, mimicking the bound transition state of the endogenous substrate but lacking the cleavable peptide bond.[14]

Peptidomimetics

The replacement of two or more residues by a non-peptidic moiety is generally referred to as peptidomimetic design, and represents the final goal of design based upon endogenous lead peptides. The rational design of such peptidomimetics is usually based upon an extensive analysis of likely bioactive conformation. This may result from a knowledge of receptor-bound structure from X-ray crystallographic or NMR data or from pharmacophores established through the procedures listed above. Conformational motifs of peptides have proved popular targets for peptidomimetic design since the prospect of designing mimics which are applicable to a range of peptides justifies the more extensive synthetic effort in building organic structures. The reverse turn in particular has been targeted as an important structural motif which can be simply replaced by semirigid organic structures.[24]

2.2. Agonists and Antagonists of Peptide Ligands

Opioid Ligands (Endorphins, Enkephalins, Dynorphins)

The discovery of endogenous ligands which competed with morphine (**1**) for opioid receptors in animals provided the first bridge between peptide and non-peptide and thus the most compelling evidence that peptidomimetic research would provide an effective route to the discovery of new drugs. The search for the endogenous ligands themselves was triggered by the establishment of discrete opioid receptors,[25,26] and of the many research groups that were able to isolate endogenous opioid agonists,[27-30] it was Hughes et al. who were able to characterize the two pentapeptides Met- and Leu-enkephalin with the aid of Edman sequencing and mass spectrometry.[31] The identification of other endogenous endorphin and dynorphin peptides soon followed, and at that time high hopes were held that these peptides might provide the lead to non-narcotic analgesic drugs, further heightened by the discernment of the opioid μ-, δ-, κ-, and σ-receptor subtypes.[32]

The apparent structural analogy between morphine and enkephalin was soon seized upon, with the common phenolic group being noted as critical to activity in both classes (Figure 1).[33] As Schiller et al. described the "hypothesis of an analogous spatial disposition of critical and identical chemical functions in the peptides and in the molecular framework of morphine."[34] The highly potent oripavines included a phenylalanine mimic and led to the estimation of a pharmacophore reliant on a phenol–phenyl separation of approximately 9-10 Å. In enkephalin, the potential for a reverse turn about the glycine residues supported this hypothesis, although subsequent studies and the inactivity of morphine overlapping tripeptides cast doubt on this postulate.[35]

The strategies for optimizing opioid peptide structure subsequently have hinged on stabilizing the critical amino-terminal sequence. A large number of linear peptides have shown improved selectivity for receptor subtypes relative to the endogenous ligands as well as the important contribution made by the potent opioid peptides isolated from amphibians, the deltorphins and dermorphin.[36] Conformationally constrained peptides have proven very successful with incorporation of short range cyclizations in linear peptides and long-range disulfide and other linkages, providing potent, receptor subtype selective ligands (**2a-d**).[37-39] Some exam-

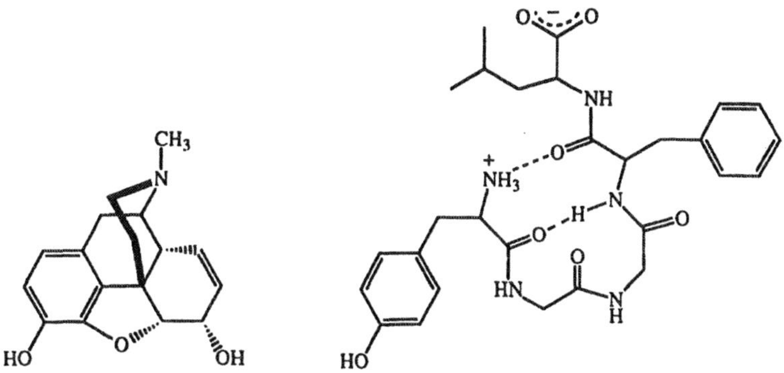

1 Morphine

Leu-Enkephalin

Figure 1. Comparison of morphine and Leu-enkephalin.

ples of modified opioid peptides, including one that is a derivative of somatostatin, are shown in Table 1.

With morphine as a model peptidomimetic, the early attempts at designing enkephalin-based peptidomimetics were quite adventurous. Belanger et al. targeted the putative β-turn stabilized by the Gly-Gly dipeptide of enkephalin, replacing this unit with bicyclic scaffolds (**3**,**4**) designed to restrain the Tyr and Phe residues in a more rigid orientation.[40] The failure to recover opioid receptor affinity, while disappointing, has not prevented the pursuit of other novel scaffolds. Krstenansky et al. had more success with another bicyclic moiety incorporating heteroatoms which yielded a peptidomimetic (**5**) of one-third the potency of morphine.[41] The β-turn structure has been targeted further by incorporation of two of the best

Table 1. Some Examples of Modified Opioid Peptides

Y	G	G	F	L	R	R	I	R	P	K	L	K	W	D	N	Q	Dynorphine A
Y	G	G	F	M	T	S	E	K	S	Q	T	P	L	V	T		α-Endorphin
Y	G	G	F	L	R	R	Q	F	K	V	V	T					Dynorphine B
Y	G	G	F	L													Leu-enkephalin
Y	G	G	F	M													Met-enkephalin
Y	a	F	G	Y	P	S	NH_2										Dermorphin
Y	a	F	D	V	V	G	NH_2										Deltorphin II
Y	a	G	mF[a]	G	ol												DAMGO (μ-selective)
Y	s	G	F	I	T												DSLET (δ-selective)
Y	d	G	F	Dp[b]	R	R	I	R	P	K	L	K					Cyclic Dyn A analogue (κ-selective)
f	C	Y	w	K	T	Pn[c]	T	NH_2									Somatostatin analogue

Notes: [a] mF = 4-MePhe.
[b] Dp = diaminopropionic acid.
[c] Pn = Penicillamine.

known dipeptide mimics, the Freidinger γ-lactam[42] (**6**) and Nagai and Sato's β-turn dipeptide (BTD) (**7**),[43] although neither were able to produce useful activity. The failure of these mimics probably illustrates their inability to meet more specific conformational requirements of the receptor.

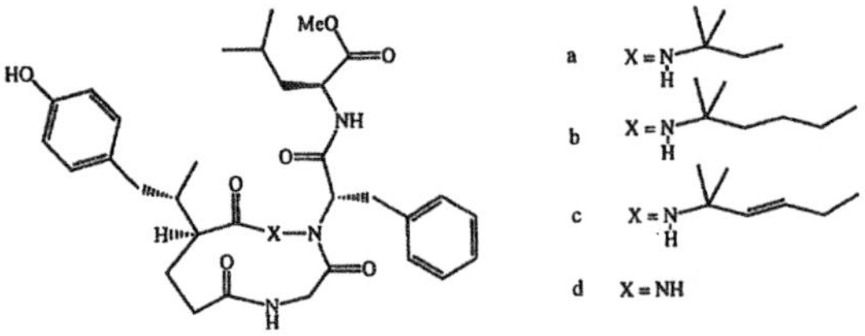

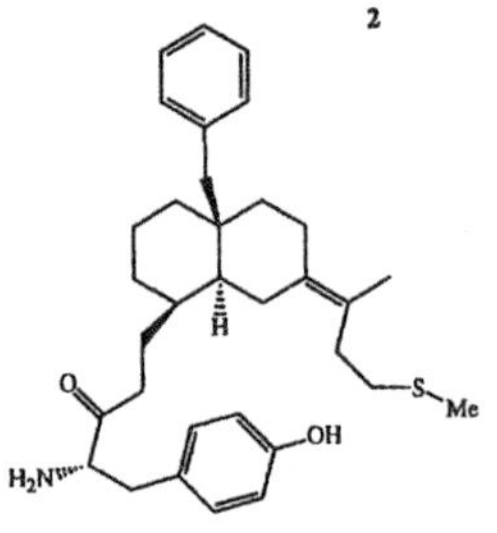

Recent reports of the novel endogenous ligand nociceptin, also known as orphanin FQ, provides an excellent example of the "state-of-the-art" in endogenous peptide discovery which uses the power of molecular biology to discover a ligand for a receptor of unknown function, but which may have a significant role in the regulation of pain perception.[44]

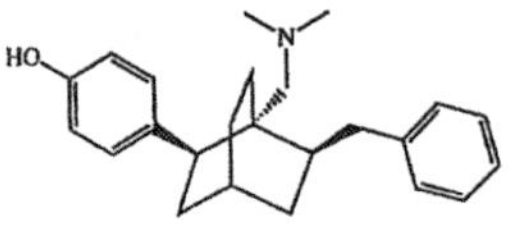

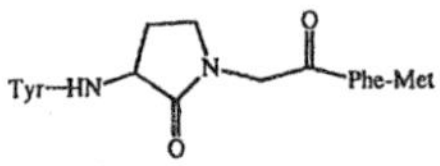

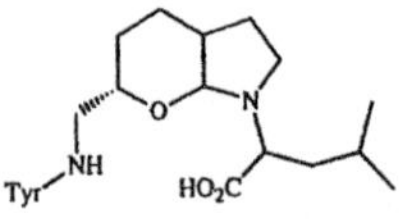

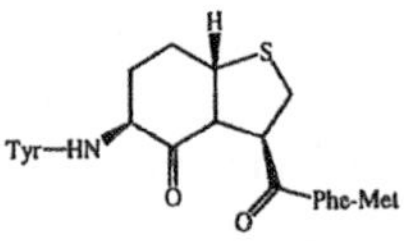

A number of groups have reported the isolation of a cDNA clone with similarities to opioid receptors and abundantly expressed in rat and mouse brain, but cells transfected with this gene expressed a receptor which was not recognized by any of the common opioid ligands.[44] Thus the hunt was on for the endogenous ligand to this orphan receptor. Two separate reports identified a heptadecapeptide with significant amino-terminal homology to the opioid peptides. Meunier et al. extracted the peptide from rat brain via a sequence of gel-filtration, cation-exchange and reversed-phase liquid chromatography to yield a peptide of M_r 1810 whose sequence was determined to be YGGFTGARKSARKLANQ.[45] The same peptide was also extracted from porcine hypothalamus and purified by cation-exchange and reversed-phase HPLC by Reinscheid et al. [46] The sequence similarities to the opioid family of peptides suggested that the peptide may induce analgesia; however *in vivo* assays of the peptide showed that it induced hyperalgesia, with treated animal models showing increased sensitivity to painful stimuli.

The antagonism of the nociceptin action would therefore appear to be a potential means of controlling pain, possibly with the elimination of opioid drug side effects. Dooley and Houghten have undertaken the only structure–activity studies reported to date.[47] A library of 34 peptides were examined comprising an alanine scan, N-terminal truncates, and C-terminal truncates. The results of these analyses showed the critical nature of the amino-terminal sequence to affinity, as well as the noncritical nature of the C-terminus to receptor-binding affinity. The similarity to the binding of opioid peptide to their receptors suggests that similar approaches to peptidomimetics will be made as for homologous dynorphin mimics.[48]

Gonadotropin-Releasing Hormone (GnRH)

Gonadotropin-releasing hormone (GnRH; Table 2) is synthesised in the hypothalamus and upon secretion binds to its receptor and stimulates the release of gonadotropins, leutinizing hormone (LH), and follicle-stimulating hormone (FSH). As such, GnRH and its analogues are of potential use in contraception, as well as in the treatment of fertility disorders (agonists) and diseases such as endometriosis, polycystic ovarian disease, and prostate carcinoma (antagonists or receptor down-regulating agonists).[49]

The presence in hypothalamic extracts of a hormones capable of stimulating the release of leutinizing hormone and follicle-stimulating hormone was established by a number of workers in the 1960s. Schally et al.[50,51] accomplished the isolation of LHRH in a 12-step process from which 100,000 pig hypothalami produced 830 μg of purified peptide, a 2 million-fold purification process. The composition of GnRH was elucidated using amino acid analysis. The sequence analysis was complicated by the presence of N-terminal pGlu and a C-terminal amide. However, by sequencing analysis of peptide fragments from thermolysin and chymotrypsin degradation, and other techniques including mass spectrometry, the correct sequence was established. Finally, synthesis of the sequence confirmed the identity of the active material and that the releasing hormones for LH and FSH were one and the same peptide.

Very soon after followed the first reports of agonists of superior potency and stability to the native hormone. First, successful modification of the C-terminal sequence (replacement of glycinylamide by an ethyl group)[52] was followed by the critical residue replacement of glycine by a range of bulky D-amino acids.[53] The group of peptides listed in Table 2 represents the currently utilized drugs for endocrine-dependent cancers, endometriosis and hirsutism.[54] While these materials are not orally active, they are well suited to use in depot formulations for subcutaneous implants and so have been very successful as drugs.

Table 2. Sequence of GnRH and Angonists Currently in Use

pGlu	His	Trp	Ser	Tyr	Gly	Leu	Arg	Pro	Gly-NH_2	GnRH
pGlu	His	Trp	Ser	Tyr	D-Leu	Leu	Arg	Pro-NHEt	—	Leuproreli 1974
pGlu	His	Trp	Ser	Tyr	D-Trp	Leu	Arg	Pro	Gly-NH_2	Triptorelin 1976
pGlu	His	Trp	Ser	Tyr	D-Ser(tBu)	Leu	Arg	Pro	aGly	Goserelin 1978
pGlu	His	Trp	Ser	Tyr	D-Ser(tBu)	Leu	Arg	Pro-NHEt	—	Buserelin 1979
pGlu	His	Trp	Ser	Tyr	D-Nap	Leu	Arg	Pro	Gly-NH_2	Nafarelin 1982
pGlu	His	Trp	Ser	Tyr	D-Trp	Leu	Arg	Pro-NHEt	—	Deslorelin
pGlu	His	Trp	Ser	Tyr	D-His(Bn)	Leu	Arg	Pro-NHEt	—	Histrelin

The dramatic and rapid success in the field of GnRH agonists was due to the concomitant and perhaps fortuitous induction of bioactive conformation and stabilization against enzymic degradation. The receptor bound conformation of GnRH is thought to include a type II' β-turn at Tyr^5-Gly^6-Leu^7-Arg^8, and this has been the focus of most analogue design.[55]

Until 1980, the underlying rationale for this premise was based upon the apparent requirement for a D-amino acid or glycine at position 6. Freidinger et al. confirmed the likelihood of a type II' β-turn being implicated in describing a γ-lactam replacement for Gly^6-Leu^7 in LHRH (GnRH) to produce an agonist 8.9-fold more potent in *in vitro* assays, and 2.4-fold more potent *in vivo* than the parent peptide.[56] The γ-lactam dipeptide mimic was designed to restrict the angle, $\varphi_1 = 120\gamma$ as required in a type-II' β-turn and was synthesized in six steps from methionine. It was the first report of a conformationally constrained bioactive peptide incorporating a non-peptide β-turn mimic, and has provided the template short-range cyclization for ever more sophisticated β-turn mimics.

The antagonists of GnRH have similarly been pursued with a large array of nonnatural substitutions leading to antagonists which have ultimately led to replacement of eight of the 10 residues of the native peptide (**8**), in the antagonists, antide and cetrorelix, only five ribosomal amino acids are retained (Figure 2).[57] Interestingly, work in this area has had to overcome the apparent side effect of drug-induced histamine release. The mechanism by which histamine release occurs is still unknown but has been reduced to low levels in the most recent antagonists, such as cetrorelix.

As with GnRH and its agonists, the lack of direct structural data on the bioactive conformation was a problem in analogue design. Some form of conformational restriction was demanded as to identify this bioactive conformation and to lead design. Macrocyclic lactamization has proved quite successful in this case, since

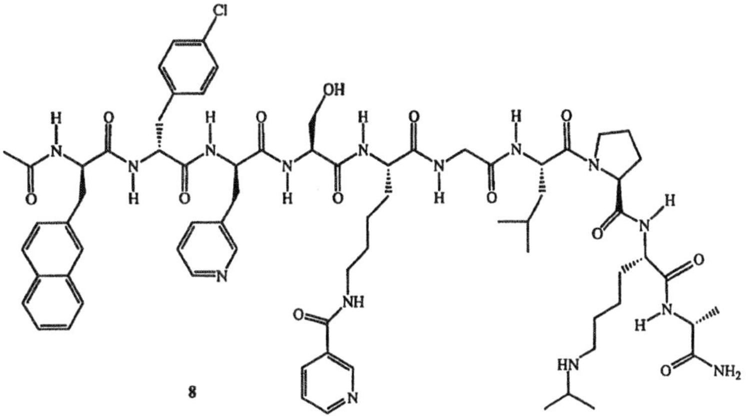

Figure 2. Linear and cyclic GnRH antagonists.

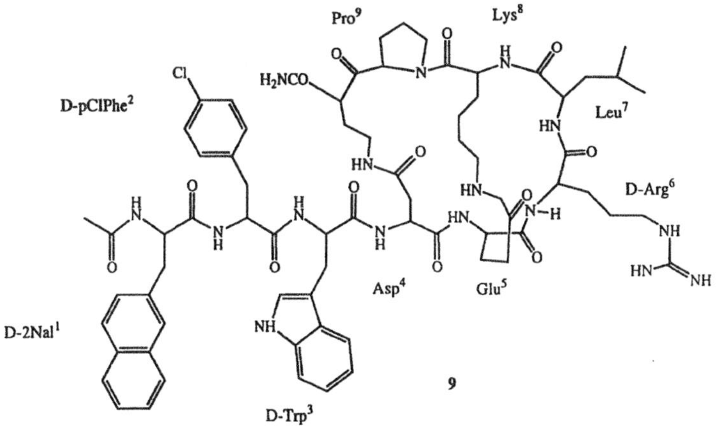

building on the first report of a head-to-tail coupled antagonist with weak activity Rivier et al. were able to synthesize cyclic peptides with links between side chains of residues 4 and 10, and these peptides retained high potency. Analysis of this peptide led to a proposed conformation for residues 4-10 with a type-II' β-turn at residues 6 and 7. Further, possibilities for further conformational restriction were suggested by the proximity of side chains of residues 5 and 8. Thus a dicyclic peptide (**9**) with two macrocyclic lactam bridges has been prepared with enhancement of activity[58] (Figure 2). With such a constrained analogue at hand, a well-refined model for the active conformation of GnRH is being established that can lead to further analogue design. It should be noted that the macrocyclisations account for four of the 10 side chains in the molecule; however the regions critical for binding as distinct from inducing binding conformations are still ambiguous.[59]

Somatostatin (Ala-Gly-Cys-Lys-Asn-Phe-Phe-Trp-Lys-Thr-Phe-Thr-Ser-Cys)

Somatostatin (**10**) is a 14-residue cyclic peptide also formed in the hypothalamus, and has been implicated in an array of actions in central and peripheral nervous systems as well as in autocrine and paracrine regulation. Current therapeutic uses of somatostatin and analogues relate to treatment of gastric ulcers, acromegalia, carcinoid tumours, and intestinal sclerodoma. Somatostatin has been applied to treatment of diabetes mellitus.[60]

The discovery of somatostatin came serendipitously out of the search for agents with growth hormone (GH)-releasing activity. In fact, hypothalamic extracts from half a million sheep obtained for the isolation of GnRH (see above) were further screened for inhibition of GH secretion and yielded 8.5 mg of a peptide identified as Somatotropin-release inhibiting factor (SRIF). Critical to the isolation was the development of a reliable *in vitro* assay, as well as characterization of the peptide by analytical methods and solid-phase synthesis of the cyclic peptide.[61]

The cyclic nature of this peptide has significantly expedited the development process of somatostatin analogues (Figure 3), and a β-turn was established as an inherent feature of bioactive analogues. In this case, replacement of L-Trp^8 by D-Trp^8 and truncation of the peptide proved significant in the development of peptidomimetics since activity could be retained and indeed enhanced several-fold in a cyclic hexapeptide, MK 678 (Seglitide, **11**) and an octapeptide, Octreotide.[62] Octreotide has found clinical use in the treatment of acromegaly and of gastrointestinal and neuroendocrine tumors. The success of this work and subsequent substitutions has allowed the establishment of models for the binding requirements of the key side chains of the tetrapeptide, Phe-D-Trp-Lys-Thr. An array of cyclic and linear somatostatin analogues have been examined over time and has led importantly to the identification of receptor subtypes with important implication for drug design.[63]

Somatostatin is implicated in a vast range of processes and as such it is unsurprising that it would bind to multiple receptor subtypes. While receptor heterogeneity was first proposed in 1984, it is only in recent years that these subtypes have been discerned and it is the array of somatostatin analogues previously synthesized that are defining their properties and occurrence and presenting new therapeutic opportunities. Ultimately, molecular cloning has established the existence of five human somatostatin receptors, sstr1-5.[60] The implications for therapeutic design are shown by the correlation of high subtype affinity to a specific physiological effect. For example, inhibition of growth hormone release correlates very precisely with affinity for sstr2 subtypes, while ligand binding to sstr5 receptors appears to be responsible for inhibition of insulin secretion. The potential thus exists for selective treatment of tumors expressing specific somatostatin receptors. Further, if non-peptide ligands which can cross the blood-brain barrier can be devised, novel treatments for neurological disorders such as epilepsy and Alzheimer's Disease have been envisaged.[60]

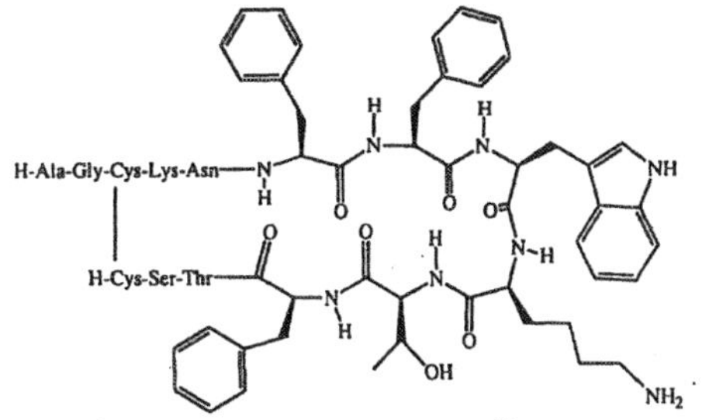

10

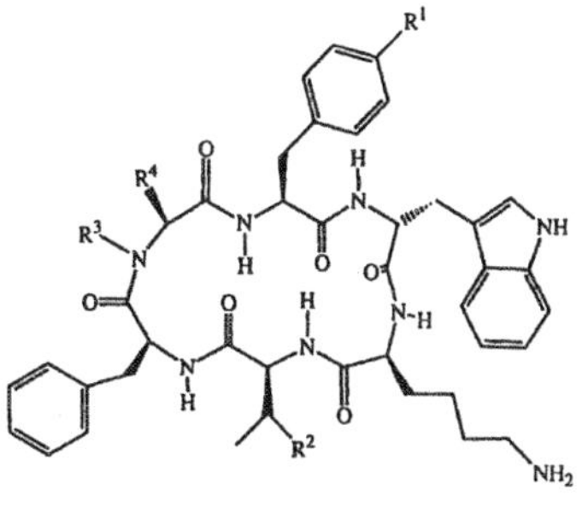

11 R^1 = OH, R^2 = R^3 = R^4 = Me
12 R^1 = H, R^2 = OH, R^3=R^4= $-(CH_2)_3-$

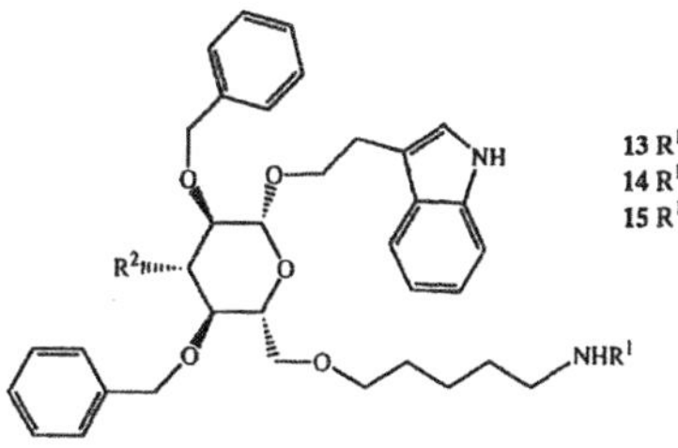

13 R^1 = R^2 = H
14 R^1 = H, R^2 = OBn
15 R^1= Ac, R^2 = OBn

Figure 3. Evolution of somatostatin analogues.

As an impressive first step in the search for non-peptide ligands Nicolaou et al. took the relatively rigid structure of Seglitide and L-363,301 (**12**) as models, and set out to completely replace the backbone by an organic peptidomimetic.[64] This they achieved spectacularly with the incorporation of the critical side chains on to a 3-deoxy-D-glucose template which yielded compound **13**, the first non-peptide somatostatin mimic, which had an IC_{50} of 1.3 μM at pituitary somatostatin receptors. Interestingly, at low concentrations the molecule behaves as an agonist, while at higher concentrations it shows antagonistic effects, and is the first reported somatostatin antagonist.[65] The wider application of the sugar nucleus is discussed in Section 4. Some other non-peptide somatostatin inhibitors have recently been extracted from marine sponges using the strategies outlined in Section 3.[66]

Substance P (Arg-Pro-Lys-Pro-Gln-Gln-Phe-Phe-Gly-Leu-Met-NH$_2$)

Substance P (SP; **16**) is an 11-amino acid peptide neurotransmitter that is widely distributed in both the central and peripheral nervous system. SP is the endogenous ligand for the neurokinin 1 (NK-1) receptor and plays an important role in several physiological processes such as neurogenic inflammation and pain transmission. Neurokinin antagonists may be useful in the treatment of disorders such as emesis and schizophrenia, and in diseases involving neurogenic inflammation, such as asthma, migraine, and rheumatoid arthritis.[67] First described in 1931, Substance P was not characterized until 1970 when the peptide was purified via a sequence of gel filtration, cation exchange, CM-cellulose chromatography, and a final paper electrophoresis step.[68] Thus from a 100-g extract of bovine hypothalami, 150 μg of purified peptide was obtained. The amino acid sequence[69] and solid-phase synthesis[70] were reported in 1971.

The minimum sequence of SP that maintains significant functional activity is the pentapeptide $SP_{7\text{-}11}$.[71] However, although these shorter C-terminal fragments maintain full agonist activity, they are less potent than SP itself. An alanine scan of SP showed that the first five amino acids and glycine-9 can be replaced by alanine without loss of activity, while replacement of Phe7, Phe8, Leu10 and Met11 results in a significant loss of potency, although all active analogues are full agonists.[72] A similar result was obtained using a D-amino acid scan, where substitution of a D-residue at Phe7, Phe8, Leu10, and Met11 decreased potency significantly.[73] Conformational analysis of D-residue analogues and comparison with predicted bioactive conformations for SP indicated that binding of SP to the NK-1 receptor occurred at five points: the guanidinium of Arg, the aromatic rings of both Phe residues, the sulfur atom of Met, and the C-terminal carboxamide.

Numerous SP analogues were synthesised by various groups with the aim of producing more selective, potent,and bioavailable NK-1 ligands. Increased selectivity for the NK-1 subtype over NK-2 and NK-3 receptor subtypes was obtained by modification of either Gly9 or the C-terminal Met-NH2. For example, the SP agonist septide (**17**)[74] is more selective than SP, and GR73632 (**18**)[75] is both more selective and more enzymatically stable. Methylation at the C-terminal NH2 gave agonist SP-OMe (**19**) which retained the potency of SP but was more selective for the NK-1 receptor.[76] Similarly, SP agonist [Sar9, Met(O_2)-11]SP (**20**) is equipotent and more selective than SP.[77] (Table 3)

Substituting D-Trp for both Phe7 and Gly9 produced SP antagonists. For example, spantide (**21**), spantide II (**22**), and related compounds are SP antagonists at the NK-1 receptor, although their selectivity is poor.[78,79] Further structure–activity studies based on these SP antagonists showed that FR113680 (**23**),[80] a tripeptide, has activity similar to longer peptides. Even the dipeptide Cam104 (**24**)[81] was found to have reasonable potency at the NK-1 receptor. These short peptides were good starting points for the design of non-peptide SP antagonists.

The fact that an 11-amino acid endogenous ligand (SP) could be reduced to two or three amino acids with virtually no loss in potency indicates that most of the

Table 3. Names and Structures and Some Bioactive Substance P Analogues

Compound	*Structure*
Septide (**17**)	pGlu-Phe-Phe-Pro-Leu-Met-NH_2
GR73632 (**18**)	NH_2-CH_2)$_4$-CO-Phe-Phe-Pro-(N-Me)-Leu-Met-NH_2
SP-OMe (**19**)	Arg-Pro-Lys-Pro-Gln-Gln-Phe-Phe-Gly-Leu-Met-OMe
[S, M(O_2)]SP (**20**)	Arg-Pro-Lys-Pro-Gln-Gln-Phe-Phe-(N-Me)-Gly-Leu-Met(O_2)-NH_2
Spantide (**21**)	(D)Arg-Pro-Lys-Pro-Gln-(D)Trp-Phe-(D)Trp-Leu-Leu-NH_2
Spantide II (**22**)	(D)Lys-Pro-β(3-pyridenyl)Ala-Pro-(,4-chloro)Phe-Asn-(SD)Trp-Phe-(D)Trp-Leu-Ne-NH_2
FR11368 (**23**)	Ac-Thr-(N-formyl)Trp-Phe-(N-Me)(N-Bzl)
Cam-104 (**24**)	Benyloxycarbonyl-Trp-Phe-NH_2
Cam-4261 (**25**)	(2-Benzofuranyl)CH_2OCO[R]α-MeTrp-NH[S]CH(CH_3)Ph
L-668,169 (**26**)	cyclo(Gln-D-Trp-(N-Me)-Phe-Gly-R-γ-lactam-Leu-Met)$_2$
GR71251 (**27**)	Arg-Pro-Lys-Pro-Gln-Gln-Phe-Ph:-N [spirolactam, O] N-Leu-Met-NH_2

amino acids in SP are not involved in binding-but have other functions such as certain transport or conformational properties. For SP, the additional amino acids probably act as Nature's device to fold the peptide into a scaffold, allowing the correct orientation of a small number of key amino acid residues to interact with the receptor.[81] This would indicate that design of a non-peptide ligand from an endogenous neuropeptide is entirely feasible.

Studies on Cam104, including the introduction of methyl groups as conformational constraints to the Trp-Phe backbone and QSAR studies, led to the discovery of Cam4261 (**25**) as a potent peptoid SP antagonist.[82] A cyclic peptide dimer with some similarity to FR113680 has been prepared and was found to be a selective NK-1 antagonist.[83] This cyclic peptide, L-668,169 (**26**) contains Gln-D-Trp-(N-Me-Phe) and a lactam. L-668,169 has a conformational restriction in the backbone as a result of the lactam group.

Other peptides designed from SP and introducing conformational constraints include the NK-1 selective antagonist GR71251 (**27**) which contains a spirolactam group.[84,85] The discovery of GR71251 originated from molecular modeling and conformational energy maps of ϕ and ψ angles in Gly9 in constrained SP analogues in order to determine low-energy conformations. From these studies, a conformationally rigid spirolactam was found which produces an antagonist conformation by apparently diverting the C-terminal residues into a nonactivating pocket of the NK-1 receptor. No agonism was apparent at the receptor at concentrations up to 20 μM.

Efforts to replace the amide backbone of SP include the use of *trans*-olefinic dipeptide or fluoroolefin dipeptide isosteres as amide backbone replacements.[86,87] In addition, the y(CH_2NH) moiety has been introduced into SP analogues at various positions.[88,89] More recently a bis-phenylalanine mimetic was incorporated into

SP.[90] However, although this SP analogue inhibits substance P endopeptidase, it does not bind to the NK-1 receptor.

Considerable effort has been extending in trying to develop a useful drug via modification of the endogenous peptide ligand SP. However, these peptidomimetic studies have not been as successful as the screening of chemical files and natural product extracts. Recently, a number of small molecule SP antagonists have been discovered using screening approaches, and some of these will be discussed in Section 3.

2.3. Enzyme Inhibitors by Peptidomimetic Design

Protease inhibitors are an attractive target for peptidomimetic design since the substrates are proteins and peptides. Proteases regulate the release and degradation of active hormones and thus inhibitors represent an alternate means to direct agonism or antagonism of inhibiting or promoting hormone action.[14] Interestingly, this means that a single hormone may represent a target for either agonist/antagonist design as the ligand, or enzyme inhibitor design as the substrate. An example of this is angiotensin for which both ACE inhibitors[91] and angiotensin II antagonists[92] have been fruitful targets.

While the synthesis of many protease inhibitors has been led mainly by discovery of non-native natural products (eg. ACE inhibitors from the snake venom peptide, teprotide),[93] renin inhibitors which block the formation of angiotensin I have been developed from the natural renin substrate, angiotensinogen.

While the 452-amino acid sequence of the natural protein substrate, angiotensinogen, was only deduced from the cloned cDNA sequence in 1984,[94] the region of the protein cleaved by renin has been long established. The best known substrate for renin was derived by tryptic hydrolysis of equine angiotensinogen in 1957,[95] and constituted the amino-terminal tetradecapeptide of the native hormone. The sequence of the N-terminal region of human angiotensinogen (**28**) was first deduced by Tewksbury et al. in 1981.[96]

The cleavage site of angiotensinogen by renin is a Leu-Val P_1-P_1'. bond and the first successful inhibitor replaced this susceptible dipeptide unit by a Phe-Phe linkage (**29**).[97] This peptide effectively competed with angiotensinogen for the enzyme-binding site but was not cleavable by renin.

The application of the transition-state analogue approach provided the next major impetus to the design of inhibitors.[98] Replacement of the siscile amide bond by an analogous unit produced potent inhibitors. These inhibitors initially included the reduce amide (**30**) and hydroxyethylene isostere of Szelke (**31**)[99] or an sp^3 hydroxymethylene isostere (**32**), the statine unit described by Boger et al.,[100] both of which are proposed to mimic the tetrahedral transition state for the hydrolysis of the peptide. A variety of isosteric replacements for this bond have been examined with varying success. With a successful replacement for the critical bond established, the improvement of the activity profile rests with modification of the remainder of the

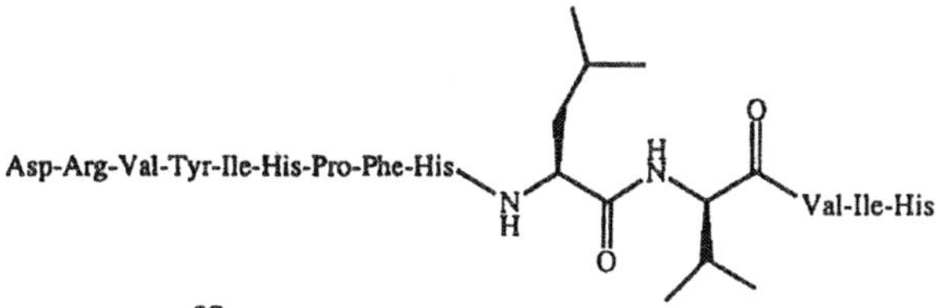

Asp-Arg-Val-Tyr-Ile-His-Pro-Phe-His, Val-Ile-His

28

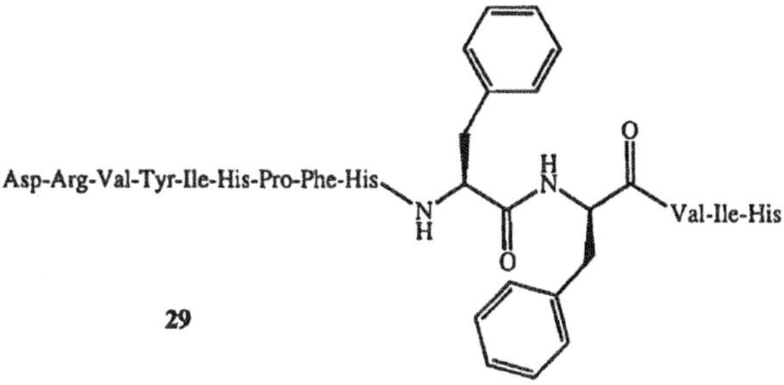

Asp-Arg-Val-Tyr-Ile-His-Pro-Phe-His, Val-Ile-His

29

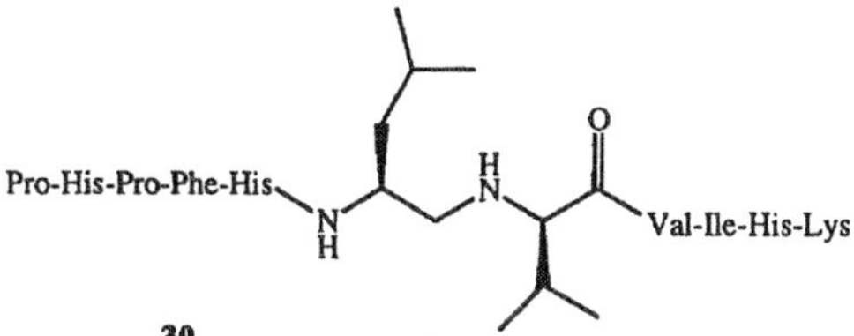

Pro-His-Pro-Phe-His, Val-Ile-His-Lys

30

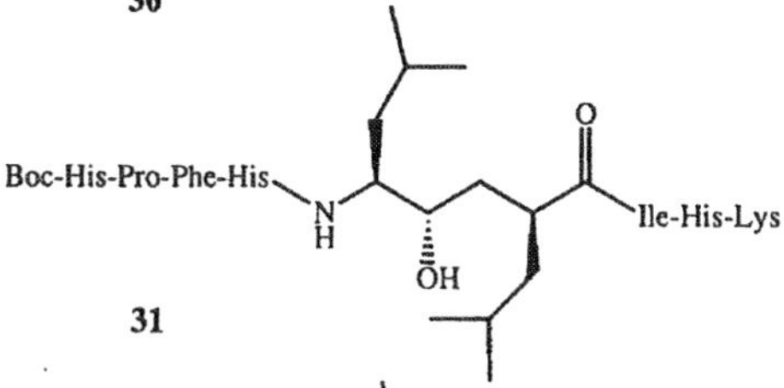

Boc-His-Pro-Phe-His, Ile-His-Lys

31

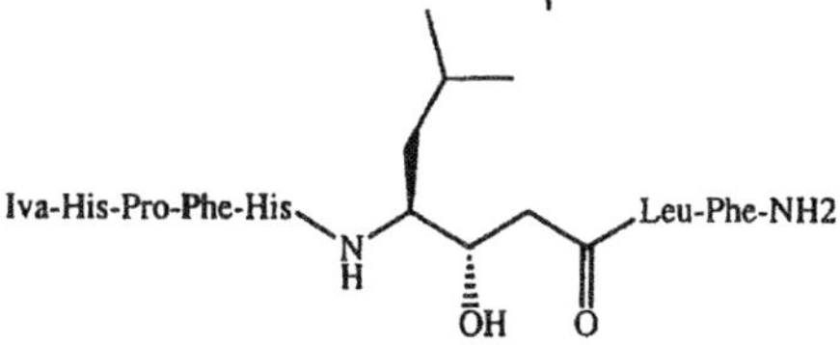

Iva-His-Pro-Phe-His, Leu-Phe-NH2

32

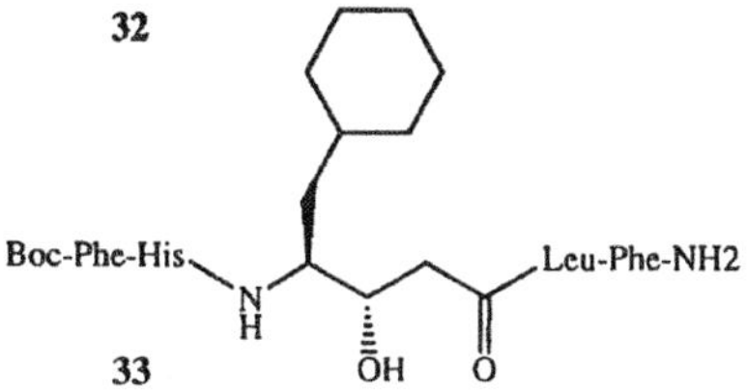

Boc-Phe-His, Leu-Phe-NH2

33

Figure 4. Transition state analogue renin inhibitors.

molecule as described for other peptide ligands, with the objectives of increasing binding affinity, resistance to other degrading enzymes, and increasing oral bioavailability. The most significant of these has been the replacement of phenylalanine at the P1 position by a cyclohexylalanine (**33**) which apparently optimizes binding to a hydrophobic pocket of the enzyme, allowing significant truncation of the required sequence such that the C-terminal side of the isosteric residue could be removed. (Figure 4).[11]

The introduction of conformational constraint has been extensively examined in renin inhibitor design. Short range cyclizations such as γ-lactams and even the 1,2,4-triazolo[4,3-a]pyrazine derivatives (**34**) have been examined as well as macrocyclic lactams (**35**) linking inhibitors' side chains shown to be in close proximity by crystallographic studies with related aspartyl proteases (Figure 5).[101] Further attempts to remove peptide character from these renin inhibitors have included the synthesis of vinylogous amides, which was unsuccessful, and mono- and bis-pyrrolinones which yielded an inhibitor with submicromolar affinity.[102] It should be noted that success in this area has also led to the design of HIV protease inhibitors.[11]

2.4. Conclusion

The examples described here represent most of the strategic approaches used in generating peptidomimetics from the endogenous peptides. In many respects the approaches are driven by the perceived importance of the peptide backbone in de-

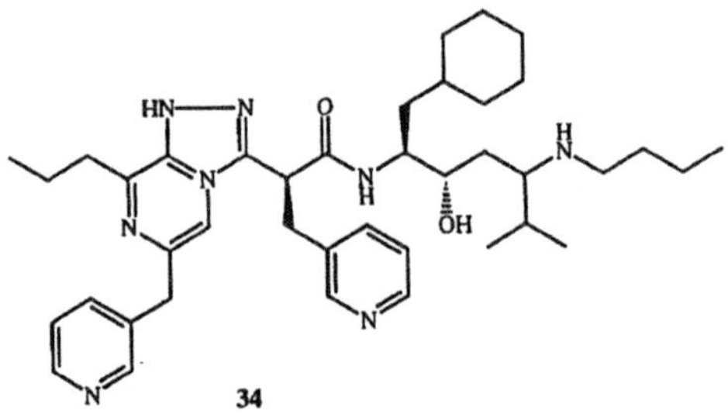

34

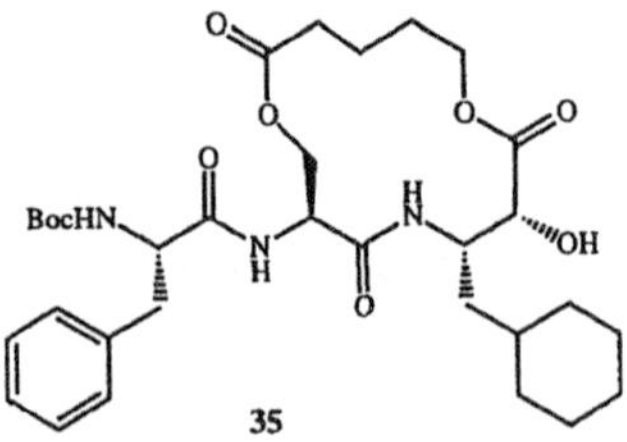

35

Figure 5. Conformationally constrained renin inhibitors.

scribing the orientation of the amino acid side chains, but significant progress is being made in looking beyond the backbone towards the positioning the key functionality of the molecule. Even within the peptide structure, analogues are being evaluated in which quite radical departures from the parent structure but retaining the proposed key elements have been made.

The modification of endogenous peptides represents the biggest body of published work in the area of peptidomimetics. However, it doesn't necessarily follow that this makes the area the most successful. It does demonstrate the safety of working with this form of analogue design with respect to gathering data from which to build up structure–function data. If a single-residue change abolishes biological activity, that is noteworthy, while in the mass screening and *de novo* design strategies, a negative result cannot be utilized and generally goes unpublished. What this process does allow is the discernment of more complex effects of endogenous ligands, such as interactions with multiple receptors or subtypes, enzymic activation, or inactivation.

3. NON-PEPTIDE NATURAL PRODUCTS THAT MIMIC ENDOGENOUS PEPTIDE LIGANDS

Over the last 10 years drug discovery via screening of natural product extracts and compound libraries has shifted from a whole cell assay approach, such as the traditional search for antibiotics from microbial culture, to a mechanistic approach. This mechanistic approach to drug discovery has developed out of our increased understanding of signal-transduction pathways. Inhibition of enzymes and antagonism of ligand–receptor binding are now the major targets for drug discovery. The advantages of a mechanistic approach is that the mode of action of a drug can be more readily evaluated and potential side effects predicted. Also, mechanism-based assays are amenable to higher throughput than whole cell assay.

Many endogenous ligands for receptors are peptides. Therefore, the aim of mechanism-based drug discovery is often to find a small non-peptide which displaces and hence mimics a peptide ligand. The first example of a non-peptide natural product discovered via screening for displacement of the binding of a peptide ligand was the isolation by Merck of asperlicin (**36**) as a cholecystokinin antagonist. Since the discovery of asperlin in 1985, a number of non-peptide natural products have been discovered as inhibitors of other peptide ligands.

Natural products which compete with the binding of endogenous peptide ligands can be considered non-peptide peptidomimetics in that they mimic the binding behavior of the endogenous peptide agonist, especially if they bind to the receptor at the same site. In some cases, conformational information for the peptide ligand has been used to guide the optimization process of the natural product lead. The following are selected examples of non-peptide natural products that mimic endogenous peptide ligands.

3.1. Cholecystokinin Antagonists

Cholecystokinin (CCK) is a 33-amino acid polypeptide that is found in various molecular forms in gastrointestinal tissue and in the central and peripheral nervous systems. CCK mediates a broad range of physiological responses by interacting with various receptors. The agent is involved in stimulation of gall bladder contraction and pancreatic exocrine secretion, is a regulator of both gut motility and appetite, and is a neurotransmitter.[103-107]

In 1985, Merck published the discovery of the natural product asperlicin (**36**) as a new selective and competitive non-peptide antagonist of CCK at peripheral CCK receptors.[108] This discovery was especially significant since it was the first example of the discovery, via a mechanism-based screen, of a non-peptide natural product which displaced a peptide ligand from its receptor. Asperlicin was found to have both *in vitro* and *in vivo* activity and to be selective for peripheral CCK receptors. However, asperlicin displayed low aqueous solubility, hindering its use as a tool for investigating the physiological and pharmacological actions of CCK.[109] In addition, asperlicin was less potent than other known CCK antagonists. Therefore, considerable effort has been expended improving asperlicin's potency and water solubility.[110]

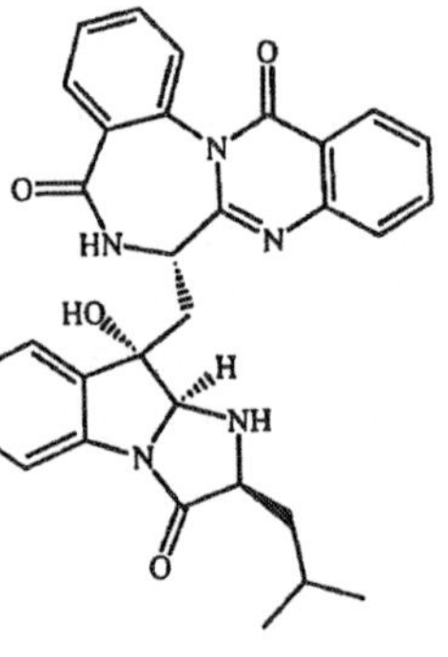

36 asperlicin

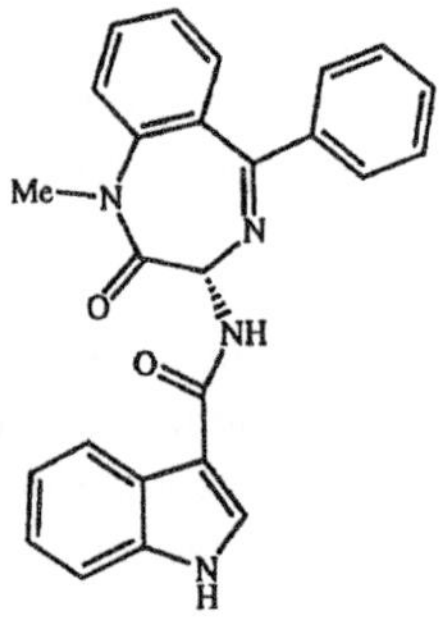

37 L-364,718

The core pharmacophore of asperlicin is the benzodiazepine substructure (Figure 6).[109] By starting with this core substructure and exploring substitution patterns based on information obtained from studying the structure of CCK and the anti-anxiety agent, diazepam, Merck synthesized L-364,718 (**37**).[109] This molecule is a subnanomolar potent antagonist at pancreatic CCK receptors and is highly specific and orally active.[111] L-364,718 is the first example of a potent peptidomimetic designed from structural comparison of an endogenous neuropeptide ligand, a known drug, and a non-peptide natural product antagonist. The design of L-354,718 indicates that asperlicin is a true peptidomimetic in that it probably binds to the same site on the CCK receptor as does CCK. Subsequently, Merck published a series of benzolactams which had similar activity to L-364,718, showing that the benzodiazepine partial structure per se is not essential for CCK antagonism.[112] Molecular modeling studies indicated that the benzodiazepine and benzolactam core structures overlap well and therefore probably interact with similar binding areas on the CCK receptor.[112]

While asperlicin is a selective CCK-A antagonist, its structure was used by a group at Lilly Research Laboratories as the starting point in the development of a selective CCK-B antagonist.[113] Substructure analysis of asperlicin revealed an embedded 4(3H)-quinazolinone ring system (Figure 6). In this analysis, asperlicin can be considered a conformationally constrained quinazolinone, or alternatively the Lilly compound **38** can be considered a conformationally flexible pharmacophore of asperlicin. The initial structure **38** obtained by comparison with asperlicin was used as a starting point for optimization studies, from which the selective CCK-B antagonist **39** was obtained. Quinazolinone **39** has an IC_{50} of 9.3 nM at mouse CCK-B receptors and represents a structurally novel series of non-peptide CCK-B receptor ligands. The development of the CCK-B selective ligand **39** from the CCK-A selective ligand asperlicin is an example of rational drug design based on the structure of a known receptor ligand. This example also illustrates how a mimetic approach, whether to an endogenous peptide ligand or to a non-peptide natural product ligand, can provide potent ligands to related receptors.

3.2. Substance P (SP) Antagonists

The neurokinins are a family of undecapeptides which share the common C-terminal amino acid sequence, Phe-X-Gly-Leu-Met-NH2. Substance P (SP) is the most well-studied of the neurokinin and is the endogenous ligand for the neurokinin-1 receptor.[114] When SP is released from the terminals of certain primary

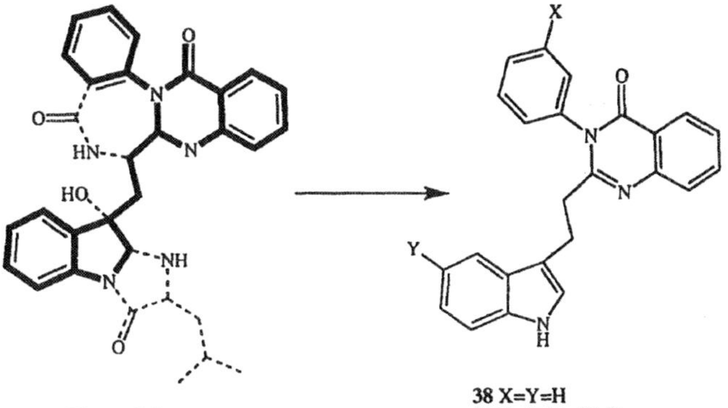

36 asperlicin

38 X=Y=H
39 X=O-iPr, Y=Br

Figure 6. Substructure analysis of asperlicin as a conformationally restricted quinazolinone.

afferent neurons following nervous stimulus it induces an excitatory effect on spinal dorsal horn neurons[115] that generates a painful stimulus.[116] In addition to this nociceptive role, SP has been shown to exist in elevated concentrations in nerves supplying sites of chronic inflammation[117] and has been shown to cause histamine release from mast cells and basophils from different species.[118] Because of this role in pain and inflammation, a competitive antagonist to SP at the NK-1 receptor is potentially useful as a non-narcotic analgesic or as an antiinflammatory agent, with potential in the treatment of diseases such as arthritis, asthma, and inflammatory bowel disease.

A number of SP inhibitors have been discoverd recently, the most well known of which is the Pfizer compound CP-96345 (**40**).[119] Although CP-96345 was discovered as the result of chemical file screening, a number of natural product inhibitors have also been found. These include two natural product cyclic peptide SP antagonists,[120,121] the non-peptide inhibitors fiscalin,[122] anthrotainin,[123] and spiroquinazoline,[124] and the symmetrical diketopiperazine dimer WIN 64821 (**41**).[125]

WIN 64821 can be considered a peptidomimetic for two reasons. The first is that

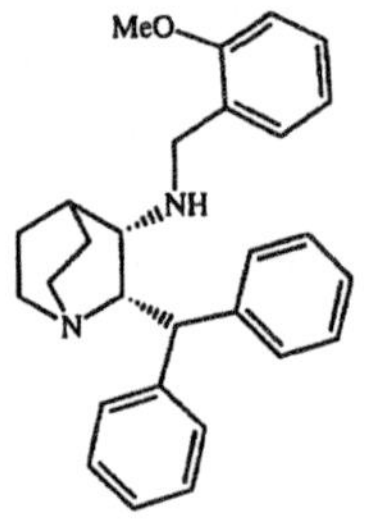

40 CP-96345

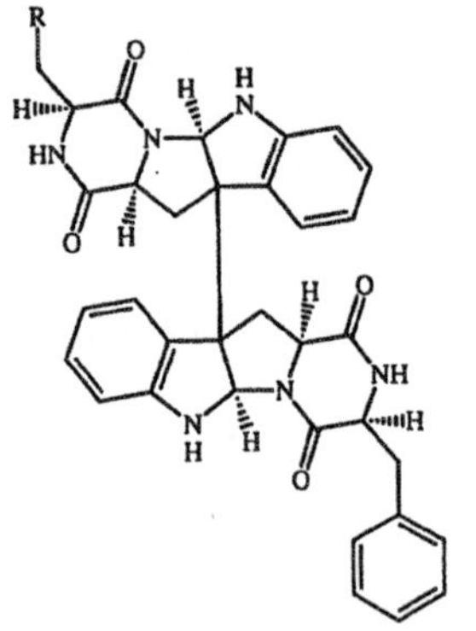

41 WIN 64821 (R=phenyl)
42 WIN 64745 (R=dimethyl)

it displaces SP from the NK-1 receptor and the second is that it is biosynthetically, but not structurally, a peptide. WIN 64821 was isolated from an *Aspergillus* sp of fungus together with a less active analogue WIN 64745 (**42**) and was found to inhibit radiolabeled SP binding in a variety of tissues, including those of human organs.[126] WIN 64821 showed competitive inhibition of SP binding with K_i values ranging from 0.24 μM in human astrocytoma cells to 7.89 μM in rat submaxillary membranes. This species difference indicates the difficulties of using a nonhuman cell line for primary screening. In NK-1 functional assays, WIN 64821 was found to be a competitive inhibitor of SP-induced contractility in guinea pig ileum (pA2 6.6), as well as an inhibitor of SP-induced $^{45}Ca^{2+}$ efflux from human astrocytoma cells (IC_{50}-0.6 μM).[126]

Although some analogues of WIN 64821 were synthesized chemically,[127] most were produced using a directed biosynthesis approach.[128] Structure–activity results indicated that both phenyl rings and one of the two indoline rings were involved in binding to the NK-1 receptor. The presence of two or three aromatic rings appears to be a consistent feature of many potent NK-1 antagonists, indicating that these compounds may bind at the same site within the receptor.

The solution structure of WIN 64821 was determined using a combination of NMR spectroscopy and molecular modeling.[125] Comparison of the low-energy structures for WIN 64821 with the low-energy structures of analogues, together with structure–activity correlations, indicated that the binding conformation for WIN 64821 is one in which the two phenyl and the one indoline rings involved in binding to the receptor lie on one side of the molecule.

A second SP antagonist which is structurally different from WIN 64821 but is consistent with the notion that aromaticity/hydrophobicity is necessary for potent binding with the NK-1 receptor is the cyclic peptide WIN 66306 (**43**).[121] The potency of this cyclic peptide is increased over 100-fold by methylation of the prenylation β-hydroxy tyrosine functionality to give WIN 67689 (**44**). This cyclic heptapeptide contains three glycines, and therefore only three side chains potentially involved in binding. The peptide exists in a single conformation in solution with a type-II β-turn centred at Pro-Gly.

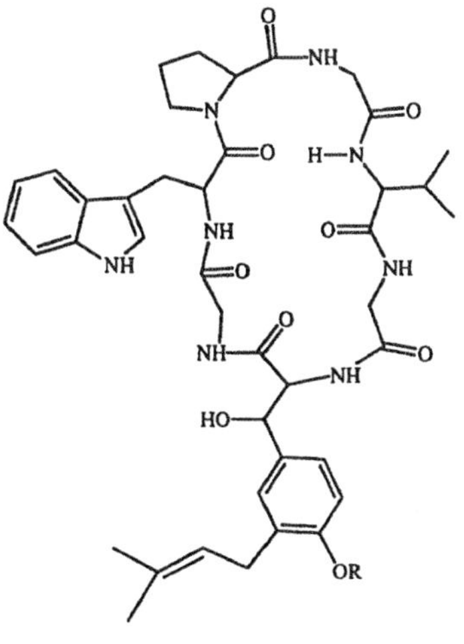

43 WIN 66306 (R=H)
44 WIN 67689 (R=Me)

WIN 66306 is an example of a peptide which appears amenable to peptidomimetic studies. That is, Pro-Gly could be replaced with a β-turn mimetic and the two glycines replaced. Alternatively, the entire peptide backbone could be replaced with a glucose scaffold in a similar manner to that performed during a peptidomimetic approach to the design of somatostatin antagonists (see Section 2.2 of this review).[65]

During the last 5 years a number of potent substance P antagonists have been obtained both from natural sources and through the screening of chemical databases. There are obvious structural similarities among some of these compounds, including similarly spaced aromatic moieties. Some efforts have been made to quantitate this similarity and produce mimetics by analyzing a number of these compounds with respect to their conformational space, and to discover bio-relevant conformations and pharmacophoric patterns. This information is then applied to a database searching to retrieve compounds that have the characteristic features of substance P receptor antagonists. Using this type of substance P mimetic strategy, a series of new antagonists was recently discovered.[129]

3.3. Paclitaxel (Taxol®)

Paclitaxel (**45**) is the first in a new class of anticancer compounds which prevent the disassembly of microtubules and thereby cause the arrest of cells at the G2/M step of the cell cycle.[130] Paclitaxel was first isolated from the bark of the Western Yew (*T. Brevifolia*) in the late 1960s but was not vigorously studied as a potential anticancer compound until the discovery in 1979 that paclitaxel promotes polymerization of the cellular protein tubulin, causing it to assemble into stable microtubules.[130] It was later shown that in addition to promoting assembly, paclitaxel also inhibited disassembly of microtubules to tubulin.[131] Although paclitaxel's low water solubility and poor yield from the bark of Western Yew were major obstacles to drug development, it entered Phase I clinical trials in 1983 on the strength of its unique anticancer mechanism. It was approved for the treatment of ovarian cancer by the U.S. Food and Drug Administration in 1992.

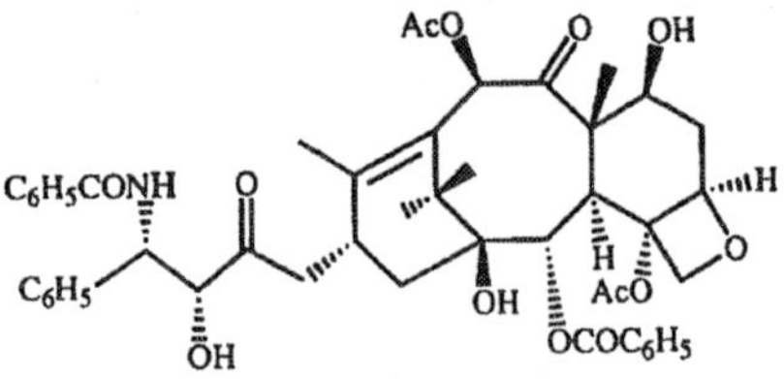

45 Taxol®

Photolabeling studies have shown the binding site for paclitaxel to be the amino acid residues 217-231 of β-tubulin.[132,133] Thus its action is due to its ability to bind within a protein-binding site. Recently, a monoclonal anti-idiotypic antibody that mimics paclitaxel was prepared.[134] This antibody (antibody 82H) not only mimics the binding of paclitaxel to microtubules, but also promotes the assembly of tubulin into microtubules. The finding that a polypeptide can mimic paclitaxel supports the hypothesis that paclitaxel is a peptidomimetic compound which displaces an endogenous peptide ligand at the β-tubulin-binding site.[134]

Paclitaxel is a good example of how the limited supply of a natural product can delay its introduction as a drug. It was initially isolated from the bark of the Western Hew in 0.007% yield[135] and tens of thousands of kilograms of bark were needed for clinical trials, raising ecological concerns. Although total synthesis of paclitaxel has been achieved,[136,137] the difficulty and expense of synthesis at present preclude this as a method of supply. Of the alternative approaches to paclitaxel production, including plant tissue culture,[138] related fungal production,[139] and partial synthesis,[140] the most successful has been partial synthesis from the paclitaxel precursor 10-deacetylbaccatin III (10-DAB; **46**). 10-DAB is available from the leaves of a number of yew species in relatively high yield,[141] and partial synthesis has the added advantage that it permits the synthesis of paclitaxel analogues as well as paclitaxel. One more potent analogue of paclitaxel has already progressed through clinical trials[142] and additional analogues are sure to follow. If paclitaxel is a natural product peptidomimetic which displaces an endogenous peptide, then characterization of such a substance could lead to the discovery of new chemical classes of compounds with paclitaxel-like activity.

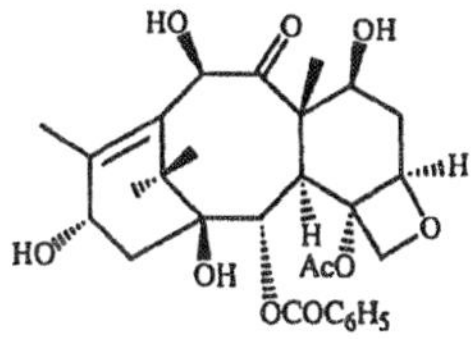

46 10-DBA

4. FROM VENOM PEPTIDES TO DRUGS

Biologically active peptides from venoms have been used both to elucidate physiological mechanisms and as starting points for the design of new therapeutic drugs. For example, small peptides isolated from the Brazilian arrowhead viper *Bothrops jaracusa*[143] were found to prolong the action of bradykinin by blocking an enzyme that also activated angiotensin's precursor. Therefore the discovery of these venom peptides led to the discovery of angiotensin-converting enzyme (ACE). Structural studies, partly based on these peptides, led to the marketing of ACE inhibitor drugs such as captopril, enalapril, and lisinopril, which are in the top 30 best selling medicines in the world. Studies toward the development of therapeutics from venoms have led to a number of potential drugs. These include drugs to control abnormal clotting, for the treatment of heart disease, as anticonvulsants, as immunosuppressants, and for the treatment of neurological disorders.

4.1. Drugs from Conus Venoms

Cone snails are marine predators that feed on marine worms, mollusks, or fish.[144] Cone snails use venom, injected via barbed radula tooth, to quickly paralyze their prey. Conus venom varies from the venom of most other animals in that it contains small structurally rigid peptides of approximately 10 to 30 amino acids in length with one to three disulfide bonds.[145] These disulfide bonds add structural constraints to the flexibility of these peptides, which tend to make conotoxins extremely receptor selective and potent.

Most conotoxins fall into three distinct sequence and receptor specific classes: (1) conotoxins that interact with acetyl choline receptors (α-conotoxins), (2) sodium channels (μ-conotoxins), or (3) calcium channels (ω-conotoxins). A number of conotoxins are being investigated as potential drugs, including one which is in clinical trials as an analgesic for intractable pain.[146] A synthetic version of ω-conotoxin MVIIA, a neuroprotective agent in a model of strokes in rats,[147] is under development by the Neurex Corporation as a neuroprotective and is in clinical trials for the treatment of head trauma.[148]

Although peptide conotoxins themselves are potential drugs, the major interest in conotoxins is that their structural rigidity makes the peptides good starting points for drug design using peptidomimetic chemistry. NMR studies on conotoxins indicates that the multiple disulfide bonds confer considerable rigidity on the conformation of the peptide.[149,150] Three major disulfide-bonded patterns exist and these are shown in Figure 7. α-Conotoxins have a two-loop framework, μ-conotoxins a three-loop framework, and ω-conotoxins a four-loop framework.

The presence of these frameworks provides two major advantages over unconstrained venoms. The first is that the more tightly a peptide segment is constrained, the less likely it is to bind to any particular target, but the tighter and more specific it will bind to its specific target. In fact, synthetic libraries of peptides built with and without constraints indicate that constrained libraries provide more potent and specific lead structures.[151,152,153] The second advantage is that the constrained conotoxins have a preferred conformation that will be close to the receptor-binding conformation and this can be determined using X-ray crystallography or NMR spectroscopy and molecular modeling. In highly constrained peptides, the amide backbone may be essentially fixed so that only the side-chain groups have significant freedom.

Conotoxin	Disulfide Framework
α-conotoxins	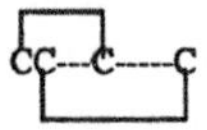 CC---C-----C
μ-conotoxins	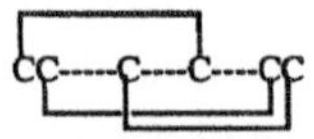 CC-----C----C----CC
ω-conotoxins	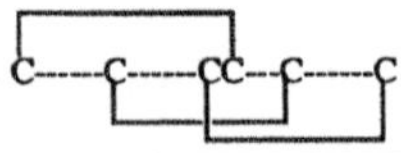 C------C------CC---C------C

Figure 7. Disulfide framework for the three common conotoxin classes.

Because it appears necessary to have only approximately six correctly positioned amino acids for potent and specific binding to macromolecular targets, it should be possible to decrease the size of bioactive conotoxins, perhaps by removing nonactive loops. However, there has been little success to date in significantly decreasing the size of an active conotoxin while maintaining biological activity. Nor has a non-peptide been developed via systematic design from a bioactive conotoxin. Several programs are underway with the aim of designing non-peptide drug leads from conotoxins, including a multimillion dollar collaboration in Australia between AMRAD and the 3D Centre at the University of Queensland.[154] A major contribution of conotoxins toward drug design is the provision of biologically relevant peptidic frameworks that can be used in conjunction with combinatorial chemistry for the production of multiple constrained peptide libraries for biological screening.

4.2. Anticoagulant Snake Venoms

Many snake venoms have anticoagulant or procoagulant activities.[155] Snake venoms often contain both proteinases and their inhibitors. The inhibitors presumably prevent autolysis and facilitate protection of the host tissues. The presence of both activators and inhibitors of blood coagulation and platelet aggregation has made snake venoms a rich source of peptides with therapeutic potential. Recently, a number of small proteins and peptides called "disintegrins" have been isolated from snake venoms as potent inhibitors of platelet aggregation.[156,157] Mechanistically these venom components act as potent antagonists of a cytoadhesive cell surface receptor, glycoprotein IIb/IIIa (GpIIb/IIIa), on platelets, preventing the interaction between platelets and fibrinogen. Disintegrins have been isolated from the venoms of the Okinawan habu (*Timeresurus gramineus*),[156] the saw-scaled viper (*Echis carinatus*),[158] the puff adder (*Bitis arietans*),[159] and the Malayan pit viper (*Agkistrodon rhodostoma*).[157] All these proteins are similar in that they contain the Arg-Gly-Asp (RGD) sequence that has been shown to be responsible for the observed competitive inhibition of platelet aggregation.[160]

Cell adhesion processes are mediated by a number of matrix-associated adhesive glycoproteins, such as fibronectin, fibrinogen, vitronectin, thrombospondin, and von-Willebrand factor, all of which contain the triad sequence, Arg-Gly-Asp (RGD).[161] This sequence is recognized by a range of receptors (called integrins), and a range of short RGD-containing peptides were shown to inhibit the adhesion process via interaction with one or more integrins. As drug targets, antagonists were potential therapeutics for a variety of diseases associated with abnormal extracellular matrix function including cardiovascular disease, cancer, osteoporosis, and inflammation. Agonists, which promote the interaction of cells and tissues with artificial matrices, are targets in organ and tissue healing and regeneration processes such as the treatment of burns and chronic wounds. Most significantly for drug design strategies, the affinity and selectivity of RGD peptides for integrins has been shown to depend upon the conformation of the RGD sequence and the induction of receptor affinity and selectivity is crucial to the development of useful therapeutics in this area.

The integrin GpIIb/IIIa on the surface of blood platelets mediates platelet aggregation upon binding by the RGD-containing peptide, fibrinogen, and is a primary process in stopping bleeding after vascular injury. However, thrombus formation can ultimately lead to the development of coronary artery thrombosis such that IIb/IIIa antagonists are important targets. The use of RGD-based peptides has therefore been examined extensively. However, the exclusion of affinity for a range of other integrins is required or the risk of serious side effects is quite high. The disintegrins are specific for the interaction of GpIIb/IIIa with fibrinogen and are highly conserved, although they vary in length. They are cystine-rich and have relatively fixed conformations close to the RGD sequence, which influences the specificity of the particular disintegrin.[162] NMR studies have aided in the designs of peptidomimetics from the disintegrins.[163]

The design of cyclic peptides and linear pseudopeptides based on the structures of disintegrins has been recently reviewed.[164,165,166] The existence of a simple continuous pharmacophore has enabled the generation of important structure–function studies and in particular the identification of conformational inputs influencing receptor selectivity.

Cyclic peptides such as G4120 and MK-852 have been successful in increasing potency and enzymic stability relative to there linear counterparts,[166] and the role of conformation to specificity is being examined using cyclic peptides where the RGD sequence is held in varying conformations by structurally rigid amino acids. An impressive selectivity of binding to GpIIb/IIIa versus GpV/IIIb was shown for a series of cyclo(Xxx-Yyy-Arg-Gly-Asp) peptides in which the variation of adjacent amino acids could switch selectivity for one receptor to the other.[167] Based on structure-activity information obtained from both venom and synthetic peptide inhibitors, SmithKline researchers developed the cyclic peptide SK&F 107260 (**47**), after considerable exploration around the RGD sequence. Based on **47**, the group designed a peptidomimetic (**48**), which incorporates a γ-turn mimetic at the aspartic acid residue.[168] Although peptidomimetic **48** is approximately 10-fold less potent than cyclic peptide **47**, the former still has significant activity in the dog platelet aggregation assay used.

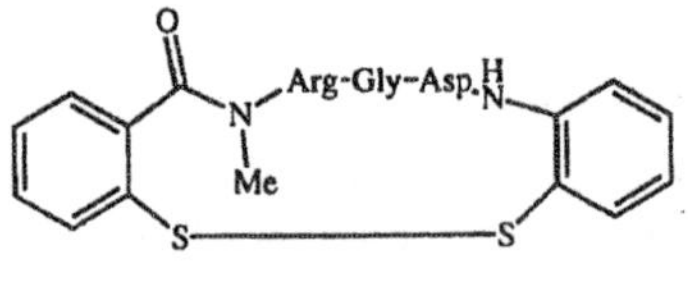

47 SK&F 107260

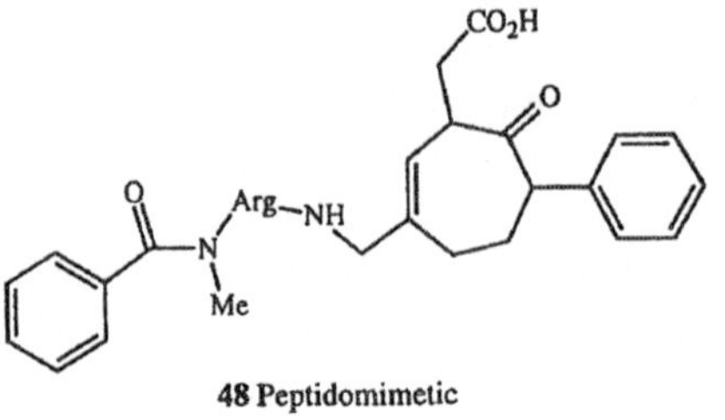

48 Peptidomimetic

Structure–bioactivity studies for disintigrins and related peptides containing the RGD sequence indicates that the Asp carboxylate distanced appropriately from the Arg guanidine is necessary for inhibition of the GpIIb/IIIa/fibrinogen interaction. Such a small pharmacophore readily lends itself to the development of non-peptides which may be orally active. Drugs in this group include Lamifiban, and Xemlofiban which has an ester prodrug moiety.[166] A number of groups have discovered small molecule inhibitors that incorporate a carboxylate and a guanidine or related positively charged surrogate appropriately distanced through the use of a scaffolding. These small molecule inhibitors include compounds developed by SmithKline Beecham (**49**),[169] Genentech (**50**),[170] Roche (**51**),[171] Glaxo (**52**),[172] and

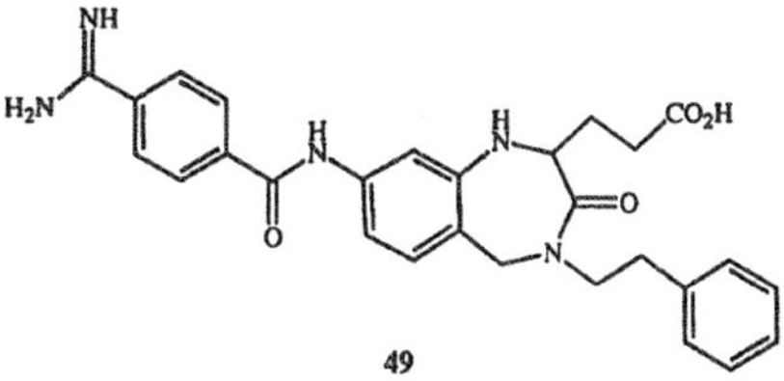

49

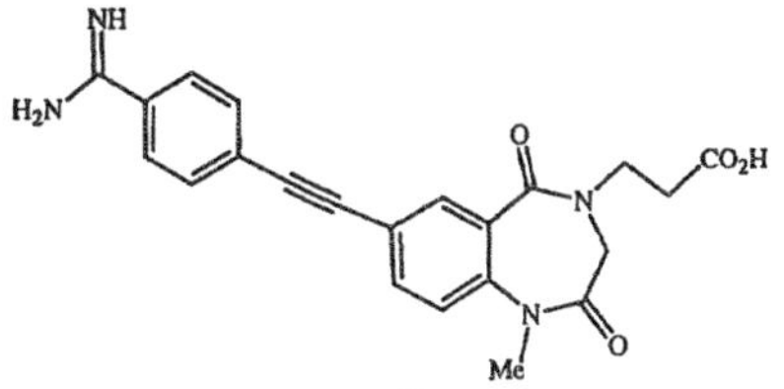

50

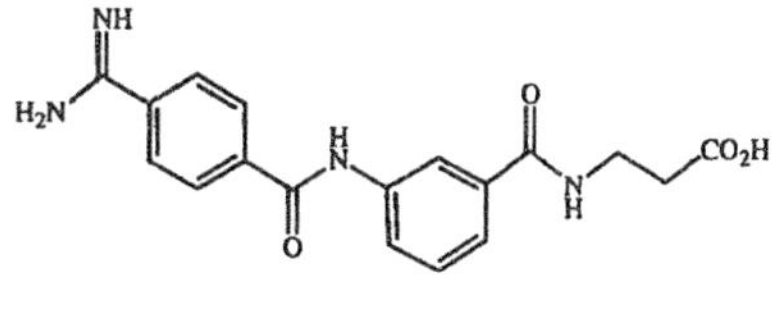

51

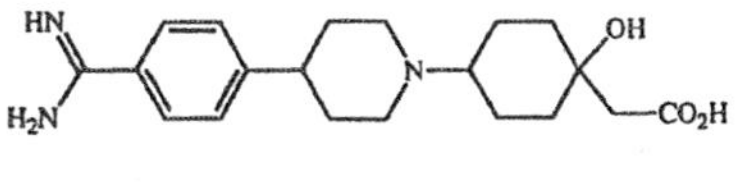

52

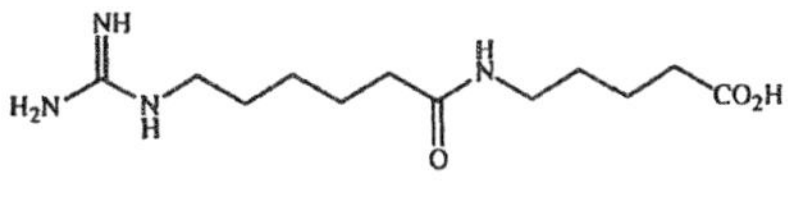

53

Figure 8. Small lmolecule inhibitors of the GpIIb/IIIa/Fibrinogen interaction, based on the RGD sequence.

the Weizmann group (**53**).[173] Structural comparison of these compounds indicates that the spacing between the guanidine and the carboxylate is similar in each case. Structure bioactivity studies on **53** indicate that for a linear structure the number of atoms present in the spacer needs to be eleven.[173]

4.3. Peptide Toxins Targeted Against Potassium Channels

Numerous venom toxins including sea anemone toxins,[174,175] spider toxins,[176-178] and snake venom toxins,[179] have been isolated that interact with various ion channels, including sodium, calcium, and potassium channels. These peptide toxins have been used as pharmacological tools for the characterization of classes and subtypes of ion channels, as labeled ligands for the development of new receptor targets, and as therapeutic leads. For example, toxins such as apamin, charybdotoxin, noxiustoxin, and dendrotoxin have been useful for establishing the roles and types of potassium channels in various tissues.[180] Potassium channels are exceptionally diverse in both variety and function and provide important therapeutic targets. It has been suggested, for example, that blockers of certain potassium channels might provide symptomatic relief in neurodegenerative disorders such as Alzheimer's disease, or as anticonvulsant drugs for epilepsy.[148]

An important group of potassium channel blockers that have potential in drug design are the dendrotoxins. Dendrotoxins were isolated from mamba venom and are small proteins (57–60 amino acids) that are cross-linked by three disulfide bonds. The dendrotoxins were discovered because of the ability of green mamba venom to augment the chemical signaling between nerves and muscles, and the purified toxins were found to enhance the evoked release of neurotransmitters by selectively blocking certain voltage-dependent potassium channels to which they bind with high affinity. The site of binding of the dendrotoxins remains unknown, although it appears that they might bind directly to the extracellular anionic sites of the potassium channels by ionic interactions.[181]

Recent three-dimensional structures of dendrotoxins indicate that the proteins are relatively structured due to the disulfide linkages and have similar conformations, containing an amphiphilic framework that exposes the proposed active region of the dendrotoxins to the anionic sites of the potassium channel receptors.[182] Structure–bioactivity studies where one or more of the three disulfide bonds were selectively reduced also indicated that the relatively rigid secondary structure was crucial for bioactivity.[183] Site-directed mutagenesis studies are currently underway to further determine the importance of specific amino acids for potassium channel specificity and binding potency.[184-186] Further structure–bioactivity studies on these peptide toxins will open the door to the design of peptidomimetic analogues, which appears to be the goal of companies such as Wyeth-Ayerst.[179,187] The design of peptidomimetics based on the dendrotoxins will be facilitated by their relatively small and rigid structures, in a similar way as for the conotoxins discussed above.

5. NATURAL PRODUCT SCAFFOLDS

As the gap is bridged between peptide and non-peptide drugs it has become increasingly apparent that the elaborate secondary and tertiary structure of peptides serves mainly as scaffolding to correctly orient key functional atoms for receptor interaction and that this task can similarly be performed by non-peptidic molecular scaffolds.[10] However, the rational design of such a scaffold is a difficult process—it requires detailed understanding of the structural bases for ligand–receptor binding, design of a precise non-peptide template, and the successful total synthesis of the compound before a new ligand can be tested. That few ligand–receptor interactions are known in fine detail, that algorithms for *de novo* design of ligands are in their infancy, and organic synthesis of novel molecules still represents a significant challenge makes this a high risk-process. Compromises are needed in this approach and easing the burden of the synthetic chemist can be achieved by biasing the design towards synthetic targets which have been studied before. In this respect, non-peptide natural products, particularly those of medicinal value are providing a significant resource in the design of novel peptidomimetics. By accident or by design, a number of natural product-derived structures have been incorporated into a diverse array of ligands for peptide receptors to replace the peptide backbone in whole or in part, and the advent of combinatorial chemistry strategies render the natural product template even greater potential in peptidomimetic discovery. Hirschmann et al. have recently described the synthesis of a derivatized steroid analogue of the integrin-binding RGD sequence which incorporated the guanidinyl and carboxyl side chains and overlaid the backbone of a cyclic peptide constrained in a type-I β-turn. One resultant molecule did compete with fibrinogen for GpIIa/IIIb receptor sites, albeit at high relative concentrations.[188] Interestingly, a steroid natural product has been recently found to be a substance P antagonist lending further weight to the potential of steroid scaffolds as peptidomimetics.[189]

5.1. β-D-Glucose

Carbohydrates in general offer great versatility as non-peptide scaffolding for peptidomimetics, not least because of the vast literature covering carbohydrate synthesis. Furthermore, the pyran ring offers significant rigidity and various points of attachment for substituents, either as side-chain mimics or for peptide backbone elaboration.

As described in Section 2.2. of this review, the glucopyransides were found to be to be mimics of the cyclic hexapeptide L-363,301 in that they bind to the somatostatin receptor,[64] mimicking the β-turn of the latter. It was discovered that these two compounds also bind weakly to the β2 adrenergic receptor. Subsequent studies showed that the glycoside **14** inhibits GRF-induced growth hormone (GH) release by cultured rat anterior pituitary cells with an IC_{50} of 3 μM, and N-acetylation of the amine yielded **15** which was highly selective for the substance P (SP) receptor, with an IC_{50} of 60 nM. That such minor chemical modification should produce such a dramatic switch in biological profile has been rationalized by the fact that at least two somatostatin receptors, as well as the SP, and the β_2-adrenergic receptors, utilise G-protein-mediated signal transduction, suggesting that the binding of **14** to these three receptors may involve similar interactions within the conserved hydrophobic domains of the receptors.

Graf von Roedern and Kessler examined the glucose scaffold in another way, viewing the 3,5-disubstitution of D-glucose as part of the peptide backbone, again with the intention of stabilizing a reverse-turn conformation.[190] They were able to prepare analogues of somatostatin with significant activity in inhibiting growth-hormone release (**54, 55**). Subsequently, this group have developed a "construction kit" of sugar amino acids as mimetics for amino acids and dipeptides.[191] Analysis of the conformation induced by these amino acids has shown them to constrain linear and turn conformations. Moreover the templates retain additional hydroxyl functionality which might be modified with peptide side-chain mimics.[192]

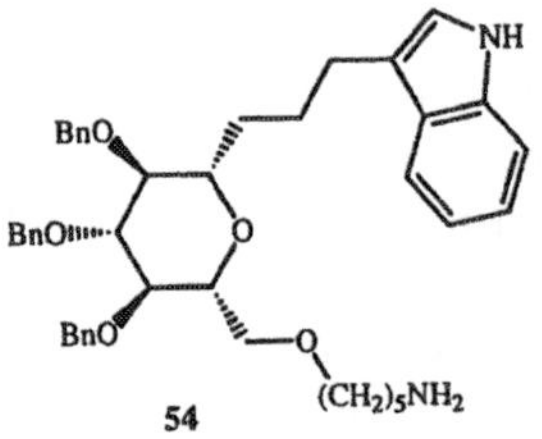

Interestingly, while many peptides and other natural products have been described which bind to oligosaccharide-binding regions of enzymes, there are few oligosaccharides that have been shown to inhibit the binding of an endogenous peptide. Attempts to define the antagonism of atrial natriuretic peptide by the oligosaccharide HS-142-1 represents the first significant step in this regard.[193,194] A recent report of oligosaccharide synthesis using solid-phase methods would appear to be an important step forward also.[195]

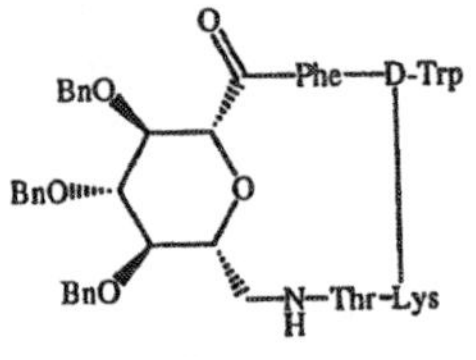

5.2. Penicillin

The antibiotic penicillin (**56**) was one of the most important discoveries of the twentieth century. A natural product, the original structure has been modified and derivatized in many ways in order to enhance biological activity and overcome resistance, etc.[196] It is against this backdrop that Nagai and Sato noted the potential for modification of the penicillin structure to give a thiazolidone δ-lactam as a potential template for peptidomimetic design. Their synthesis of the β-turn dipeptide (BTD; **57**) has been one of the most widely applied forms of conformational stabilization and has been used in analogue studies of Gramicidin S,[197] enkephalin,[198] somatostatin,[199] LHRH,[200] GRF,[201] HIV protease,[202] and tendamistat.[203] It should be noted that only the Gramicidin S analogue, LRF analogue, and HIV protease analogue have shown potent activity. A number of variants of this structure have been examined including the use of a γ-lactam as described by Subasinghe et al.[204] The syntheses of these structures follows that laid out in the synthesis of penicillin analogues.[205,206]

The crystal structure of the BTD isomer was identified using the CAVEAT program of Bartlett as overlapping a dipeptide region of Cyclosporin A, and upon incorporation into CsA yielded a peptide of increased affinity for cyclophilin and calcineurin.[207]

A penicillin-G (**56**) derived peptidomimetic has been described by Humber et al. and relates to the synthesis of HIV-1 protease inhibitors.[208] In a screening program, crude samples of a penicillin dimer (**58**) were found to have inhibitory activity. Ultimately it was found that the lactam ring opened derivative (**59**) (a side product of purification) was a potent inhibitor of HIV-1 protease. This led to the preparation of a series of related more potent amides from readily accessible penicillins.[209,210]

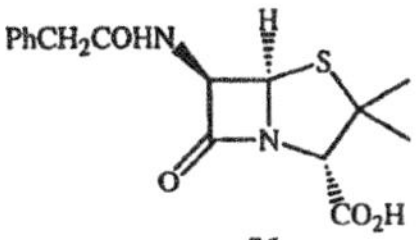

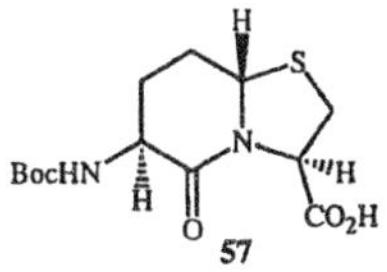

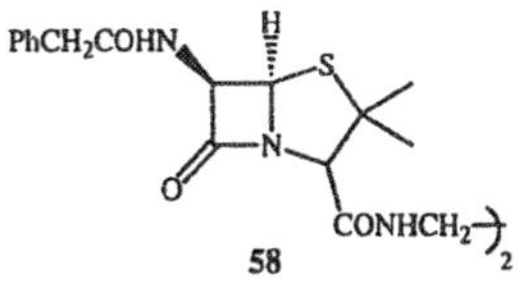

5.3. Benzodiazepines

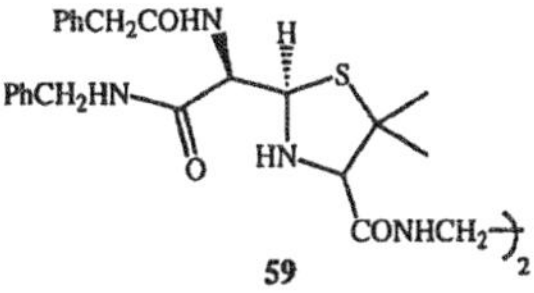

The benzodiazepines have an extraordinary diversity of action, in addition to their classical anxiolytic activity as found with diazepam. They have been found with CCK antagonists activity (see Section 3), opioid antagonists, platelet-activating factor antagonists, HIV transactivator Tat antagonists, GpIIbIIIa inhibitors, reverse transcriptase inhibitors, and *ras* farnesyltransferase inhibitors.[211] That many natural product representatives of this class exist justifies the inclusion of this heterocycle in this section, and there is a massive body of medicinal chemistry literature that demonstrates the utility of benzodiazepines as template scaffolding in peptidomimetic design.

First, the existence of benzodiazepine peptidomimetics, such as the CCK antagonists, opioid ligand tifluadom, and farnesyl transferase inhibitor, show the successful application of benzodiazepine scaffolds. Further, the potential of the benzodiazepines as idealized β turn mimics has been proposed and has been tested by Ripka et al. who have constructed a range of peptides related to Gramicidin S that contain benzodiazepines similar to **60**.

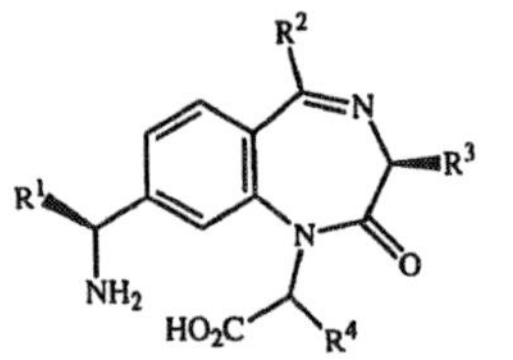

60a $R^1 = H, R^2 = Ph, R^3 = Ph, R^4 = H$
60b $R^1 = (CH_3)_2CHCH_2, R^2 = Ph, R^3 = Ph, R^4 = H$

The broader application of benzodiazepines as a scaffold for peptidomimetic discovery has been demonstrated by the solid-phase synthesis of benzodiazepines by Bunin and Ellman.[212,213] In the first report,[192] benzodiazepine derivatives were synthesized on polyethylene pins using the method developed by Geysen, and evaluated for CCK_A affinity.[214] The subsequent evaluation of an expanded library of 11,200 structurally diverse derivatives against a range of therapeutic targets is currently in progress.[211]

5.4. Tropanes

The tropane nucleus also lends itself to application as a peptidomimetic template and many of the known tropane natural products and synthetic derivatives lend themselves to ready adaptation into peptidomimetic moieties. In particular, tropanes such as **61** seem well-suited to either use as turn-inducing amino acid replacements[215] or as scaffolding for the attachment of diverse side chains, particularly using combinatorial approaches.[216] The access to optically active or racemic precursors particularly enhances the range of structures that can be readily derived with relatively generic synthetic methods. Oligotropanes of this type also represent completely novel molecules, and the incorporation of extra chemical functionality provides wide scope for the generation of molecular diversity.

Amino acid derivatives of the tropane alkaloid natural products[217] have been evaluated for potential application as peptidomimetics. These molecules contain the structural elements desirable in a peptidomimetic–amino and carbonyl groups held in an relatively rigid orientation by a cyclic structure. The structural diversity within this class of heterocycle allows for a large number of conformations to be surveyed, while at the same time retaining a relatively simple structure, well-precedented approaches to their synthesis and incorporation into peptides via solid-phase synthesis.

It has been demonstrated that synthesis of Fmoc-protected tropane-based amino acid derivatives can be achieved by elaboration of established synthetic routes to reported tropane alkaloid natural products in optically pure or racemic form from cheap, readily available precursors. The range of protected amino acids synthesized include the simple tropane–carboxylic acid nucleus as well as unsaturated, aryl-substituted and small-ring variants, all prepared from a common precursor molecule, tropinone.

The recent description of solid-phase organic synthesis (SPOS) on a tropane nucleus provides a potential strategy for the synthesis of many primary target amino acids and a whole new array of substituted tropanes which may be non-peptide peptidomimetics or developed into monomer amino acids might be accessed by application of SPOS.[216]

Selected peptides containing the racemic nortropane-2α-carboxylic acid residue ($Ntc^{2\alpha}$; **61**) have been synthesized and evaluated in biological assays for hypoglycaemic activity and platelet aggregation inhibition, and the results of these assays indicate the promise that tropane ligands hold for structure–function studies of bioactive peptides. Molecular dynamics studies have shown that tropane amino acids generally restrict the resulting peptide to a very narrow range of conformations; for example peptides containing the residue-$Ntc^{2\beta}$ (**62**) overlay a type-II' β-turn, which should make them particularly useful in structure–function studies of peptides.

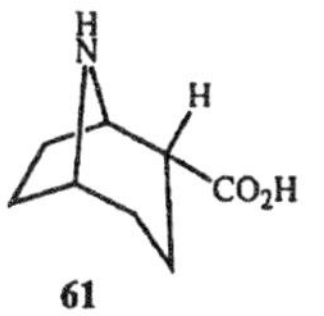

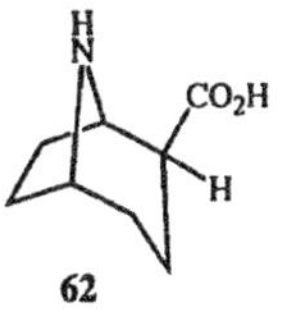

Against this backdrop of successful application of organic scaffolds in peptidomimetic design and the potential application of combinatorial chemistry to provide a large database of ligands to test, other natural products to act as molecular templates can be examined. Natural products provide a ready resource creating a diversity of three-dimensional structures readily accessed and evaluated. At present, one of the major challenges to many similar approaches is the adaptation of classical solution-phase chemistry to solid-phase synthesis of these products.

6. CONCLUSION

Historically, small molecules and peptides have been developed along parallel paths in drug research with developments in one area occurring independently of the other. The field of peptidomimetic research now has as much to do with small molecule design as it does with peptides, and in this context natural product research has reemerged as a major source of biological leads alongside the new field of combinatorial chemistry. Indeed there may be good cause to revisit many natural product sources to examine their interactions with peptide binding sites.[218]

Extracts of living organisms represent the predominant source of drug leads with about 80% of all drugs having natural product origins, yet less than 1% of the 60–100 million species have been screened in this way. Major drug companies are seeking to further explore "pharmaceutical bioprospecting". For example, Merck's agreement with Costa Rican biodiversty institute, INBio., gives Merck access to biological samples for drug development. Issues of conserving biodiversity because of the potential drugs that might disappear with species are also being examined.[220]

As far as peptidomimetic discovery goes, the combination of endogenous peptides, non-mammalian peptides, and non-peptide natural products provides several angles of attack which can provide the chemical diversity to truly understand the molecular basis of specific biological activities, and examples of overlap between these strategies are evident in this chapter. Synthetic compound libraries, either from compound databanks or combinatorial synthesis, provide a further source of diversity.

Combinatorial chemistry represents a cheaper source of molecular diversity than natural products but is limited to that which can be prepared in a laboratory. In general, this means that the molecules currently lack much of the structural complexity often characteristic of natural products. However, it seems possible that the natural product chemistry might provide solutions in this context, and that complex natural products might act as templates upon which to perform combinatorial syntheses. In this way, the structural and conformational diversity of the natural product world complements the functional group diversity of combinatorial chemistry.

7. REFERENCES

1. Dutta, A. S.; *Adv. Drug. Res.* **1991**, *21*, 145.
2. Veber, D. F. In *Peptides: Proceedings of the Twelfth American Peptide Symposium;* Smith, J. A.; Rivier, J. E., Eds.; ESCOM, Leiden, 1992, pp 3-14.
3. Liskamp, R. M. J. *Recl. Trav. Chim. Pays-Bas* **1994**, *113*, 1.
4. Olsen, G. L.; Bolin, D. R.; Bonner, M. P.; Bos, M.; Cook, C. M.; Fry, D. C.; Graves, B. J.; Hatada, M.; Hill, D. E.; Kahn, M.; Madison, V. S.; Rusieck, V. K.; Sarabu, R.; Sepinwall, J.; Vincent, G. P.; Voss, M. E. *J. Med. Chem.* **1993**, *36*, 3039.
5. Fauchere, J.-L. *Adv. Drug Res.* **1986**, *15*, 29.
6. Fauchere, J.-L; Thurieau, C. *Adv. Drug Res.* **1992**, *23*, 128.
7. Farmer, P. S.; Ariens, E. J. *Trends Pharmacol. Sci.* **1982**, *5*, 362.
8. Giannis, A.; Kolter, T. *Angew. Chem.* **1993**, *32*, 1244.
9. Ball, J. B.; Alewood, P. F. *J. Mol. Recogn.* **1990**, *3*, 55
10. Farmer, P. S. In *Drug Design;* Ariens, E., Ed.; Academic Press, 1980, Vol. X, p 119.
11. Wiley, R.A.; Rich, D. H. *Med. Res. Rev.* **1993**, *13*, 327.
12. Marshall, G. R. *Tetrahedron* **1993**, *17*, 3547.
13. Moore, G. J. *Trends Pharm. Sci.* **1994**, *15*, 124.
14. Gante, J. *Angew. Chem. Int. Ed. Engl.* **1994**, *33*, 1699.

15. Rizo, J.; Gierasch, L. M. *Annu. Rev. Biochem.* **1992**, *61*, 387.
16. von Mering, J.; Minkowski, O. *Arch. Exp. Pathol. Pharmakol.* **1889**, *26*, 371.
17. Abel, J. J. *Proc. Natl. Acad. Sci. USA* **1926** *12*, 132.
18. Sanger, F. *Science* **1959**, 129, 1340.
19. Geysen, H. M.; Meloen, R. H.; Barteling, S. J. *Proc. Natl. Acad. Sci. USA* **1984**, *81*, 3998.
20. Gallop, M.; Barrett, R. W.; Dower, W. J.; Fodor, S. P. A.; Gordon, E. M. *J. Med. Chem.* **1994**, *37*, 1233.
21. Hruby, V. J. *Life Sci.* **1982**, *31*, 189.
22. Toniolo, C. *Biopolymers* **1989**, *28*, 247.
23. Houston, M. E. Jr.; Campell, A. P.; Lix, B.; Kay, C. M.; Sykes, B. D.; Hodges, R. S. *Biochemistry* **1996**, *35*, 10041.
24. Rose, G. D.; Gierasch, L. M.; Smith, J. A. *Adv. Protein Chem.* **1985**, *37*, 1.
25. Tenerius, L. *Acta Pharmacol. Toxicol.* **1973**, *32*, 317.
26. Pert, C. B.; Snyder S. H. *Science* **1973**, *179*, 1011.
27. Terenius, L.; Wahlstrom, A. *Acta Physiol Scand.* **1975**, *94*, 74.
28. Hughes, J.; Smith, T.; Morgan, B.; Fothergill, L. *Life Sci.* **1975**, *16*, 1753.
29. Pasternak, G. W.; Goodman, R.; Snyder, S. H. *Life Sci.* **1975**, *16*, 1765.
30. Teschemacher, H.; Opheim, K. E.; Cox, B. M.; Goldstein, A. *Life Sci.* **1975** *16*, 1771.
31. Hughes, J.; Smith, T. W.; Kosterlitz, H. W.; Fothergill, L. A.; Morgan, B. A.; Morris, H. R. *Nature* **1975**, *258*, 577.
32. Martin, W. R. *J. Pharmacol. Exp. Ther.* **1976**, *197*, 517.
33. Bradbury, A. F.; Smyth, D. G.; Snell, C. R. *Nature* **1976**, *260*, 165.
34. Schiller, P. W.; Yam, C. F.; Lis, M. *Biochemistry* **1977**, *16*, 1831.
35. Gorin, F. A.; Marshall, G. R. *Proc. Natl. Acad. Sci. USA* **1977**, *74*, 5179.
36. Erspamer, V.; Melchiorri, P.; Falconieri-Erspamer, G.; Negri, L.; Corsi, R.; Severini, C.; Burra, D.; Simmaco, M.; Kreil, G. *Proc. Natl. Acad. Sci. USA* **1989**, *86*, 5188.
37. Schiller, P. W.; Di Maio, J. *Proc. Natl. Acad. Sci. USA* **1980**, *77*, 7162.
38. Gardner, B.; Nakanishi, H.; Kahn, M. *Tetrahedron* **1993**, *49*, 3433.
39. Bartosz-Bechowski, H.; Davis, P.; Zalewska, T.; Slaninova, J.; Porreca, F.; Yamamura, H. I.; Hruby, V. J. *J. Med. Chem.* **1994**, *37*, 146.
40. Belanger, P. C.; Dufresne, C.; Scheigetz, J.; Young, R. N.; Springer, J. P.; Dmitrienko, G. I. *Can. J. Chem.* **1982**, *60*,1019.
41. Krstenansky, J. L.; Baranowski, R. L.; Currie, B. L. *Biochem. Biophys. Res. Commun.* **1982**, *109*, 1368.
42. Freidinger, R. M., In *Peptides: Proceedings of the 7th American Peptide Symposium;* Rich D. H., Ed.; Pierce Chemical Co., Rockford, Il, 1981.
43. Nagai, U.; Sato, K.; Nakamura, R.; Kato, R. *Tetrahedron* **1993**, 49, 3577.
44. Rowe, P. M. *Lancet* **1996**, *347*, 606.
45. Meunier, J.-C.; Mollereau, C.; Toll, L.; Suadeau, C.; Moisand, C.; Alvinerie, P.; Butour, J.-L.; Guillemot, J.-C.; Ferrara, P.; Monsarrat, B.; Mazarguil, H.; Vassart, G.; Parmentier, M.; Costentin, J. *Nature* **1995**, *377*, 532.
46. Reinscheid, R.; Nothaker, H.-P.; Bourson, A.; Ardati, A.; Henningson, R. A.; Bunzow, J. R.; Grandy, D. K.; Langen, H.; Monsma Jr., F. J.; Civelli, O. *Science* **1995**, *270*, 792.
47. Dooley C. T.; Houghten R. A. *Life Sci.* **1996**, *59*, PL23.
48. Lung, F. D.; Collins, N.; Stropova, D.; Davis, P.; Yamamura, H. I.; Porreca, F.; Hruby V. J. *J. Med. Chem.* **1996**, *39*, 1136.
49. Karten, M.; Rivier, J. *Endocrin. Rev.* **1986**, *7*, 44.
50. Schally, A.V.; Arimura, A.; Baba, Y.; Nair, R. M. G.; Matsuo, N.; Redding, T.W.; Debeljuk, L.; White, W. F. *Biochem. Biophys. Res. Commun.* **1971**, *43*, 393.
51. Matsuo, N.; Baba, Y.; Nair, R. M. G.; Arimura, A.; Schally, A. V. *Biochem. Biophys. Res. Commun.* **1971**, *43*, 1334.

52. Fujino, M.; Lobayashi, S.; Obayashi, M.; Shinigawa, S.; Fukuda, T. *Biochem. Biophys. Res. Commun.* **1972**, *49*, 863.
53. Vilchez-Martinez, J. A.; Coy, D.H.; Arimura, A.; Coy, E. J.; Hirotsu, Y.; Schally, A. V. *Biochem. Biophys. Res. Commun.* **1974**, *59*, 1226.
54. Martindale, W. *The Extra Pharmacopoeia, 30th Edition*; Reynolds, J. E. F., Ed.; The Pharmaceutical Press, London, 1993, p 948.
55. Momany, F. A., *J. Am. Chem. Soc.* **1976**, *98*, 2990.
56. Freidinger, R. M.; Veber, D. F.; Perlow, D. S.; Brooks, J. R.; Saperstein, R. *Science* **1980**, *210*, 656.
57. Ljungquist, A.; Feng, D. M.; Tang, P.-F. L.; Kubota, M.; Okamoto, Y.; Zhang, Y.; Bowers, C. Y.; Hook, W. A.; Folkers, K. *Biochem. Biophys. Res. Commun.* **1987**, *148*, 849.
58. Rivier, J. E.; Rivier, C.; Vale, W.; Koerber, S.; Corrigan, A.; Porter, J.; Gierasch, L. M.; Hagler, A. T., In *Peptides: Chemistry, Structure and Biology*; Rivier, J. E.; Marshall, G. R., Eds.; ESCOM, Leiden, The Netherlands, 1990, p 33.
59. Bienstock, R. J.; Rizo, J.; Koeber, S. C.; Rivier, J. E.; Hagler, A. T.; Gierasch, L. M. *J. Med. Chem.* **1992**, *36*, 3265.
60. Reisine, T.; Bell, G. I. *Neuroscience* **1995**, *67*, 777.
61. Brazeau, P.; Vale, W.; Burgus, R.; Ling, N.; Butcher, M.; Rivier, J.; Guillemin, R. *Science* **1973**, *179*, 77.
62. Veber, D. F., Design and Discovery in the Development of Peptide Analogues, In *Peptides: Chemistry and Biology, Proceedings of the Twelfth American Peptide Symposium;* Escom, 1992, p 3.
63. Raynor, K.; Murphy, W. A.; Coy, D. H.; Taylor, J. E.; Moreau, J.-P.; Yasuda, K.; Bell, G. I.; Reisine, T. *Molec. Pharmacol.* **1993**, *43*, 838.
64. Nicolaou, K.C.; Salvino, J. M.; Raynor, K.; Pietranico, S.; Reisine, T.; Freidinger, R. M.; Hirschmann, R., In *Peptides: Chemistry, Structure and Biology;* Escom, Leiden, The Netherlands, 1990, p 881.
65. Hirschmann, R.; Nicolaou, K. C.; Pietranico, S.; Salvino, J.; Leahy, E. M.; Sprengeler, P. A.; Furst, G.; Smith III, A. B. *J. Am. Chem. Soc.* **1992**, *114*, 9217.
66. Vassas, A.; Bourdy, G.; Paillard, J. J.; Lavayre, J.; Pays, M.; Quirion, J. C.; Debitus, C. *Planta Med.* **1996**, *62*, 28.
67. Maggi, C. A.; Patacchini, R.; P. Rovero, P.; Giachetti, A. *J. Auton. Pharmacol.* **1993**, *13*, 23.
68. Chang, M. M.; Leeman, S. E. *J. Biol. Chem.* **1970**, *245*, 4784.
69. Chang, M. M.; Leeman, S. E.; Niall, H. D. *Nature - New Biol.* **1971**, *232*, 86.
70. Tregear, G. W.; Niall, H. D.; Potts, J. T. Jr.; Leeman, S. E.; Chang M. M. *Nature - New Biol.* **1971**, *232*, 87.
71. Bury, R. W.; Mashford, M. L. *J. Med. Chem.* **1976**, *19*, 854.
72. Couture, R.; Fournier, A.; Magnan, J.; St. Pierre, S.; Regoli, D. *Can. J. Physiol. Pharmacol.* **1979**, *57*, 1427.
73. Convert, O.; Duplaa, H.; Lavielle, S.; Chassaing, G. *Neuropeptides* **1991**, *19*, 259.
74. Laufer, R.; Gilon, C.; Chorev, M.; Selinger, Z. *J. Med. Chem.* **1986**, *29*, 1284.
75. Hagan, R. M.; Ireland, S. J.; Jordan, C. C.; Bailey, F.; Stephens-Smith, M.; Ward, P. *Br. J. Pharmacol.* **1989**, *98*, 717P
76. Cascieri, M. A.; Goldenberg, M. M.; Liang, T. *Mol. Pharmacol.* **1981**, *20*, 457.
77. Drapeau, G.; D'Orleans-Juste, P.; Dion, S.; Rhaleb, N-E.; Rouissi, N-E.; Regoli, D. *Neuropeptides* **1987**, *10*, 43.
78. Folkers, K.; Hakanson, R.; Horig, J.; Jie-Cheng, X.; Leander, S. *Br. J. Pharmacol.* **1984**, *83*, 449.
79. Folkers, K.; Feng, D-M.; Asano, N.; Hakanson, R.; Weisenfeld-Hallin, Z.; Leander, S. *Proc. Natl. Acad. Sci. USA* **1990**, *87*, 4833.
80. Hagiwara, D.; Miyake, H.; Morimoto, H.; Murai, M.; Fujii, T.; Matsuo, M. J. *Pharmacobio.-Dyn.* **1991**, *14*, S-104.

81. Horwell, D. C. *Trends Biotech.* **1995**, *13*, 132.
82. Boyle S. *Biomed. Chem.* **1994**, *2*, 357.
83. McKnight, A. T.; Maguire, J. J.; Williams, B. J.; Foster, A. C.; Trigett, R.; Iversen, L. L. *Regul. Pept.* **1988**, *22*, 127.
84. Ward, P.; Ewan, G. B.; Jordan, C. C.; Ireland, S. J.; Hagan, R. M.; Brown, J. R. *J. Med. Chem.* **1990**, *33*, 1838.
85. Glaxo Group Ltd: EP-A-360390 (28 Mar 1990).
86. Cox, M. T.; Gormley, J. J.; Hayward, C. F.; Petter, N. N. *J. Chem. Soc., Chem. Commun.* **1980**, 800.
87. Allmendinger, T; Felder, E.; Hungerbuhler, E. *Tetrahedron Lett.* **1990**, *31*, 7301.
88. Zacharia, S.; Rossowski, W.J.; Jiang, N.-Y.; Hrbas, P.; Ertan, A.; Coy, D. H. *Eur. J. Pharmacol.* **1991**, *203*, 353.
89. Jukic, D.; Mayer, M.; Schmitt, P.; Drapeau, G.; Regoli, D.; Michelot, R. *Eur. J. Med. Chem.* **1991**, *26*, 921.
90. Jenmalm, A.; Luthman, K.; Lindeberg, G.; Nyberg, F.; Terenius, L.; Hacksell, U. *Bioorg. Med. Chem. Lett.* **1992**, *2*, 1693.
91. Waeber, B.; Burnier, M.; Brunner, H. R. In *Medicinal Chemistry of the Renin-Angiotensin System;* Timmermans, P. B. M. W. M.; Wexler, R. R., Eds.; Elsevier, Lausanne, 1994, p 27.
92. Brunner, H. R.; Nussberger, J.; Burnier, M.; Waeber, B. *Clin. Exp. Hypertension* **1993**, *15*, 1221.
93. Ferreira, S. H.; Bartelt, D. C.; Greene, L. J. *Biochemistry* **1970**, *9*, 2583.
94. Kageyama, R.; Ohkubo, H.; Nakanishi, S. *Biochemistry* **1984**, *23*, 3603.
95. Skeggs, Jr., L. T.; Kahn, J. R.; Lentz, K. E.; Shumway, N. P. *J. Exp. Med.* **1957**, *106*, 439
96. Tewksbury, D. A.; Dart, R. A.; Travis, J. *Biochem. Biophys. Res. Commun.* **1981**, *99*, 1311.
97. Burton, J.; Poulsen, K.; Haber, E. *Biochemistry* **1975**, *14*, 3892.
98. Szelke, M.; Jones, D. M.; Atrash, B.; Hallett, A.; Leckie, B. J. In *Peptides: Structure and Function, Proceedings of Eighth American Peptide Symposium;* Hruby, V.J., Rich, D. H., Eds.; Pierce Chemical Co., Rockford Il, 1983, p 579.
99. Szelke, M.; Leckie, B.; Hallett, A.; Jones, D. M.; Sueiras, J.; Atrash, B.; Lever, A. F. *Nature* **1982**, *299*, 555.
100. Boger, J.; Lohr, N. S.; Ulm, E. H.; Poe, M.; Blaine, E. H.; Fanelli, G. M.; Lin, T.-Y.; Payne, L. S.; Schorn, T. W.; LaMont, B. I.; Vassil, T. C.; Stabilito, I. I.; Veber, D. F.; Rich, D. H.; Bopari, D. F. *Nature* **1983**, *303*, 81.
101. Catanzaro, D. F.; Sealey, J. E. In *Medicinal Chemistry of the Renin-Angiotensin System*; Timmermans, P. B. M. W. M.; Wexler, R. R., Eds.; Elsevier, Lausanne, 1994, p 65.
102. Smith III, A. B.; Akaishi, R.; Jones, D. R.; Keenan, M. C.; Guzman, M. C.; Holcomb, R. C.; Sprengeler, P. A.; Wood, J. L.; Holloway, M. K.; Hirschmann, R. *Biopolymers* **1995**, *37*, 29.
103. Mutt, V.; Jorpes, J. E. *Eur. J. Biochem.* **1968**, *6*, 156.
104. Rehfeld, J. F. *J. Neurochem.* **1985**, *44*, 1.
105. Williams, J. A. *Biomed. Res.* **1982**, *3*, 107.
106. Morley, J. E. *Life Sci.* **1982**, *30*, 479.
107. Beinfeld, M. C. *Neuropeptides* **1983**, *3*, 411.
108. Chang, R. S. L.; Lotti, V. J.; Monaghan, R. L.; Birnbaum, J.; Stapley, E. O.; Goetz, M. A.; Albers-Schonberg, G.A.; Patchett, A.; Liesch, J. M.; Hensens O. D.; Springer, J. P. *Science* **1985**, *230*, 177.
109. Evans, B. E.; Bock, M. G.; Rittle, K. E.; DiPardo, R. M.; Whitter, W. L.; Veber, D. R.; Anderson, P. S.; Freidinger, R. L. *Proc. Natl. Acad. Sci. USA* **1986**, *83*, 4918.
110. Bock, M. G.; DiPardo, R. M.; Rittle, K. E.; Evans, B. E.; Freidinger, R. M.; Veber, D. F.; Chang, R. S. L.; Chen, T.; Keegan, M. E.; Lotti, V. J. *J. Med. Chem.* **1986**, *29*, 1941.
111. Chang R. S. L.; Lotti, V. J. *Proc. Natl. Acad. Sci. USA* **1986**, *86*, 4923.
112. Parsons, W. H.; Patchett, A. A.; Holloway, M. K.; Smith, G. M.; Davidson, J. L.; Lotti, V. J.; Chang, R. S. L. *J. Med. Chem.* **1989**, *32*, 1681.

113. Yu, M. J.; McCowan, J. R.; Mason, N. R.; Deeter, J. B.; Mendelsohn, L. G. *J. Med. Chem.* **1992**, *35*, 2434.
114. Von Euler, U. S.; Gaddum, J. H. *J. Physiol.* **1931**, *72*, 157.
115. Henry, J. L. *Brain Res.* **1976**, *114*, 439.
116. Otsuka, M.; Konishi, S. *Trends. Neurosci.* **1983**, *6*, 317.
117. Lambeck F.; Holzer, P. *Naunyn-Schmeideberg's Arch. Pharmacol.* **1979**, *310*, 175.
118. Ali, H.; Leung, K.; Pearce, F.; Hayes, N.; Foreman, J. *Int. Archs. Allergy Appl. Immun.* **1986**, *79*, 413.
119. Snider, R. M.; Constantine, J. W.; Lowe, J. A.; Longo, K. P.; Lebel, W. S.; Woody, H. A.; Drozda, S. E.; Desai, M. C; Vinick, F. J.; Spencer, R. W.; Hess, H-J. *Science* **1991**, *251*, 435.
120. Morimoto, H.; Murai, M.; Maeda, Y.; Yamaoka, M.; Nishikawa, M.; Kiyotoh, S.; Fujii, T. *J. Pharmacol. Exp. Ther.* **1992**, *262*, 398.
121. Barrow, C. J.; Doleman, M. S.; Bobko, M. A.; Cooper, R. *J. Med. Chem.* **1994**, *37*, 356.
122. Wong, S-M.; Musza, L.; Kydd, G. C.; Kullnig, R.; Gillum A. M.; Cooper, R. *J. Antibiotics* **1993**, *46*, 811.
123. Wong, S-M.; Kullnig, R.; Dedinas, J.; Appell, K. C.; Kydd, G. C.; Gillum, A. M.; Cooper R.; Moore, R. *J. Antibiotics* **1993**, *46*, 214.
124. Barrow, C. J.; Sun, H. H. *J. Nat. Prod.* **1994**, *57*, 471.
125. Barrow, C. J.; Ping, C.; Snyder, J. K.; Sedlock, D. M.; Sun H. H.; Cooper, R. *J. Org. Chem.* **1993**, *58*, 6016.
126. Oleynek, J. J.; Sedlock, D. M.; Barrow, C. J.; Appell, K. C.; Casiano, R.; Haycock, D.; Ward, S. J.; Kaplita, P.; Gillum, A. M. *J. Antibiotics* **1994**, *47*, 399.
127. Barrow, C. J.; Musza L. L.; Cooper, R. *Bioorg. Med. Chem. Lett.* **1995**, *5*, 377.
128. Popp, J. L.; Musza, L. L.; Barrow, C. J.; Rudewicz, P. J.; Houck, D. R. *J. Antibiotics* **1994**, *47*, 411.
129. Goldstein, S.; Neuwels, M.; Moureau, F.; Berckmans, D.; Lassoie, M-A.; Differding, E.; Houssin, R.; Henichart, J-P. *Lett. Pept. Science* **1995**, *2*, 125.
130. Schiff, P. B.; Fant, J.; Horwitz, S. B. *Nature* **1979**, *277*, 665.
131. Horwitz, S.; Lothstein, L.; Manfredi, J. J.; Mellado, W.; Parness, J.; Roy, S. N.; Schiff, P. B.; Sorbara, L.; Zeheb, R. *Ann. N. Y. Acad. Sci.* **1986**, *466*, 733.
132. Rao, S.; Krauss, N. E.; Heerding, J. M.; Swindell, C. S.; Ringel, I.; Orr, G. A.; Horwitz, S. B. *J. Biol. Chem.* **1994**, *269*, 3134.
133. Rao, S.; Orr, G. A.; Chaudhary, A. G.; Kingston, D. G.; Horwitz, S. B. *J. Biol. Chem.* **1995**, *270*, 20235.
134. Leu, J-G.; Chen, B-X.; Dianmanduros, A. W.; Erlanger, B. F. *Proc. Natl. Acad. Sci. USA* **1994**, *91*, 10690.
135. Cragg, G. M.; Schepatz, S. A.; Suffness, M.; Grever, M. R. *J. Nat. Prod*.**1993**, *56*, 1657.
136. Holton, R. A.; Somoza, C.; Kim, H-B.; Liang, F.; Biediger, R. J.; Boatman, P. D.; Shindo, M.; Smith, C. C.; Kin S.; Nadizadeh, H.; Suzuki, Y.; Tao, C.; Vu, Ph.; Tang, S.; Zhang, P.; Murthi, K. K.; Gentile, L. N.; Jyanwei H. L. *J. Am. Chem. Soc.* **1994**, *116*, 1597.
137. Nicolaou, K. C.; Yang Z.; Liu J. J.; Ueno, H.; Nantermet, P. G.; Guy, R. K.; Claiborne, C. F.; Renaud, J.; Couladouros, E. A.; Paulvannan, K. *Nature* **1994**, *367*, 630.
138. Christen, A. A.; Bland, J.; Gibson D. M. *Proc. Am. Assoc. Cancer Res.* **1989**, *30*, 566.
139. Stierle, A.; Strobel, G.; Stierle, D. *Science* **1993**, *260*, 214.
140. Kingston, D. I. *Trends Biotec.* **1994**, *12*, 222.
141. Denis, J-N.; Greene, A. E.; Guenard, D.; Gueritte-Voegerlein, R.; Mangatal, F. L.; Potier, P. *J. Am. Chem. Soc.* **1988**, *110*, 5917.
142. Guenard, D.; Gueritte-Voegelein, F.; Potier, P. *Acc. Chem. Res.* **1993**, *26*, 160.
143. Cushman, D. W. In *Ezyme Inhibitors and Drugs;* Sandler, M., Ed.; London, Macmillan, 1980, p 231.
144. Olivera, B. M.; Gray, W. R.; Zeikus, R.; McIntosh, J. M.; Varga, J.; Riverier, J.; de Santos, V.; Cruz, L. J. *Science* **1985**, *230*, 1338.

145. Gray, W. R.; Olivera, B. M.; Cruz, L. J. *Ann. Rev. Biochem.* **1988,** *57*, 665.
146. Olivera, B. M.; Hillyard, D. R.; March, M.; Yoshikami, D. *Trends Biotech.* **1995,** *13*, 422.
147. Valentino, K. *Proc. Natl. Acad. Sci. USA* **1993,** *90*, 7894.
148. Harvey, A. L. *Chemistry & Industry* **1995**, 914.
149. Davis, J. H.; Bradley, E. K.; Miljanich, G. P.; Nadasdi, L.; Ramachandran, J.; Basus, V. *Biochemistry* **1993,** *32*, 7396.
150. Pallaghy, P. K.; Duggan, B. M.; Pennigton, M. W.; Norton, R. S. *J. Mol. Biol.* **1993,** *234*, 405.
151. Ladner, R. C. *Trends Biotech.* **1995,** *13*, 426.
152. Hoess, R. H.; Mack, A. J.; Walton, H.; Reilly, T. M. *J. Immunol.* **1994,** *153*, 724.
153. Leatherbarrow, R. J.; Salcinski, H. I. *Biochemistry* **1991,** *30*, 10717.
154. Lewis, R. J.; Bingham, J-P.; Jones, A.; Alewood, P. F.; Andrews, P. R. *Australian Biotechnol.* **1994,** *4*, 298
155. Kornalik F. In *Snake Toxins;* Harvey, A. L., Ed.; New York, Pergamon Press, 1991, p 322.
156. Gould, R. J.; Polokoff, M. A.; Friedman, P. A.; Huang, T-F; Holt, J. C.; Cook, J. J.; Niewiarowski, S. *Proc. Soc. Exp. Bio. Med.* **1990,** *195*, 168.
157. Dennis, M. S.; Henzel, W. J.; Pitti, R. M.; Lipari, M. T.; Napier, M. A.; Deisher, T. A.; Bunting, S.; Lazarus, R. A. *Proc. Natl. Acad. Sci. USA* **1990,** *87*, 2471.
158. Gan, Z. R.; Gould, R. J.; Jacobs, J. W.; Friedman, P. A.; Polokoff, M. A. *J. Biol. Chem.* **1988,** *263*, 18827.
159. Shebuski, R. J.; Ramjit, D. R.; Bencen, G. H.; Polokoff, M. A. *J. Biol. Chem.* **1989,** *264*, 21550.
160. Ruoslahti, E.; Pierschbacher, M. D. *Science* **1984,** *238*, 491.
161. Craig, W. S.; Cheng, S.; Mullen, D. G.; Blevitt, J.; Pierschbacher, M. D. *Biopolymers* **1995** *37*, 157.
162. Scarborough, R. M.; Rose, J. W.; Naughton, M. A.; Phillips, D. R.; Nannizzi, L.; Arfsten, A.; Campbell, A. M.; Charo I. F. *J. Biol. Chem.* **1993,** *268*, 1058.
163. Alder, M.; Lazarus, R. A.; Dennis, M. S.; Wagner G. *Science* **1991,** *253*, 445.
164. Barker, P. L.; Webb II, R. B. *Adv. Med. Chem.* **1995,** *3*, 57.
165. Ojima, I.; Chakravarty, S.; Dong, Q. *Bioorg. Med. Chem. Lett.* **1995,** *3*, 337.
166. Lefkovits, J.; Plow, E. F.; Topol, E. J. *N. Eng. J. Med.* **1995,** *332*, 1553.
167. Bach, A. C., II; Espina, J. R.; Jackson, S. A.; Stouten, P. F. W.; Duke, J. L.; Mousa, S. A.; DeGrado, W. F. *J. Am. Chem. Soc.* **1996,** *118*, 293.
168. Callahan, J. F.; Bean, J. W.; Burgess, J. L.; Eggleston, D. S.; Hwang, H. S.; Kopple, K. D.; Koster, P. F.; Nichols, A.; Peishoff, C. E.; Samanen, J. M.; Vasko, J. A.; Wong, A.; Huffman W. F. *J. Med. Chem.* **1992,** *35*, 3970.
169. Ku, T. W.; Ali, F. E.; Barton, L. S.; Bean, J. W.; Bondinell, W. E.; Burgess, J. L.; Callahan, J. G.; Calvo, R. R.; Chen, L.; Eggleston, D. S.; Gleason, J. G.; Huffman, W. F.; Hwang, S. M.; Jakas, D. R.; Karash, C. B.; Keenan, R. M.; Kopple, K. D.; Miller, W. H.; Newlander, K. A.; Nichols, A.; Parker, M. F.; Peishoff, C. E.; Samanen, J. M.; Uzinskas, I.; Venslavsky, J. W. *J. Am. Chem. Soc.* **1993,** *115*, 8861.
170. McDowell, R. S.; Blackburn, B. K.; Gadek, T. R.; McGee, L. R.; Rawson, T.; Reynolds, M. E.; Robarge, K. D.; Somers, T. C.; Thorsett, E. D.; Tischler, M.; Webb II, R. R.; Venuti M. C. *J. Am. Chem. Soc.* **1994,** *116*, 5069.
171. Alig, L.; Edenhofer, R.; Muller, M.; Trzeciak, A.; Weller, T. U.S. patent 5,039,805. Hoffmann-LaRoche, August 13, 1991.
172. Porter, B.; Eldred, C. D.; Kelly H. A. European patent application EP 0 537 980 A1.
173. Greenspoon, N.; Hershkoviz, R.; Alon, R.; Varon, D.; Shenkman, B.; Marx, G.; Federman, S.; Kapustina, G.; Lider, O. *Biochemistry* **1993,** *32*, 1001.
174. Kem, W. R. In *The Biology of Nematocysts;* Academic Press, 1988, pp 375-405. Honerjager, P. *Rev. Physiol. Biochem. Pharmacol.* **1982,** *92*, 2.
175. Hellberg, S.; Kem, W. R. *Int. J. Peptide Protein Res.* **1990,** *36*, 440.
176. Geren, C. R. *J. Toxicol. -Toxin Rev.* **1986,** *5*, 161.

177. Jackson, H.; Usherwood, R. N. R. *Trends Neurosci.* **1988,** *11*, 278.
178. Jackson, H.; Parks, T. N. *Ann. Rev. Neurosci.* **1989,** *12*, 405.
179. Harvey, A. L. *Chemistry & Industry* **1995,** 914.
180. Castle, N. A.; Haylett, D. G.; Jenkinson, D. H. *Trends Neurosci.* **1989,** *12*, 59.
181. Swaminathan, P.; Hariharan, M.; Murali, R.; Singh C. U. *J. Med. Chem.* **1996,** *39*, 2141.
182. Skarzynski ,T. *J. Mol. Biol.* **1992,** *224*, 671.
183. Hollecker, M.; Marshall, D. L.; Harvey A. L. *Br. J. Pharmacol.* **1993,** *110*, 790.
184. Smith, L. A.; Lafaye, P. J.; LaPenotiere, H. F.; Spain, T.; Dolly, J. O. *Biochemistry* **1993,** *32*, 5692.
185. Smith, L. A.; Olsen, M. A.; Lafaye, P. J.; Dolly, J. O. *Toxicon* **1995,** *33*, 459.
186. Danse, J. M.; Rowan, E. G.; Gasparini, S.; Ducancel, F.; Vatanpour, H.; Young, L. C.; Poorheidari, G.; Lajeunesse, E.; Drevet, P.; Menez, R.; Pinkasfeld, S.; Boulain, J-C.; Harvey, A. L.; Menez, A. *FEBS Lett.* **1994,** *356*, 153.
187. Swaminathan, P.; Hariharn, M.; Murali, R.; Singh, C. U. *J. Med. Chem.* **1996,** *39*, 2141.
188. Hirschmann, R.; Sprengler, P. A.; Kawasaki, T.; Leahy, J. W.; Shakespeare, W. C.; Smith III, A. B. *Tetrahedron* **1993,** *49*, 3665.
189. Venapalli, B. R.; Aimone, L. D.; Apell, K. C.; Bell, M. R.; Dority, J. A.; Goswami, R.; Hall, P. L.; Kumar, V.; Lawrence, K. B.; Logan, M. E.; Scensny, P. M.; Seelye, J. A.; Tomczuk, B. E.; Yanni, J. M. *J. Med. Chem.* **1992,** *35*, 374.
190. Graf von Roedern, E.; Kessler, H. *Angew. Chem. Int. Ed. Engl.* **1994,** *33*, 670.
191. Graf von Roedern, E.; Lohof, E.; Hessler, G.; Hoffmann, M.; Kessler, H. *J. Am. Chem. Soc.* **1996,** *118*, 10156.
192. McDevitt, J. P.; Lansbury, Jr, P. T. *J. Am. Chem. Soc.* **1996,** *118*, 3818.
193. Qiu Y.; Nakahara Y.; Ogawa T. *Biosc., Biotech. Biochem.* **1996,** *60*, 986.
194. Sano T.; Imura R.; Morishita, Y.; Matsuda, Y.; Yamada, K. *Life Sci.* **1992,** *51*, 1445.
195. Adinolfi, M.; Barone, G.; De Napoli, L.; Iadonisi, A.; Piccialli, G. *Tetrahedron Lett.* **1996**. *37*, 5007.
196. Nathwani, D.; Wood, M. J. *Drugs* **1993,** *45*, 866.
197. Nagai, U.; Sato, K. *J. Chem. Soc., Perkin Trans. I* **1986,** 1231.
198. Nagai, U.; Sato, K. *Peptides: Structure and Function, Proceedings of the 9th American Peptide Symposium;* Deber, C. M.; Hruby, V. J.; Kopple, K. D., Eds.; Pierce Chemical Co., Rockford, IL, 1985, p 465.
199. Nagai, U.; Sato, K, *Japanese Patent 63,280,099* (1988), *Chem. Abstr. 111*, 23952m.
200. Nagai, U.; Sato, K, Nakamura, R.; Kato, R. *Tetrahedron* **1993,** *49*, 3577.
201. Sato, K.; Hotta, M.; Dong, M.-H. *Int. J. Peptide Protein Res.* **1991,** *38*, 340.
202. Baca, M., Alewood, P. F., Kent, S. B. H. *Protein Sci.* **1993,** *2*, 1085.
203. Etzkorn, F.; Guo, T.; Lipton, M.; Goldberg, S. D.; Bartlett, P. A. *J. Am. Chem. Soc.* **1994,** *116*, 10412.
204. Subasinghe, N. L.; Bontems, R. J.; McIntee, E.; Mishra, R. K.; Johnson, R. L. *J. Med. Chem.* **1993,** *36*, 2356.
205. Sheehan, J. C.; Henry-Logan, K. R. *J. Am. Chem. Soc.* **1959,** *81*, 3089.
206. Baldwin, J. E.; Freeman, R. T.; Lowe, C.; Schofield, C. J.; Lee, E. A. *Tetrahedron* **1989** *45*, 4537.
207. Alberg, D. G.; Schreiber, S. L. *Science* **262**, 248.
208. Humber, D. C.; Cammack, N.; Coates, J. A. V.; Cobley, K. N.; Orr, D. C.; Storer, R.; Weingarten, G. G.; Weir, M. P. *J. Med. Chem.* **1992,** *35*, 3081.
209. Humber D. C.; Bamford M. J.; Bethell R. C.; Cammack N.; Cobley K.; Evans D. N.; Gray N. M.; Hann M. M.; Orr D. C.; Saunders J. *J. Med. Chem.* **1993,** *36*, 3120.
210. Storer, R.; Cammack, N.; Cobley, K.; Evans, D.; Hann, M.; Humber, D.; Mistry, A.; Orr, D.; Weingarten, G.; Wonnacott, A. In *Perspectives in Medicinal Chemistry;* Testa, B.; Kyburz, E.; Fuhrer, W.; Giger, R., Eds.; Verlag Helvetica Chimica Acta, Basel. 1993, pp 61-72.
211. Bunin, B. A.; Plunkett, M. J.; Ellman, J. A. Synthesis and evaluation of 1,4-benzodiazepine libraries, In: *Methods in Enzymology. Combinatorial Chemistry;* Abelson, J. N., Ed.; San Diego, Academic Press Inc., 1996, p 448.

212. Plunkett, M.J.; Ellman, J. A. *J. Am. Chem. Soc.* **1995**, *117*, 3306.
213. Bunin, B.A.; Ellman, J. A. *J. Am. Chem. Soc.* **1992**, *114*, 10997.
214. Bunin, B. A.; Plunkett, M. J.; Ellman, J. A. *Proc. Natl. Acad. Sci. USA* **1994**, *91*, 4708.
215. Thompson, P. E.; Hearn, M. T. W. *Tetradedron Lett.* **1997**, *38*, 2907-2910.
216. Koh, J. S.; Ellman, J. A. *J. Org. Chem.*, **1996**, *61*, 4494.
217. Fodor, G.; Dharanipragada, R. *Nat. Prod. Rep.* **1993**, *10*, 199.
218. Williamson, E. M.; Okpako, D. T.; Evans, F. J. *Selection, Preparation and Pharmacological Evaluation of Plant Material;* Wiley, Chichester, 1996.
219. Farnsworth, N. R. *Ciba Foundation Symposium* **1994**, *185*, 42.
220. Kleiner, K. *New Scientist* **1995**, 5.

INDEX

J

A

I

P

R

E

S

S

Advances in Biophysical Chemistry

Edited by **C. Allen Bush,** *Department of Chemistry and Biochemistry, The University of Maryland, Baltimore County*

Volume 1, 1990, 247 pp. $112.50/£72.00
ISBN 1-55938-159-0

CONTENTS: Preface. Stable-Isotope Assisted Protein NMR Spectroscopy in Solution, *Brian J. Stockman and John L. Markley.* ^{31}P and ^{1}H Two-Dimensional NMR and NOESY-Distance Restrained Molecular Dynamics Methodologies for Defining Sequence-Specific Variations in Duplex Oligonucleotides, *David G. Gorenstein, Robert P. Meadows, James T. Metz, Edward Nikonowcz and Carol Beth Post.* NMR Study of B- and Z-DNA Hairpins of d[(CG) $_3T_4(CG)_3$] in Solution, *Satoshi Ikuta and Yu-Sen Wang.* Molecular Dynamics Simulations of Carbohydrate Molecules, *J.W. Brady, Cornell University.* Diversity in the Structure of Hemes, *Russell Timkovich and Laureano L. Bondoc.*

Volume 2, 1991, 180 pp. $112.50/£72.00
ISBN 1-55938-396-8

CONTENTS: Preface, *C. Allen Bush.* Methods in Macromolecular Crystallography, *Andrew J. Howard and Thomas L. Poulos.* Circular Dichroism and Conformation of Unordered Polypeptides, *Robert W. Woody.* Luminescence Studies with Horse Liver Alcohol Dehydrogenase: Information on the Structure, Dynamics, Transitions and Interactions of this Enzyme. Surface-Enhanced Resonance Raman Scattering (SERRS) Spectroscopy: A Probe of Biomolecular Structure and Bonding at Surfaces, *Therese M. Cotton, Jae-Ho Kim and Randall E. Holt.* Three-Dimensional Conformations of Complex Carbohydrates, *C. Allen Bush and Perseveranda Cagas.* Index.

Volume 3, 1993, 263 pp. $112.50/£72.00
ISBN 1-55938-425-5

CONTENTS: Introduction to the Series: An Editor's Foreword, *Albert Padwa.* Preface, *C. Allen Bush.* Raman Spectroscopy of Nucleic Acids and Their Complexes. *George J. Thomas, Jr. and Masamichi Tsuboi.* Oligosaccharide Conformation in Protein/Carbohydrate Complexes, *Anne Imberty, Yves Bourne, Christian Cambillau and Serge Perez. Geometric Requirements of Proton Transfers, Steve Scheiner.* Structural Dynamics of Calcium-Binding Proteins, *Robert F. Steiner.* Determination of the Chemical Structure of Complex Polysaccharides, *C. Abeygunawardana and C. Allen Bush.* Index.

Printed and bound by CPI Group (UK) Ltd, Croydon, CR0 4YY
08/05/2025
01865011-0001